A SURVEY OF GEOMETRY

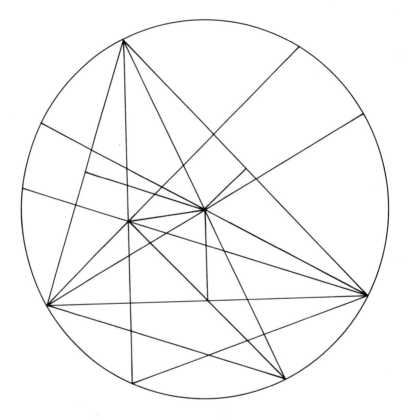

A Survey of Geometry
Revised Edition

HOWARD EVES
University of Maine

Allyn and Bacon, Inc.   Boston

To Diane

We started rambling through the hills together
Then we decided to ramble through life together

Contents

3. ELEMENTARY TRANSFORMATIONS

99

4. EUCLIDEAN CONSTRUCTIONS

154

5. DISSECTION THEORY

194

6. PROJECTIVE GEOMETRY

240

7. NON-EUCLIDEAN GEOMETRY

282

Preface to the
Revised Edition

Since writing the preface of the first edition of this work, the gloomy plight there described of beginning collegiate geometry has brightened considerably. The pendulum seems definitely to be swinging back and a goodly amount of excellent textual material is appearing.

The widespread use of the first volume of our work has called forth the present revision. While this revised edition is essentially like the first edition, opportunity has been taken for making a great number of small improvements and a few larger changes. Among the larger changes has been the rewriting of certain sections, such as those on isometries, similarities, the method of loci, and metamathematics. Another pronounced change has been the splitting of many former sections into shorter ones, so that any section taken in class can be comfortably covered in one 50-minute class session. A third major change has been the amplification, by a factor of three or four, of the hints and suggestions for the solution of the problems.

The book is very rich and one cannot hope to cover all the sections in even a year course meeting three hours a week. A number of sections were not written for class coverage, anyway, but rather for the interested student seeking material for a term paper or for "junior" research. Among these sections are 2.7, 3.11, 4.4, 4.7, 4.8, 5.5, 5.6, 5.8, 6.5, 6.6, 7.7, 7.8, 8.6, 8.7, 8.8, 8.9. Other sections, a selection of which might be similarly treated, are 1.5, 1.6,

1.7, 4.5, 4.6, 5.4, 5.7, 6.9, 7.6, 8.5. The instructor may, of course, wish to comment in class on the contents (without going into proofs) of some of the omitted sections. All in all, there is a good deal of room for an instructor to exercise personal preference and to vary the course from year to year.

Preface to the
First Edition

The undergraduate college textbook situation in geometry today is much like what it was in algebra just prior to 1940.

Before 1940 there was scarcely any textual material available in English for introducing college undergraduate mathematics students to the vast and rich area of so-called modern algebra. True, there were a few books in the area, but these either were written for the graduate student or tended to concentrate on only certain facets of modern algebra and did not constitute anything like a true survey of the subject. Then, in 1941, appeared the magnificent Birkhoff-MacLane *A Survey of Modern Algebra,* and the undergraduate mathematics student was furnished a textbook that gave him a glimpse into most of the important areas of modern algebra. Here was a text designed for a single course in which the student was able to gain a good idea of the nature of such topics as group theory, vector spaces, matrix algebra, linear groups, determinants, Boolean algebra, transfinite arithmetic, rings and ideals, algebraic number fields, and Galois theory — topics any one of which can easily be extended into a full course of its own. Oriented by this textbook in the entire area of modern algebra, a student interested in algebra can see where his preferences lie and can go on to more detailed courses on selected topics of his choosing.

Today we have a dozen or more good undergraduate textbooks of modern algebra, and introductory courses in modern algebra based on these textbooks are now offered in practically every college and university of the country — a situation quite different from that of some twenty or more years ago, when only a few institutions gave such courses, and these, not from easily-available textbooks, but from hard-to-obtain sets of notes.

There are some excellent textbooks available today on specialized areas of geometry, such as college geometry, projective geometry, non-Euclidean geometry, analytic geometry, differential geometry, topology, and so on, and there are some so-called introductions to higher geometry, which, however, largely concentrate on projective geometry, or on the foundations of geometry, or on advanced methods in analytic geometry, or on some other particular field of geometric study. But there is a real dearth of anything like a true survey of the subject that can be used in the undergraduate classroom.

What we need is a text that might do for geometry what the Birkhoff-MacLane text of 1941 did for algebra. The time for such a venture in America seems ripe, for after several decades of general neglect, there appears to be stirring a revived interest in the venerable subject of geometry. Without wanting in any way to appear presumptuous, the present text is humbly offered from this viewpoint.

In writing the book, much trouble was encountered in deciding what to include and what to leave out, and easily five times the material of the final work was written before ruthless weeding and pruning reduced the thing to a sensible size. Replies to a letter to a number of eminent geometers and teachers seemed to indicate that there should be chapters on historical origins, modern elementary (or college) geometry, elementary transformation theory, Euclidean constructions, the dissection of areas and volumes, projective geometry, non-Euclidean geometry, the foundations of geometry, analytic geometry, the Erlanger programm, limit operations in geometry, plane curves, differential geometry, combinatorial topology, *n*-dimensional geometry, and abstract spaces. There was no great unanimity of opinion as to just what topics each chapter should cover, but there was a unanimous request that the book be truly *geometric* in spirit, and not merely a treatment of some algebra couched in geometric terminology. A number of the correspondents expressed the hope that the book would lay some emphasis on geometric *method*, and would not be just a compendium of geometric facts.

Finally, a development in some sixteen chapters with six or seven subsections per chapter, all sewed together with a generous and varied supply of problems, was decided upon. Because, to satisfy the varying desires of different instructors, more material was included than could normally be covered, and because geometrical exposition necessarily requires more space than most other mathematical exposition, it seemed wise to split the complete work into two separately bound parts, or volumes, each part to supply an instructor with ample material for a three-hour one-semester course. It was decided to make the later chapters of each part somewhat independent of one another, thus allowing the instructor to choose from them and to cover as many sections of them as desired. It also seemed wise to make the first part of the work synthetic and relatively elementary, and to reserve analytical considerations and most of the more advanced and abstract portions of the subject for the second part.

It is perhaps pertinent to make a comment or two here about the problems of the text. There is a distinction between what may be called a *problem* and what may be considered an *exercise*. The latter serves to drill a student in some technique or procedure, and requires little, if any, original thought. Thus, after a student beginning algebra has encountered the quadratic formula, he should undoubtedly be given a set of exercises in the form of specific quadratic equations to be solved by the newly acquired tool. The working of these exercises will help clinch his grasp of the formula and will assure his ability to use the formula. An exercise, then, can always be done with reasonable dispatch and with a minimum of creative thinking. In contrast to an exercise, a problem, if it is a good one for its level, should require thought on the part of the student. The student must devise strategic attacks, some of which may fail, others of which may partially or completely carry him through. He may need to look up some procedure or some associated material in texts, so that he can push his plan through. Having successfully solved a problem, the student should reconsider it to see if he can devise a different and perhaps better solution. He should look for further deductions, generalizations, applications, and allied results. In short, he should live with the thing for a time, and examine it carefully in all lights. To be suitable, a problem must be such that the student cannot solve it immediately. One does not complain about a problem being too difficult, but rather too easy.

It is impossible to overstate the importance of problems in mathematics. It is by means of problems that mathematics develops and actually lifts itself by its own bootstraps. Every research article, every doctoral thesis, every new discovery in mathematics, results from an attempt to solve some problem. The posing of appropriate problems, then, appears to be a very suitable way to introduce the promising student to mathematical research. And, it is worth noting, the more problems one plays with, the more problems one may be able to propose on one's own. The ability to propose significant problems is one requirement to be a creative mathematician.

There are relatively few exercises in this work, but many and assorted problems of varying degrees of difficulty. Problems are an integral part of geometry, for one *learns* geometry chiefly by *doing* it. There is a lot of very interesting geometry introduced through the problems, and a student will miss much if he does not dip into this material. Some of the problems are not at all easy for a beginner, and, to allay frustration and alleviate insomnia, a collection of suggestions for the solution of many of the problems appears towards the end of each volume.

A short bibliography will be found at the conclusion of each chapter, and an enterprising student may care to start a deeper study of some material of the chapter with one of these references.

Acknowledgments

It is a pleasure to thank The Cambridge University Press for kindly giving me permission to quote from its admirable publication, T. L. Heath, *The Thirteen Books of Euclid's Elements* (1926); the proof of Euclid's first proposition (in Section 8.1) and Euclid's first principles and the statements of the propositions of Book I of his *Elements* (in Appendix 1) are taken from this source. Thanks are also extended to Prentice-Hall, Inc., for allowing me (in Figure 6.3b) to reproduce a figure from Carrol V. Newsom and Howard Eves, *Introduction to College Mathematics*, 2d ed. (1954). Helpful material, particularly for parts of Chapters 1 and 8, was garnered from appropriate places in Howard Eves, *An Introduction to the History of Mathematics* (1969), and Howard Eves and Carroll V. Newsom, *An Introduction to the Foundations and Fundamental Concepts of Mathematics* (1965), two publications of Holt, Rinehart, and Winston, Inc.; I thank my good friend and coauthor of the second work for allowing this use of material from our book. Special thanks go to the administrative officers and Board of Trustees of the University of Maine for granting me a semester of leave; without this leave it is doubtful that the first edition of this work could even have been begun. Dr. Joseph M. Murray, then Dean of the College of Arts and Sciences, and Professor Spofford H. Kimball, then Chairman of the Department of Mathematics and Astronomy, at the University of Maine, did all they could to encourage and make possible the task of writing the first version of this work; I am very grateful to these two persons. I am also very grateful to the many users of the first edition who wrote warm and helpful comments. And, of course, deep thanks also go to the efficient staff of Allyn and Bacon, Inc., for the very excellent and understanding assistance received throughout the preparation of the work.

Howard Eves

All that a man has to say or do that can possibly concern mankind is, in some shape or other, to tell the story of his love . . .

HENRY DAVID THOREAU

A SURVEY OF GEOMETRY

1

The Fountainhead

This first chapter is introductory in nature and is designed to serve two purposes. In the first place, it describes the source from which essentially all geometrical investigations of the modern era have arisen. This makes possible, in the succeeding chapters, an analysis of the origins of the fundamental ideas there introduced—an analysis without which neither a true understanding nor a genuine appreciation of those ideas is possible.

In the second place, this first chapter serves somewhat in the nature of a review for the reader who has been too long away from his elementary geometry. Before discussing the advances of the modern era, the reader is given a chance to recall and revive some of the basic concepts and terminology that were considered in his high school geometry course, and lacking which it would be rather foolhardy to proceed. This ability to feel one's way around again can best be acquired by working a fair sample of the problems that appear at the ends of the various sections of the chapter.

1.1 THE EARLIEST GEOMETRY

The first geometrical considerations of man are unquestionably very ancient, and would seem to have their origin in simple observations stemming from human ability to recognize physical form and to compare shapes and sizes.

There were innumerable circumstances in the life of even the most primitive man that would lead to a certain amount of subconscious geometric discovery. The notion of distance was undoubtedly one of the first geometrical concepts to be developed. The estimation of the time needed to make a journey led very early to the realization that the straight line constitutes the shortest path from one point to another; indeed, most animals seem instinctively to realize this. The need to bound land led to the notion of simple geometric figures, such as rectangles, squares, and triangles. In fact, it seems natural, when fencing a piece of land, first to fix the corners and then to join these by straight lines. Other simple geometrical concepts, such as the notion of vertical, of parallel, and of perpendicular, would have been suggested by the construction of walls and dwellings.

Many observations in the daily life of early man must have led to the conception of curves, surfaces, and solids. Instances of circles were numerous —for example, the periphery of the sun or the moon, the rainbow, the seed heads of many flowers, and the cross-section of a log. Shadows cast by sun or lamp would reveal circles and conic sections. A thrown stone describes a parabola; an unstretched cord hangs in a catenary curve; a wound rope lies in a spiral; spider webs illustrate regular polygons. The growth-rings of a tree, the swelling circles caused by a pebble cast into a pond, and figures on certain shells suggest the idea of families of curves. Many fruits and pebbles are spherical, and bubbles on water are hemispherical; some bird eggs are approximately ellipsoids of revolution; a ring is a torus; tree trunks are circular cylinders; conical shapes are frequently seen in nature. Early potters made many surfaces and solids of revolution. The bodies of men and animals, most leaves and flowers, and certain shells and crystals illustrate the notion of symmetry. The idea of volume arises immediately in the consideration of receptacles to hold liquids and other simple commodities.

Examples like the above can be multiplied almost indefinitely. Physical forms which possess an ordered character, contrasting as they do with the haphazard and unorganized shapes of most bodies, necessarily attract the attention of a reflective mind—and some elementary geometric concepts are brought to light. Such geometry might, for want of a better name, be called *subconscious geometry*. This subconscious geometry was employed by very early man in the making of decorative ornaments and patterns, and it is probably quite correct to say that early art did much to prepare the way for later geometric development. The evolution of subconscious geometry in little children is well known and easily observed.

Now, in the beginning, man considered only concrete geometrical problems, which presented themselves individually and with no observed interconnections. When human intelligence was able to extract from a concrete geometrical relationship a general abstract relationship containing the former as a particular case, geometry became a science. In this capacity, geometry has the advantage of ordering practical problems into sets such that the problems in a set can be solved by the same general procedure. One thus arrives at the notion of a geometrical law or rule. For example, comparing the lengths of circular courses with their diameters would lead, over a period of time, to the geometrical law that the ratio of circumference to diameter is a constant.

There is no evidence which permits us to estimate the number of centuries that passed before man was able to raise geometry to the status of a science, but all the writers of antiquity who concerned themselves with this matter unanimously agree upon the Nile valley of ancient Egypt as the place where subconscious geometry first became *scientific geometry*. The famous Greek historian Herodotus (*ca.* 485 B.C.–*ca.* 425 B.C.) has stated the thesis in this wise:

> They said also that this king [Sesostris] divided the land among all Egyptians so as to give each one a quadrangle of equal size and to draw from each his revenues, by imposing a tax to be levied yearly. But every one from whose part the river tore away anything, had to go to him and notify what had happened. He then sent the overseers, who had to measure out by how much the land had become smaller, in order that the owner might pay on what was left, in proportion to the entire tax imposed. In this way, it appears to me, geometry originated, which passed thence to Hellas.

Thus the traditional account finds in early Egyptian surveying practices the beginnings of geometry as a science; indeed, the word "geometry" means "measurement of the earth." While we cannot be certain of this origin, it does seem safe to assume that scientific geometry arose from practical necessity, appearing several thousand years before our era in certain areas of the ancient orient as a science to assist in engineering, agriculture, business, and religious ritual. There is historical evidence that this occurred not only along the Nile River of Egypt, but also in other great river basins, such as the Tigris and Euphrates of Mesopotamia, the Indus and Ganges of south-central Asia, and the Hwang Ho and the Yangtze of eastern Asia. These river basins cradled advanced forms of society known for their engineering prowess in marsh drainage, irrigation, flood control, and the erection of great edifices and structures. Such projects required the development of much practical geometry.

1.2 THE EMPIRICAL NATURE OF PRE-HELLENIC GEOMETRY

As far back as history allows us to grope into the past, we still find present a sizeable body of material that can be called practical, or scientific, geometry.

The earliest existing records of man's activity in the field of geometry are some inscribed baked clay tablets unearthed in Mesopotamia and believed to date, in part at least, from Sumerian times of about 3000 B.C. There are other generous supplies of Babylonian cuneiform tablets coming from later periods, such as the First Babylonian Dynasty of King Hammurabi's era, the New Babylonian Empire of Nebuchadnezzar, and the following Persian and Selucidan eras. From these tablets we see that ancient Babylonian geometry is intimately related to practical mensuration. Numerous concrete examples show that the Babylonians of 2000 to 1600 B.C. were familiar with the general rules for computing the area of a rectangle, the areas of right and isosceles triangles (and perhaps the general triangle), the area of the special trapezoid having one side perpendicular to the parallel sides, the volume of a rectangular parallelepiped, and, more generally, the volume of a right prism with special trapezoidal base. The circumference of a circle was taken as three times the diameter, and the area as one-twelfth the square of the circumference (both correct for $\pi = 3$), and the volume of a right circular cylinder was then obtained by finding the product of the base and the altitude. The volume of a frustum of a cone or of a square pyramid appears incorrectly as the product of the altitude and half the sum of the bases. There also seems to be evidence that the ancient Babylonians used the incorrect formula

$$K = (a + c)(b + d)/4$$

for the area of a quadrilateral having a, b, c, d for consecutive sides. These peoples knew that corresponding sides of two similar right triangles are proportional, that the altitude through the vertex of an isosceles triangle bisects the base, and that an angle inscribed in a semicircle is a right angle. The Pythagorean Theorem was also known, even as far back as approximately 2000 B.C. There is a recently discovered tablet in which $3\frac{1}{8}$ is used as an estimate for π.

Our chief sources of information concerning ancient Egyptian geometry are the Moscow and Rhind papyri, mathematical texts containing 25 and 85 problems respectively, and dating from approximately 1850 B.C. and 1650 B.C. There is also, in the Berlin Museum, the oldest extant astronomical or surveying instrument—a combination plumb line and sight rod—which comes from the ancient Egypt of about 1950 B.C. The Berlin Museum also possesses an Egyptian sundial dating from about 1500 B.C., and which is the oldest sundial in existence. These instruments reveal, of course, a knowledge at the times of some associated practical geometry. One should also point out that the great pyramid of Gizeh, whose very careful construction certainly involved some practical geometry, was erected about 2900 B.C.

Twenty-six of the 110 problems in the Moscow and Rhind papyri are geometric. Most of these problems stem from mensuration formulas needed for computing land areas and granary volumes. The area of a circle is taken as equal to that of the square on $\frac{8}{9}$ of the diameter, and the volume of a right cylinder as the product of the area of the base by the length of the altitude. Recent investigations seem to show that the ancient Egyptians knew that the area of any triangle is given by half the product of base and altitude. Some of the problems seem to concern themselves with the cotangent of the dihedral angle between the base and a face of a pyramid, and others show an acquaintance with the elementary theory of similar figures. Although there is no documentary evidence that the ancient Egyptians were aware of the Pythagorean Theorem, early Egyptian surveyors realized that a triangle having sides of lengths 3, 4, and 5 units is a right triangle. It is curious that the incorrect formula

$$K = (a + c)(b + d)/4$$

for the area of an arbitrary quadrilateral with successive sides of lengths a, b, c, d appears in an inscription found in the tomb of Ptolemy XI, who died in 51 B.C.

Very remarkable is the existence in the Moscow papyrus of a numerical example of the correct formula for the volume of a frustum of a square pyramid,

$$V = h(a^2 + ab + b^2)/3,$$

where h is the altitude and a and b are the lengths of the sides of the two square bases. No other unquestionably genuine example of this formula has been found in pre-Hellenic mathematics, and since its proof demands some form of integral calculus, its discovery must certainly be regarded as an extraordinary piece of induction. E. T. Bell has aptly referred to this early Egyptian achievement as the "greatest Egyptian pyramid."

Very likely, mathematical accomplishments similar to those of ancient Egypt and Babylonia also occurred in ancient India and China, but we know very little indeed with any degree of certainty about those accomplishments. The ancient Egyptians recorded their work on stone and papyrus, the latter fortunately resisting the ages because of Egypt's unusually dry climate, and the Babylonians used imperishable baked clay tablets. In contrast to the use of these media, the early Indians and Chinese used very perishable writing materials like bark bast and bamboo. Thus it has come to pass that we have a fair quantity of definite information, obtained from primary sources, about the mathematics of ancient Egypt and Babylonia, while we know very little about the study in ancient India and China.

It is interesting to note that in all pre-Hellenic mathematics we do not find a single instance of what we today call a logical demonstration. In place of a general argument there is merely a step-by-step description of some process applied to particular numerical cases. Beyond some very simple considerations, the mathematical relations employed by the early Egyptians and Babylonians resulted essentially from "trial and error" methods, with the result that many of their formulas are incorrect. In other words, pre-Hellenic mathematics was little more than a practically workable empiricism —a collection of rule-of-thumb procedures that gave results of sufficient acceptability for the simple needs of those early civilizations. Mathematics, and geometry in particular, appears as a laboratory study.

Empirical reasoning may be described as the formulation of conclusions based upon experience and observation; no real understanding is involved, and the logical element does not appear. Empirical reasoning often entails stodgy fiddling with special cases, observation of coincidences and the frequent employment of analogy, experience at good guessing, considerable experimentation, and flashes of intuition.

In spite of the empirical nature of pre-Hellenic mathematics, with its complete neglect of proof and the seemingly little attention paid to the difference between exact and approximate truth, one is nevertheless struck by the extent and diversity of the problems successfully attacked. Apparently a great deal of elementary mathematical truth can be discovered by empirical methods when supplemented by extensive experimentation carried on patiently over a long period of time.

PROBLEMS

1. Show that the ancient Babylonian formula

$$K = (a + c)(b + d)/4,$$

for the area of a quadrilateral having a, b, c, d for consecutive sides, gives too large an answer for all nonrectangular quadrilaterals.

2. Interpret the following, found on a Babylonian tablet believed to date from about 2600 B.C.:

"60 is the circumference, 2 is the perpendicular, find the chord."

"Thou, double 2 and get 4, dost thou not see? Take 4 from 20, thou gettest 16. Square 20, thou gettest 400. Square 16, thou gettest 256. Take 256 from 400, thou gettest 144. Find the square root of 144. 12, the square root, is the chord. Such is the procedure."

3. Given a and b as the legs of a right triangle, Babylonian geometers approximated the hypotenuse c by the formula $c = a + (b^2/2a)$. Justify this approximation by using the Pythagorean relation and the binomial theorem.

4. In the Rhind papyrus the area of a circle is taken as equal to that of a square on $\frac{8}{9}$ of the circle's diameter. Show that this is equivalent to taking $\pi = (\frac{4}{3})^4 = 3.1604 \ldots$.

5. Solve the following two problems found in the Moscow papyrus:
(a) The area of a rectangle is 12, and the width is $\frac{3}{4}$ of the length, what are the dimensions?
(b) One leg of a right triangle is $2\frac{1}{2}$ times the other, the area is 20, what are the dimensions?

6. (a) In the Moscow papyrus we find the following numerical example: "If you are told: A truncated pyramid of 6 for the vertical height by 4 on the base by 2 on the top. You are to square this 4, result 16. You are to double 4, result 8. You are to square 2, result 4. You are to add the 16, the 8, and the 4, result 28. You are to take one third of 6, result 2. You are to take 28 twice, result 56. See, it is 56. You will find it right."

Show that this illustrates the general formula

$$V = h(a^2 + ab + b^2)/3,$$

giving the volume of a frustum of a square pyramid in terms of the height h and the sides a and b of the bases. (This problem constitutes the so-called "greatest Egyptian pyramid.")

(b) Assuming the familiar formula for the volume of any pyramid (volume equals one third the product of base and altitude), show that the volume of a frustum of a pyramid is given by the product of the height of the frustum and the heronian mean of the bases of the frustum. (If m and n are two positive numbers, then $H = (m + \sqrt{mn} + n)/3$ is called the *heronian mean* of the two numbers.)

7. The *Śulvasūtras*, ancient Hindu religious writings dating from about 500 B.C., are of interest in the history of mathematics because they embody certain geometrical rules for the construction of altars that show an acquaintance with the Pythagorean Theorem. Among the rules furnished there appear empirical solutions of the circle-squaring problem which are equivalent to taking $d = (2 + \sqrt{2})s/3$ and $s = 13d/15$, where d is the diameter of the circle and s is the side of the equivalent square. These formulas are equivalent to taking what values for π?

8. The Hindu mathematician Āryabhata wrote early in the sixth century A.D. His work is a poem of 33 couplets called the *Ganita*. Following are translations of two of the couplets:

(1) The area of a triangle is the product of the altitude and half the base; half of the product of this area and the height is the volume of the solid of six edges.

(2) Half the circumference multiplied by half the diameter gives the area of the circle; this area multiplied by its own square root gives the volume of the sphere.

Show that, in each of these couplets, Āryabhata is correct in two dimensions but wrong in three. We note that Hindu mathematics remained empirical long after the Greeks had introduced the deductive feature.

9. (a) An early Chinese work, dating probably from the second century B.C., which had considerable influence on the development of mathematics in China was the *K'ui-ch'ang Suan-Shu*, or *Arithmetic in Nine Sections*. In this work we find the empirical formula $s(c + s)/2$ for the area of a circular segment of chord c and depth s. Obtain a correct formula in terms of these quantities.

(b) Solve the following problem found in the Chinese *Arithmetic in Nine Sections*:

There grows in the middle of a circular pond 10 feet in diameter a reed which projects one foot out of the water. When it is drawn down it just reaches the edge of the pond. How deep is the water?

10. (a) There are reports that ancient Egyptian surveyors laid out right angles by constructing 3–4–5 triangles with a rope divided into 12 equal parts by 11 knots. Show how this can be done.

(b) Since there is no documentary evidence to the effect that the Egyptians were aware of even a particular case of the Pythagorean Theorem, the following purely academic problem arises: Show, without using the Pythagorean Theorem, its converse, or any of its consequences, that the 3–4–5 triangle is a right triangle. Solve this problem by means of Figure 1.2a, which appears in the *Chóu-pei,* the oldest known Chinese mathematical work, which may date back to the second millennium B.C.

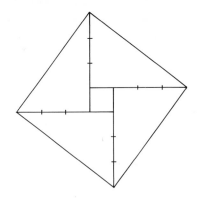

Figure 1.2a

1.3 THE GREEK CONTRIBUTION OF MATERIAL AXIOMATICS

The economic and political changes of the last centuries of the second millennium B.C. caused the power of Egypt and Babylonia to wane, new peoples came to the fore, and it happened that the further development of geometry passed over to the Greeks. The extent of the debt of Greek geometry to ancient oriental geometry is difficult to estimate, nor is the path of transmission from the one to the other yet satisfactorily uncovered. That the debt is considerably greater than formerly believed has become evident with the twentieth-century researches on Babylonian and Egyptian records. Early Greek writers themselves expressed respect for the wisdom of the East and this wisdom was available to anyone who could travel to Egypt and Babylonia.

But, whatever the strength of the historical connection between Greek and ancient oriental geometry, the Greeks transformed the subject into something vastly different from the set of empirical conclusions worked out by their predecessors. The Greeks insisted that geometric fact must be established, not by empirical procedures, but by deductive reasoning; geometrical conclusions must be arrived at by logical demonstration rather than by trial-and-error experimentation. Geometrical truth is to be attained in the study room rather than in the laboratory. In short, the Greeks transformed the empirical, or scientific, geometry of the ancient Egyptians and Babylonians into what we might call *deductive*, *demonstrative*, or *systematic geometry*.

It is disappointing that, unlike the study of ancient Egyptian and Babylonian geometry, there exist virtually no primary sources for the study of early Greek geometry. We are forced to rely upon manuscripts and accounts that are dated several hundred years after the original treatments had been written. In spite of this situation, however, scholars of classicism have been able to build up a rather consistent, though somewhat hypothetical, account of the history of early Greek geometry. This work required amazing ingenuity and patience; it was carried through by painstaking comparisons of derived texts and by examination of countless literary fragments and scattered remarks made by later authors, philosophers, and commentators.

Our principal source of information concerning very early Greek geometry is the so-called *Eudemian Summary* of Proclus. This summary constitutes a few pages of Proclus' *Commentary on Euclid, Book I*, and is a very brief outline of the development of Greek geometry from the earliest times to Euclid. Although Proclus lived in the fifth century A.D., a good thousand years after the inception of Greek geometry, he still had access to a number of historical and critical works which are now lost to us except for the fragments and allusions preserved by him and others. Among these lost works is what was apparently a full history of Greek geometry, covering the period prior to 335 B.C., written by Eudemus, a pupil of Aristotle. The *Eudemian Summary* is so named because it is admittedly based upon this earlier work.

According to the *Eudemian Summary*, Greek geometry appears to have started in an essential way with the work of Thales of Miletus in the first half of the sixth century B.C. This versatile genius, declared to be one of the " seven wise men " of antiquity, was a worthy founder of systematic geometry, and is the first known individual with whom the use of deductive methods in geometry is associated. Thales, the summary tells us, sojourned for a time in Egypt and brought back geometry with him to Greece, where he began to apply to the subject the deductive procedures of Greek philosophy. He is credited with a number of very elementary geometrical results, the value of which is not to be measured by their content but rather by the belief that he supported them with a certain amount of logical reasoning instead of intuition and experiment. For the first time a student of geometry was committed to a form of deductive reasoning, partial and incomplete though it may have been. Moreover, the fact that the first deductive thinking was done in the field of geometry, instead of algebra for instance, inaugurated a tradition in mathematics which was maintained until very recent times.

The next outstanding Greek mathematician mentioned in the *Eudemian Summary* is Pythagoras, who is claimed to have continued the systemization of geometry that was begun some fifty years earlier by Thales. Pythagoras was born about 572 B.C., on the island of Samos, one of the Aegean islands near Thales' home city of Miletus, and it is quite possible that he studied under the older man. It seems that Pythagoras then visited Egypt and perhaps traveled even more extensively about the ancient orient. When, on returning home, he found Ionia under Persian dominion, he decided to migrate to the Greek seaport of Crotona in southern Italy. Here he founded the celebrated Pythagorean school, a brotherhood knit together with secret and cabalistic rites and observances, and committed to the study of philosophy, mathematics, and natural science.

In spite of the mystical nature of much of Pythagorean study, the members of the society contributed, during the two hundred or so years following the founding of their organization, a good deal of sound mathematics. Thus, in geometry, they developed the properties of parallel lines and used them to prove that the sum of the angles of any triangle is equal to two right angles. They contributed in a noteworthy manner to Greek geometrical algebra, and they developed a fairly complete theory of proportion, though it was limited only to commensurable magnitudes, and used it to deduce properties of similar figures. They were aware of the existence of at least three of the regular polyhedral solids, and they discovered the incommensurability of a side and a diagonal of a square. Although much of this information was already known to the Babylonians of earlier times, the deductive aspect of mathematics is thought to have been considerably exploited and advanced in this work of the Pythagoreans. Chains of propositions in which successive propositions were derived from earlier ones in the chain began to emerge. As the chains lengthened, and some were tied to others, the bold idea of developing all of geometry in one long chain suggested itself. It is claimed in the *Eudemian Summary* that a Pythagorean, Hippocrates of Chios, was the first to attempt, with at least partial success, a logical presentation of geometry in the form of a single chain of propositions based upon a few initial definitions and assumptions. Better attempts were made by Leon, Theudius, and others. And then, about 300 B.C., Euclid produced his epoch-making effort, the *Elements*, a single deductive chain of 465 propositions neatly and beautifully comprising plane and solid geometry, number theory, and Greek geometrical algebra. From its very first appearance this work was accorded the highest respect, and it so quickly and so completely superseded all previous efforts of the same nature that now no trace remains of the earlier efforts. The effect of this single work on the future development of geometry has been immense and is difficult to overstate, as will be amply attested in subsequent chapters of our book. In the next section we shall consider in some detail the contents of this magnificent work; in the remainder of this section we comment on its remarkable form.

At some time between Thales in 600 B.C. and Euclid in 300 B.C. was developed the notion of a logical discourse as a sequence of statements obtained by deductive reasoning from a set of initial statements assumed at the outset of the discourse. Certainly, if one is going to present an argument by deductive procedure, any statement of the argument will have to be derived from some previous statement or statements of the argument, and such a previous statement must itself be derived from some still more previous statement or statements. Clearly this cannot be continued backward indefinitely, nor should one resort to illogical circularity by deriving a statement B from a statement A, and then later deriving statement A from statement B. The way out of this difficulty, the Greeks felt, is to set down, at the start of the argument, a collection of primary statements whose truths are acceptable to the reader, and then to proceed, by purely deductive reasoning, to derive all the other statements of the discourse.

Now both the primary and derived statements of the discourse are statements about the technical matter of the discourse, and hence involve special or technical terms. These terms need to be defined. Since technical terms must be defined by means of other technical terms, and these other technical terms by means of still others, one is faced with a difficulty similar to that encountered with the statements of the discourse. In order to get started, and to avoid circularity of definition where term y is defined by means of term x, and then later term x by means of term y, one is again forced to set down at the very start of the discourse a collection of basic technical terms whose intended meanings should be made clear to the reader. All subsequent technical terms of the discourse must then be defined by means of technical terms already introduced.

An argument which is carried out according to the above plan is today said to be developed by *material axiomatics*. Certainly the most outstanding contribution of the early Greeks to mathematics was the formulation of the pattern of material axiomatics and the insistence that mathematics be systematized according to this pattern. Euclid's *Elements* is the earliest extensively developed example of the use of the pattern that has come down to us. In more recent times, as we shall see in a later chapter, the pattern of material axiomatics has been very significantly generalized to yield a more abstract form of argument known as *formal axiomatics*. For the time being we content ourselves by summarizing the pattern of material axiomatics.

Pattern of Material Axiomatics

(A) Initial explanations of certain basic technical terms of the discourse are given, the intention being to suggest to the reader what is to be meant by these basic terms.

(B) Certain primary statements concerning the basic terms, and which are felt to be acceptable as true on the basis of the properties suggested by the initial explanations, are listed. These primary statements are called the *axioms*, or the *postulates*, of the discourse.

(C) All other technical terms of the discourse are defined by means of previously introduced terms.

(D) All other statements of the discourse are logically deduced from previously accepted or established statements. These derived statements are called the *theorems* of the discourse.

PROBLEMS

1. There are two versions of how Thales, when in Egypt, evoked admiration by calculating the height of a pyramid by shadows. The earlier account, given by Hieronymus, a pupil of Aristotle, says that Thales determined the height of the pyramid by measuring the shadow it cast at the moment a man's shadow was equal to his height. The later version, given by Plutarch, says that he set

up a stick and then made use of similar triangles. Both versions fail to mention the difficulty, in either case, of obtaining the length of the shadow of the pyramid —that is, the distance from the apex of the shadow to the center of the base of the pyramid.

Devise a method, based on similar triangles and independent of latitude and time of year, for determining the height of a pyramid *from two shadow observations.*

2. (a) The early Pythagoreans are credited with the origination of the so-called *figurate numbers.* These numbers, considered as the number of dots in certain geometrical configurations, represent a link between geometry and arithmetic. Figures 1.3a, 1.3b, and 1.3c account for the geometrical nomenclature of

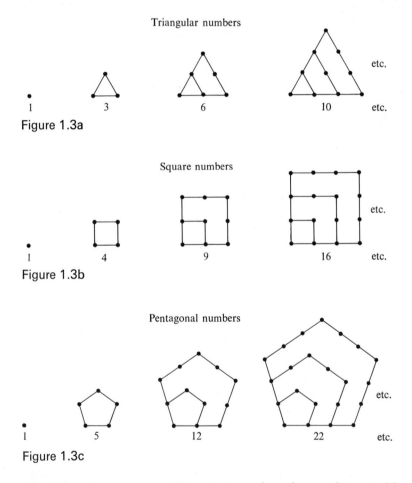

Triangular numbers

1   3   6   10   etc.   etc.

Figure 1.3a

Square numbers

1   4   9   16   etc.   etc.

Figure 1.3b

Pentagonal numbers

1   5   12   22   etc.   etc.

Figure 1.3c

triangular numbers, square numbers, pentagonal numbers, and so on. Many interesting theorems concerning figurate numbers can be established in purely geometric fashion. Show that Figures 1.3d, 1.3e, 1.3f, respectively, prove the theorems:

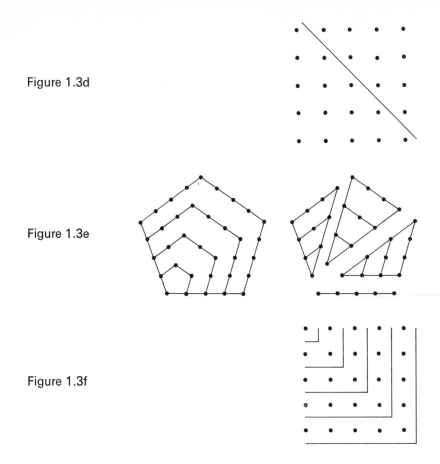

Figure 1.3d

Figure 1.3e

Figure 1.3f

I. *Any square number is the sum of two successive triangular numbers.* II. *The nth pentagonal number is equal to* n *plus three times the* (n–1)*th triangular number.* III. *The sum of any number of consecutive odd integers, starting with 1, is a perfect square.*

(b) Prove, geometrically, that *eight times any triangular number, plus 1, is a square number.*

(c) Find algebraic expressions for the nth triangular, square, and pentagonal numbers, and use these expressions to obtain algebraic proofs of the theorems in parts (a) and (b).

(d) An *oblong number* is the number of dots in a rectangular array of dots having one more column than rows. Show both geometrically and algebraically that: I. *The sum of the first* n *positive even integers is an oblong number.* II. *Any oblong number is twice a triangular number.*

3. (a) Tradition is unanimous in ascribing to Pythagoras the independent discovery of the theorem on the right triangle which now universally bears his name— that the square on the hypotenuse of a right triangle is equal to the sum of the squares on the legs. We have noted that this theorem was known to the Babylonians of Hammurabi's time, over a thousand years earlier, but the first

general proof of the theorem may have been given by Pythagoras. There has been much conjecture as to the proof Pythagoras might have offered, and it is generally felt that it probably was a dissection type of proof such as is suggested by Figure 1.3g. Supply the proof.

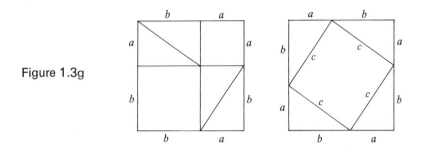

Figure 1.3g

(Since Pythagoras' times, many different proofs of the Pythagorean Theorem have been supplied. In the second edition of his book *The Pythagorean Proposition*, E. S. Loomis has collected and classified 370 demonstrations of this famous theorem.)

(b) State and prove the converse of the Pythagorean Theorem.

4. Show that there can be no more than five regular polyhedra.

5. (a) Prove that $\sqrt{2}$ is not a rational number.
(b) Show that a side and a diagonal of a square are incommensurable (that is, have no common unit of measure).
(c) Prove that the straight line through the points $(0,0)$ and $(1,\sqrt{2})$ of a rectangular Cartesian coordinate system passes through no point, other than $(0,0)$, of the coordinate lattice.
(d) Show how the coordinate lattice may be used to find rational approximations of $\sqrt{2}$.

6. Assuming the equality of alternate interior angles formed by a transversal cutting a pair of parallel lines, prove the following:
(a) The sum of the angles of a triangle is equal to a straight angle.
(b) The sum of the interior angles of a convex polygon of n sides is equal to $n - 2$ straight angles.

7. Assuming the area of a rectangle is given by the product of its two dimensions, establish the following chain of theorems:
(a) The area of a parallelogram is equal to the product of its base and altitude.
(b) The area of a triangle is equal to half the product of any side and the altitude on that side.
(c) The area of a right triangle is equal to half the product of its two legs.
(d) The area of a triangle is equal to half the product of its perimeter and the radius of its inscribed circle.
(e) The area of a trapezoid is equal to the product of its altitude and half the sum of its bases.

(f) The area of a regular polygon is equal to half the product of its perimeter and its apothem.

(g) The area of a circle is equal to half the product of its circumference and its radius.

8. Assuming (1) A central angle of a circle is measured by its intercepted arc, (2) The sum of the angles of a triangle is equal to a straight angle, (3) The base angles of an isosceles triangle are equal, (4) A tangent to a circle is perpendicular to the radius drawn to the point of contact, establish the following chain of theorems:

(a) An exterior angle of a triangle is equal to the sum of the two remote interior angles.

(b) An inscribed angle in a circle is measured by one half its intercepted arc.

(c) An angle inscribed in a semicircle is a right angle.

(d) An angle formed by two intersecting chords in a circle is measured by one half the sum of the two intercepted arcs.

(e) An angle formed by two intersecting secants of a circle is measured by one half the difference of the two intercepted arcs.

(f) An angle formed by a tangent to a circle and a chord through the point of contact is measured by one half the intercepted arc.

(g) An angle formed by a tangent and an intersecting secant of a circle is measured by one half the difference of the two intercepted arcs.

(h) An angle formed by two intersecting tangents to a circle is measured by one half the difference of the two intercepted arcs.

9. As a simple example of a discourse conducted by material axiomatics, consider a certain (finite and nonempty) collection S of people and certain clubs formed among these people, a club being a (nonempty) set of people organized for some common purpose. Our basic terms are thus the *collection S of people* and the *clubs* to which these people belong. About these people and their clubs we assume:

POSTULATE 1. *Every person of S is a member of at least one club.*

POSTULATE 2. *For every pair of people of S there is one and only one club to which both belong.*

DEFINITION. Two clubs having no members in common are called *conjugate clubs.*

POSTULATE 3. *For every club there is one and only one conjugate club.*

From these postulates deduce the following theorems:

THEOREM 1. *Every person of S is a member of at least two clubs.*

THEOREM 2. *Every club contains at least two members.*

THEOREM 3. *S contains at least four people.*

THEOREM 4. *There exist at least six clubs.*

The persevering student may care to try to establish the following much more difficult theorem.

THEOREM 5. *No club contains more than two members.*

10. Using the same basic terms as in Problem 9, let us assume:

POSTULATE 1. *Any two distinct clubs have one and only one member in common.*

POSTULATE 2. *Every person in* S *belongs to two and only two clubs.*
POSTULATE 3. *There are exactly four clubs.*

From these postulates deduce the following theorems:

THEOREM 1. *There are exactly six people in* S.
THEOREM 2. *There are exactly three people in each club.*
THEOREM 3. *For each person in* S *there is exactly one other person in* S *not in the same club.*

1.4 EUCLID'S *ELEMENTS*

Whoever even casually pages through a copy of Euclid's *Elements*, is bound to realize that, notwithstanding certain imperfections, he is examining one of the foremost works ever compiled. This treatise by Euclid is rightfully regarded as the first great landmark in the history of mathematical thought and organization. No work, except the Bible, has been more widely used, edited, or studied. For more than two millennia it has dominated all teaching of geometry, and over a thousand editions of it have appeared since the first one was printed in 1482. As the prototype of the modern mathematical method, its impact on the development of mathematics has been enormous, and a surprising number of important subsequent developments in geometry owe their origin and inspiration to some part or feature of this great work.

It is no detraction that Euclid's work is largely a compilation of works of predecessors, for its chief merit lies precisely in the consummate skill with which the propositions were selected and arranged in a logical sequence presumably following from a small handful of initial assumptions. Nor is it a detraction that the searchlight of modern criticism has revealed certain defects in the structure of the work; it would be very remarkable indeed if such an early and colossal attempt by the axiomatic method should be free of blemishes.

No copy of Euclid's *Elements* actually dating from the author's time has been found. The modern editions of the *Elements* are based upon a revision prepared by Theon of Alexandria almost 700 years after the original work had been written. It was not until the beginning of the nineteenth century that an older copy, showing only minor differences from Theon's recension, was discovered in the Vatican library. A careful study of citations and commentary by early writers indicates that the initial definitions, axioms, and postulates of the original treatise differed some from the revisions, but that the propositions and their proofs have largely remained as Euclid wrote them.

In the thirteen books that comprise Euclid's *Elements* there is a total of 465 propositions. Contrary to popular impression, many of these propositions are concerned, not with geometry, but with number theory and with Greek (geometrical) algebra.

Book I commences, of course, with the necessary preliminary definitions, postulates, and axioms. Though today mathematicians use the words "axiom" and "postulate" synonymously, most of the early Greeks made a distinction, the distinction adopted by Euclid perhaps being that an axiom is an initial assumption common to all studies, whereas a postulate is an initial assumption pertaining to the study at hand. The 48 propositions of Book I fall into three groups. The first 26 deal mainly with properties of triangles and include the three well-known congruence theorems. Propositions I 27* through I 32 establish the theory of parallels and prove that the sum of the angles of a triangle is equal to two right angles. The remaining propositions of the book deal with parallelograms, triangles, and squares, with special reference to area relations. Proposition I 47 is the Pythagorean Theorem, and the final proposition, I 48, is the converse of the Pythagorean Theorem. The material of this book was developed by the early Pythagoreans.

Book II deals with the transformation of areas and the Greek geometrical algebra of the Pythagorean school. It is in this book that we find the geometrical equivalents of a number of algebraic identities. At the end of the book are two propositions which establish the generalization of the Pythagorean Theorem that we today refer to as the "law of cosines."

Book III contains those familiar theorems about circles, chords, tangents, and the measurement of associated angles which we find in our high school geometry texts.

In Book IV are found discussions of the Pythagorean constructions, with straightedge and compasses, of regular polygons of three, four, five, six, and fifteen sides.

Book V gives a masterly exposition of the theory of proportion as originated by Eudoxus. It was this theory, which is applicable to incommensurable as well as commensurable magnitudes, that resolved a "logical scandal" created by the Pythagorean discovery of irrational numbers. Prior to the discovery of irrational numbers it was intuitively felt that any two line segments are commensurable, and the Pythagorean treatment of proportion was built on this false premise. The Eudoxian theory of proportion later provided a foundation, developed by Richard Dedekind in the late nineteenth century, for the real number system of analysis. Present-day high school geometry tests do not employ the Eudoxian theory, but rather the earlier Pythagorean theory completed by some elementary limit theory.

Book VI applies the Eudoxian theory of proportion to plane geometry. Here we find the fundamental theorems on similar triangles and constructions giving third, fourth, and mean proportionals. We also find a geometrical solution of quadratic equations, and the proposition that the internal bisector of an angle of a triangle divides the opposite side into segments proportional to the other two sides. There probably is no theorem in this book that was not known to the early Pythagoreans, but the pre-Eudoxian proofs of many of them were at fault since they were based upon an incomplete theory of proportion.

* By I 27 is meant Proposition 27 of Book I.

Books VII, VIII, and IX, containing a total of 102 propositions, deal with elementary number theory. In these books are many beautiful theorems about the natural numbers, but, since we are here concerned only with geometry, we forgo any discussion of them.

Book X deals with irrationals, that is, with line segments which are incommensurable with respect to some given line segment. Many scholars regard this book as perhaps the most remarkable in the *Elements*. Much of the subject matter of this book is believed due to Theaetetus, but the extraordinary completeness, elaborate classification, and finish are usually credited to Euclid. It taxes one's credulity to realize that the results of this book were arrived at by rhetorical reasoning unassisted by any convenient algebraic notation.

The remaining three books, XI, XII, and XIII, concern themselves with solid geometry, covering most of the material, with the exception of much of that on spheres, commonly found in high school texts today. The definitions, the theorems about lines and planes in space, and theorems concerning parallelepipeds are found in Book XI. Volumes are cleverly treated in Book XII, and constructions of the five regular polyhedra are given in Book XIII.

The traditional American high school texts in plane and solid geometry contain material on rectilinear figures, circles, proportion and similar figures, regular polygons, lines and planes in space, volumes of solids, and the sphere. Except for most of the work on spheres, this is largely the material of Euclid's Books I, III, IV, VI, XI, and XII. The material in current high school texts concerning the measurement of the circle and the sphere, and the material dealing with spherical triangles, is of later origin and is not found in the *Elements*.

The reader will find listed in Appendix I Euclid's initial definitions, postulates, and axioms, and the statements of the propositions of his Book I. This material will be referred to later.

PROBLEMS

1. By the "elements" of a deductive study the Greeks meant the leading, or key, theorems which are of wide and general use in the subject. Their function has been compared to that of the letters of the alphabet in relation to language; as a matter of fact, letters are called by the same name in Greek. The selection of the theorems to be taken as the elements of the subject requires the exercise of some judgement. If you were to choose two of the following theorems for "elements" of a course in plane geometry, which would you choose?
   (1) The three altitudes of a triangle, produced if necessary, meet in a point.
   (2) The sum of the three angles of a triangle is equal to two right angles.
   (3) An angle inscribed in a circle is measured by half its intercepted arc.
   (4) The tangents drawn from any point on the common chord produced of two given intersecting circles are equal in length.

2. A geometry teacher is going to present the topic of parallelograms to her class. After defining *parallelogram,* what theorems about parallelograms should the teacher offer as the "elements" of the subject?

3. Preparatory to teaching the topic on similar figures, a geometry teacher gives a lesson or two on the theory of proportion. What theorems should she select for the "elements" of the treatment, and in what order should she arrange them?

4. Consider the following four statements, called, respectively, the *direct* statement, the *converse* statement, the *inverse* statement, and the *contrapositive* statement: (1) All *a* are *b*. (2) All *b* are *a*. (3) All non-*a* are non-*b*. (4) All non-*b* are non-*a*.
   (a) Show that the direct and contrapositive statements are equivalent.
   (b) Show that the converse and inverse statements are equivalent.
   (c) Taking "All parallelograms are quadrilaterals," as the direct statement, give the converse, inverse, and contrapositive statements.

5. (a) Prove the theorem: *If a triangle is isosceles, then the bisectors of its base angles are equal.*
   (b) Try to prove the converse of the theorem in part (a). (This converse is known as the *Steiner-Lehmus Theorem* and is not so easy to establish.)

6. (a) How does the modern definition of a circle differ from Euclid's definition?
   (b) How many of the forty-eight propositions of Euclid's Book I are constructions?
   (c) A construction can be regarded as an existence theorem. Explain this.

7. Show that the "law of cosines" is a generalization of the Pythagorean Theorem.

8. Indicate how each of the following algebraic identities might be established geometrically, assuming *a, b, c, d* are positive quantities.
   (a) $(a + b)^2 = a^2 + 2ab + b^2$
   (b) $(a - b)^2 = a^2 - 2ab + b^2$, $a > b$
   (c) $a^2 - b^2 = (a + b)(a - b)$, $a > b$
   (d) $a(b + c) = ab + ac$
   (e) $(a + b)^2 = (a - b)^2 + 4ab$, $a > b$
   (f) $(a + b)(c + d) = ac + bc + ad + bd$

9. (a) Let *r* and *s* denote the roots of the quadratic equation

$$x^2 - px + q^2 = 0,$$

   where *p* and *q* are positive numbers. Show that $r + s = p$, $rs = q^2$, and *r* and *s* are both positive if $q \leq p/2$.
   (b) To solve the quadratic equation of part (a) geometrically for real roots, we must find line segments *r* and *s* from given line segments *p* and *q*. That is, we must construct a rectangle equivalent to a given square and having the sum of its base and altitude equal to a given line segment. Devise a suitable construction based on Figure 1.4a, and show geometrically that for real roots to exist we must have $q \leq p/2$.
   (c) Let *r* and *s* denote the roots of the quadratic equation

$$x^2 - px - q^2 = 0,$$

   where *p* and *q* are positive numbers. Show that $r + s = p$, $rs = -q^2$, the roots are real, and the numerically larger one is positive while the other is negative.

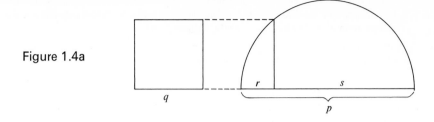

Figure 1.4a

(d) To solve the quadratic equation of part (c) geometrically, we must find line segments r and s from given line segments p and q. That is, we must construct a rectangle equivalent to a given square and having the difference of its base and altitude equal to a given line segment. Devise a suitable construction based on Figure 1.4b.

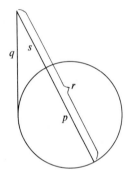

Figure 1.4b

(e) Devise constructions for geometrically solving for real roots the quadratic equations $x^2 + px + q^2 = 0$ and $x^2 + px - q^2 = 0$, where p and q are positive numbers.

(f) Given a unit segment, geometrically solve the quadratic equation

$$x^2 - 7x + 10 = 0.$$

(g) Given a unit segment, geometrically solve the quadratic equation

$$x^2 - 4x - 21 = 0.$$

(h) With straightedge and compass, divide a segment m into two parts such that the difference of their squares shall be equal to their product.

(i) Show that, in part (h), the longer segment is a mean proportional between the shorter segment and the whole segment. (The line segment is said to be divided in *extreme and mean ratio,* or in *golden section.*)

10. Important in Greek geometrical algebra is the Pythagorean theory of transforming an area from one rectilinear shape into another rectilinear shape. The Pythagorean solution of the basic problem of constructing a square equal in area to that of a given polygon may be found in Propositions 42, 44, 45 of Book I and Proposition 14 of Book II of Euclid's *Elements.* A simpler solution, probably also known to the Pythagoreans, is the following. Consider any

polygon $ABCD \cdots$ with a projecting vertex B (see Figure 1.4c). Draw BR parallel to AC to cut DC in R. Then, since triangles ABC and ARC have a common base AC and equal altitudes on this common base, these triangles have equal areas. It follows that polygons $ABCD \cdots$ and $ARD \cdots$ have equal areas. But the derived polygon has one fewer sides than the given polygon. By a repetition of this process we finally obtain a triangle having the same area as the given polygon. Now if b is any side of this triangle and h the altitude on b, the side of an equivalent square is given by $\sqrt{(bh)/2}$, that is, by the mean proportional between b and $h/2$. Since this mean proportional is easily constructed with straightedge and compass, the entire problem can be carried out with these tools.

Figure 1.4c

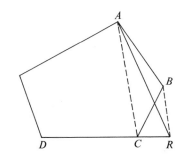

Many interesting area problems can be solved by the above simple process of drawing parallel lines.

(a) Draw an irregular hexagon and then construct, with straightedge and compass, a square having the same area.

(b) With straightedge and compass divide a quadrilateral $ABCD$ into three equivalent parts by straight lines drawn through vertex A.

(c) Bisect a trapezoid by a line drawn from a point P in the smaller base.

(d) Transform triangle ABC so that the angle A is not altered, but the side opposite angle A becomes parallel to a given line MN.

(e) Transform a given triangle into an isosceles triangle having a given vertex angle.

1.5 THE GEOMETRICAL CONTRIBUTIONS OF EUCLID AND ARCHIMEDES

Euclid, Archimedes, and Apollonius mark the apogee of Greek geometry, and it is no exaggeration to say that almost every significant subsequent geometrical development, right up to and including the present time, finds its origin in some work of these three great scholars. Since a genuine understanding of fundamental ideas is scarcely possible without some analysis of origins, it is pertinent to our study to consider, at least briefly, the geometrical accomplishments of these three men.

Of Euclid and his *Elements* we have already written at some length. Very little is known about the life of Euclid except that he was perhaps the first professor of mathematics at the famed University of Alexandria, and the father of the illustrious and long-lived Alexandrian School of Mathematics. Even his dates and his birthplace are not known, but it seems probable that he received his mathematical training in the Platonic school at Athens.

Though Euclid's *Elements* is by far his most influential work, indeed the most influential single work in geometry in the entire history of the subject, he did write several other geometrical treatises, some of which have survived to the present day. One of the latter, entitled the *Data*, is concerned with material of the first six books of the *Elements*. A *datum* may be defined as a set of parts of a figure such that if all but one are given, then that remaining one is determined. Thus the parts A, a, R of a triangle, where A is one angle, a the opposite side, and R the circumradius, constitute a datum, for, given any two of these parts, the third is thereby determined. This is clear either geometrically or from the relation $a = 2R \sin A$. It is apparent that a collection of data of this sort could be useful in the analysis which precedes the discovery of a construction or a proof, and this is undoubtedly the purpose of the work.

Another work in geometry by Euclid, which has come down to us through an Arabian translation, is the book *On Divisions*. Here we find construction problems requiring the division of a figure by a restricted straight line so that the parts will have areas in a prescribed ratio. An example is the problem of dividing a given triangle into two equal areas by a line drawn through a given point in the plane of the triangle.

Other geometrical works of Euclid that are now lost to us, and are known only from later commentaries, are the *Pseudaria*, or book of geometrical fallacies, *Porisms*, a relatively deep work about which there has been considerable speculation, *Conics*, a treatise in four books which was later completed and then added to by Apollonius, and *Surface Loci*, perhaps a treatise on surfaces of double curvature but about which nothing really certain is known. These works tend to show that Euclid delved considerably deeper into geometry than just the material of the *Elements*.

Euclid's other works concern applied mathematics, and two of these are extant: the *Phaenomena*, dealing with the spherical geometry required for observational astronomy, and the *Optics*, an elementary treatise on perspective. Euclid is supposed also to have written a work on the *Elements of Music*.

One of the very greatest mathematicians of all time, and certainly the greatest of antiquity, was Archimedes, a native of the Greek city of Syracuse on the island of Sicily. He was born about 287 B.C. and died during the Roman pillage of Syracuse in 212 B.C. There is a report that he spent time in Egypt, in all likelihood at the University of Alexandria, for he numbered among his friends Conon, Dositheus, and Eratosthenes; the first two were successors of Euclid, the last was a librarian, at the University. Many of Archimedes' mathematical discoveries were communicated in letters to these men.

The works of Archimedes are not compilations of achievements of predecessors, but are highly original creations. They are masterpieces of mathematical exposition and resemble to a remarkable extent, because of their high finish, economy of presentation, and rigor in demonstration, the articles found in present-day research journals. Some ten treatises have come down to us and there are various traces of lost works. Probably the most important contribution made to mathematics in these works is Archimedes' anticipation of some of the methods of the integral calculus.

Three of Archimedes' extant works are devoted to plane geometry. They are *Measurement of a Circle*, *Quadrature of the Parabola*, and *On Spirals*. It was in the first of these that Archimedes inaugurated the classical method of computing π. To simplify matters, suppose we choose a circle with unit diameter. Then the length of the circumference of the circle is π. Now the length of the circumference of a circle lies between the perimeter of any inscribed polygon and that of any circumscribed polygon. Since it is a simple matter to compute the perimeters of the regular inscribed and circumscribed six-sided polygons, we easily obtain bounds for π. Now there are formulas which tell us how, from the perimeters of given regular inscribed and circumscribed polygons, we may obtain the perimeters of the regular inscribed and circumscribed polygons having twice the number of sides. By successive applications of this process, starting with the regular inscribed and circumscribed six-sided polygons, we can compute the perimeters of the regular inscribed and circumscribed polygons of 12, 24, 48, and 96 sides, in this way obtaining ever closer bounds for π. This is essentially what Archimedes did, finally obtaining the fact that π lies between $\frac{223}{71}$ and $\frac{22}{7}$, or that, to two decimal places, π is given by 3.14. This procedure of Archimedes was the start in the long history of securing ever more accurate approximations for the number π, reaching in 1967 the fantastic accuracy of 500,000 decimal places.

In the *Quadrature of the Parabola*, which contains 24 propositions, it is shown that the area of a parabolic segment is $\frac{4}{3}$ that of the inscribed triangle having the same base and having its opposite vertex at the point where the tangent is parallel to the base. The summation of a convergent geometric series is involved. The work *On Spirals* contains 28 propositions devoted to properties of the curve which is now known as the spiral of Archimedes and which has $r = k\theta$ for a polar equation. In particular, the area enclosed by the curve and two radii vectors is found essentially as would be done today as a calculus exercise.

There are allusions to lost works on plane geometry by Archimedes, and there is reason to believe that some of the theorems of these works have been preserved in the *Liber assumptorum*, or *Book of Lemmas*, a collection which has reached us through the Arabic. One Arabian writer claims that Archimedes was the discoverer of the celebrated formula

$$K = \sqrt{s(s-a)(s-b)(s-c)}$$

for the area of a triangle in terms of its three sides. This formula is found in a later work of Heron of Alexandria.

Two of Archimedes' extant works are devoted to geometry of three dimensions, namely *On the Sphere and Cylinder* and *On Conoids and Spheroids*. In the first of these, written in two books and containing a total of 60 propositions, appear theorems giving the areas of a sphere and of a zone of one base and volumes of a sphere and of a segment of one base. In Book II appears the problem of dividing a sphere by a plane into two segments whose volumes shall be in a given ratio. This problem leads to a cubic equation whose solution is not given in the text as it has come down to us, but was found by Eutocius in an Archimedean fragment. There is a discussion concerning the conditions under which the cubic may have a real and positive root. Similar considerations do not appear again in mathematics for over a thousand years. The treatise *On Conoids and Spheroids* contains 40 propositions, which are concerned chiefly with an investigation of the volumes of quadrics of revolution. In this work we find a derivation of the formula $A = \pi ab$ for the area of an ellipse having semiaxes a and b. Pappus has ascribed to Archimedes 13 semiregular polyhedra, but unfortunately Archimedes' own account of them is lost.*

There is a geometrical assumption explicitly stated by Archimedes in his work *On the Sphere and Cylinder* which deserves special mention; it is one of the five geometrical postulates assumed, in addition to those of Euclid, at the start of Book I of the work and it has become known as the *Postulate of Archimedes*. A simple statement of the postulate is as follows: *Given two unequal line segments, there is always some finite multiple of the shorter one which is longer than the other*. In some modern treatments of geometry this postulate serves as part of the postulational basis for introducing the concept of continuity. It is a matter of interest, to be looked into later, that in the nineteenth and twentieth centuries geometric systems were constructed which deny the Archimedean postulate, thus giving rise to so-called non-Archimedean geometries. Although named after Archimedes, it is only fair to point out that this postulate had been considered earlier by Eudoxus.

There are two extant treatises by Archimedes on applied mathematics, *On the Equilibrium of Planes* and *On Floating Bodies*. It is interesting that in these works on mechanics, Archimedes employed the axiomatic method. The physical postulates that must be assumed in addition to the axioms and postulates of geometry are first laid down, and the properties then carefully deduced. It was not until the sixteenth-century work of Simon Stevin that the science of statics and the theory of hydrostatics were appreciably advanced beyond the points reached by Archimedes.

Archimedes wrote two related essays on arithmetic, but these, being foreign to geometry, will not be considered here.

One of the most thrilling discoveries of modern times in the history of

* Construction patterns for the Archimedean solids, and for many other polyhedral solids, can be found in Miles C. Hartley, *Patterns of Polyhedra*, rev. ed. (Ann Arbor, Mich.: Edwards Brothers, 1957).

mathematics was the discovery by Heiberg, in Constantinople, as late as 1906, of Archimedes' long lost treatise entitled *Method*. This work is in the form of a letter addressed to Eratosthenes and is important because of the information it furnishes concerning a "method" which Archimedes used to discover many of his theorems. Although the "method" can today be made rigorous by modern integration processes of the calculus, Archimedes used the "method" only to discover results, which he then established rigorously by his extension of the Eudoxian method of exhaustion.

PROBLEMS

1. Let A, B, C denote the angles of a triangle; a, b, c the opposite sides; h_a, h_b, h_c the altitudes on these sides; m_a, m_b, m_c the medians to these sides; t_a, t_b, t_c the angle bisectors drawn to these sides; R and r the circumradius and inradius; b_a and c_a the projections of b and c on side a; and r_a the radius of the circle touching side a and sides b and c produced. Show that each of the following constitutes a datum:

   (a) A, B, C

   (b) a/b, b/c, c/a

   (c) b, A, h_c

   (d) $b + c$, A, $h_b + h_c$

   (e) $b - c$, A, $h_c - h_b$

   (f) h_a, t_a, $B - C$

   (g) h_a, m_a, $b_a - c_a$

   (h) R, $B - C$, $b_a - c_a$

   (i) R, $r_a - r$, a

   (j) h_a, r, r_a

2. Construct a triangle given (for notation see Problem 1 above):

   (a) a, A, $h_b + h_c$

   (b) $a - b$, $h_b + h_c$, A

   (c) R, r, h_a

3. (a) Complete the details of the following solution (essentially found in Euclid's work *On Divisions*) of the problem of constructing a straight line GH passing through a given point D within triangle ABC, cutting sides BA and BC in G and H respectively, and such that triangles GBH and ABC have the same area (see Figure 1.5a).

   Draw DE parallel to CB to cut AB in E. Denote the lengths of DE and EB by h and k respectively, and that of GB by x. Then $x(BH) = ac$. But BH/h

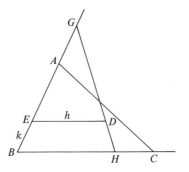

Figure 1.5a

$= x/(x - k)$. Eliminating BH we obtain $x^2 - mx + mk = 0$, where $m = ac/h$, etc., by Problem 9(b) Section 1.4.

(b) Solve the following problem, which is Proposition 28 in Euclid's work *On Divisions*: In Figure 1.5b, bisect the area $ABEC$ by a straight line drawn through the midpoint E of the circular arc BC.

Figure 1.5b

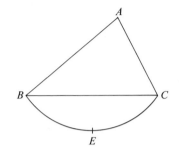

4. (a) If a is the side of a regular polygon inscribed in a circle of radius r, show that

$$b = [2r^2 - r(4r^2 - a^2)^{1/2}]^{1/2}$$

is the side of a regular inscribed polygon having twice the number of sides.

(b) If a is the side of a regular polygon circumscribed about a circle of radius r, find a formula for the side b of a regular circumscribed polygon having twice the number of sides.

5. Let p_k and a_k denote the perimeter and the area of a regular k-gon inscribed in a given circle C, and let P_k and A_k denote the perimeter and the area of a regular k-gon circumscribed about the circle C. Show that

(a) $P_{2n} = 2p_n P_n/(p_n + P_n)$ (b) $p_{2n} = (p_n P_{2n})^{1/2}$
(c) $a_{2n} = (a_n A_n)^{1/2}$ (d) $A_{2n} = 2a_{2n} A_n/(a_{2n} + A_n)$

6. Try to develop, by the synthetic methods of high school geometry, the following chain of theorems concerning the parabola.

DEFINITIONS. A *parabola* is the locus of a point that moves in a plane so that its distance from a fixed point in the plane, called the *focus* of the parabola, is always equal to its distance from a fixed line in the plane, called the *directrix* of the parabola. The midpoint of the perpendicular from the focus to the directrix is called the *vertex* of the parabola; the line through the focus and perpendicular to the directrix is called the *axis* of the parabola; the length of the chord of the parabola through the focus and perpendicular to the axis is called the *latus rectum* of the parabola. If P is any point on the parabola, F the focus, V the vertex, and M the foot of the perpendicular from P on the axis, then the lengths FP, VM, MP are respectively called the *focal radius*, the *abscissa*, and the *ordinate* of the point P.

THEOREM 1. *The vertex of a parabola lies on the parabola.*

THEOREM 2. *The latus rectum is equal to* 4(VF).

THEOREM 3. *The ordinate at any point on the parabola is the mean proportional between the latus rectum and the abscissa of the point.*

THEOREM 4. *Every point on the concave side of the parabola is nearer to the focus than to the directrix; every point on the convex side of the parabola is farther from the focus than from the directrix.*

THEOREM 5. *If P is a point on the parabola, then the bisector of the angle between* PF *and the perpendicular from* P *to the directrix is the tangent to the parabola at the point* P.

THEOREM 6. *If the tangent to the parabola at point P cuts the axis in point* T *and the normal to the parabola at* P *cuts the axis in point* N, *then* TF = FP, TV = VM, *and* FN = *half the latus rectum.*

THEOREM 7. *The foot of the perpendicular from* F *on the tangent to the parabola at a point* P *on the curve lies on the tangent to the parabola at* V.

THEOREM 8. *The line joining the focus to the intersection of two tangents makes equal angles with the focal radii drawn to the points of contact.*

THEOREM 9. *The tangents drawn through the ends of a focal chord intersect on the directrix.*

THEOREM 10. *If two tangents are drawn from a point to a parabola, the line through the point and parallel to the axis bisects the chord of contact.*

THEOREM 11. *If two tangents are drawn from a point to a parabola, and through the point the line parallel to the axis is drawn to cut the parabola in* S, *then the tangent at* S *is parallel to the chord of contact.*

THEOREM 12. *The locus of the midpoints of a family of parallel chords is a straight line parallel to the axis of the parabola.*

THEOREM 13. *The area of a parabolic segment made by a chord is two-thirds the area of the triangle formed by the chord and the tangents drawn through the ends of the chord.*

7. The *Liber assumptorum*, or *Book of Lemmas*, contains some elegant geometrical theorems credited to Archimedes. Among them are some properties of the "arbelos" or "shoemaker's knife." Let A, C, B be three points on a straight line, C lying between A and B. Semicircles are drawn on the same side of the line and having AC, CB, AB as diameters. The "arbelos" is the figure bounded by these three semicircles. At C erect a perpendicular to AB to cut the largest semicircle in G. Let the common external tangent to the two smaller semicircles touch these curves at T and W. Denote AC, CB, AB by $2r_1$, $2r_2$, $2r$. Establish the following elementary properties of the arbelos:
(a) GC and TW are equal and bisect each other.
(b) The area of the arbelos equals the area of the circle on GC as diameter.
(c) The lines GA and GB pass through T and W.
The arbelos has many properties not so easily established. For example it is alleged that Archimedes showed that the circles inscribed in the curvilinear triangles ACG and BCG are equal, the diameter of each being $r_1 r_2 / r$. The smallest circle that is tangent to and circumscribes these two circles is equal to the circle on GC, and therefore equal in area to the arbelos. Consider, in the arbelos, a chain of circles $c_1, c_2, \ldots$, all tangent to the semicircles on AB and AC, where c_1 is also tangent to the semicircle on BC, c_2 to c_1, and so on. Then, if r_n

represents the radius of c_n and h_n the distance of its center from ACB, we have $h_n = 2nr_n$. This last proposition is found in Book IV of Pappus' *Collection* and is there referred to as an "ancient proposition." Later, in Section 3.8 we shall give a singularly elegant demonstration of this proposition.

8. Cicero has related that when serving as Roman quaestor in Sicily he found and repaired Archimedes' neglected tomb, upon which was engraved a sphere inscribed in a cylinder. This device commemorates Archimedes' favorite work, *On the Sphere and Cylinder.* Verify the following two results established by Archimedes in this work:

(a) The volume of the sphere is $\frac{2}{3}$ that of the circumscribed cylinder.

(b) The area of the sphere is $\frac{2}{3}$ of the total area of the circumscribed cylinder.

(c) Define *spherical zone* (of one and two bases), *spherical segment* (of one and two bases), and *spherical sector.*

(d) Assuming the theorem: *The area of a spherical zone is equal to the product of the circumference of a great circle and the altitude of the zone,* obtain the familiar formula for the area of a sphere and establish the theorem: *The area of a spherical zone of one base is equal to that of a circle whose radius is the chord of the generating arc.*

Assuming that the volume of a spherical sector is given by one-third the product of the area of its base and the radius of the sphere, obtain the following results:

(e) The volume of a spherical segment of one base, cut from a sphere of radius R, having h as altitude and a as the radius of its base, is given by

$$V = \pi h^2(R - h/3) = \pi h(3a^2 + h^2)/6.$$

(f) The volume of a spherical segment of two bases, having h as altitude and a and b as the radii of its bases, is given by

$$V = \pi h(3a^2 + 3b^2 + h^2)/6.$$

(g) The spherical segment of part (f) is equivalent to the sum of a sphere of radius $h/2$ and two cylinders whose altitudes are each $h/2$ and whose radii are a and b, respectively.

1.6   APOLLONIUS AND LATER GREEK GEOMETERS

The third mathematical giant of Greek antiquity was Apollonius, who was born about 262 B.C. in Perga in southern Asia Minor. As a young man he went to Alexandria, studied under the successors of Euclid, and then spent most of the remainder of his life at the University. He died sometime around 200 B.C.

Although Apollonius was an astronomer of note and although he wrote on a variety of mathematical subjects, his chief bid to fame rests on his extraordinary and monumental *Conic Sections*, a work which earned him the title, among his contemporaries, of the "Great Geometer." Apollonius' *Conic Sections*, in eight books and containing about 400 propositions, is a thorough investigation of these curves, and completely superseded all earlier works on the subject. Only the first seven of the eight books have come down to us, the first four in Greek and the following three from a ninth-century

Arabic translation. The first four books, of which I, II, and III are pre-sumably founded on Euclid's earlier effort, deal with the general elementary theory of conics, while the later books are devoted to more specialized investigations.

Prior to Apollonius, the Greeks derived the conic sections from three types of cones of revolution, according as the vertex angle of the cone was less than, equal to, or greater than a right angle. By cutting each of three such cones with a plane perpendicular to an element of the cone an ellipse, parabola, and hyperbola respectively result (only one branch of the hyperbola appearing). Apollonius, on the other hand, in Book I of his treatise, obtains all the conic sections in the now familiar way from *one* arbitrary right or oblique circular *double* cone.

The names *ellipse*, *parabola*, and *hyperbola* were supplied by Apollonius, and were borrowed from the early Pythagorean terminology of application of areas. When the Pythagoreans applied a rectangle to a line segment (that is, placed the base of the rectangle along the line segment, with one end of the base coinciding with one end of the segment) they said they had a case of "ellipsis," "parabole," or "hyperbole" according as the base of the applied rectangle fell short of the line segment, exactly coincided with it, or exceeded it. Now let AB (see Figure 1.6a) be the principal axis of a conic,

Figure 1.6a

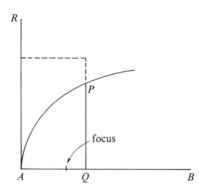

P any point on the conic, and Q the foot of the perpendicular to AB. At A, which is a vertex of the conic, draw a perpendicular to AB and mark off on it a distance AR equal to what we now call the *latus rectum*, or *parameter* p, of the conic. Apply, to the segment AR, a rectangle having AQ for one side and an area equal to $(PQ)^2$. According as the application falls short of, coincides with, or exceeds the segment AR, Apollonius calls the conic an *ellipse*, a *parabola*, or a *hyperbola*. In other words, if we consider the curve referred to a Cartesian coordinate system having its x and y axes along AB and AR respectively and if we designate the coordinates of P by x and y, then the curve is an ellipse, parabola, or hyperbola according as $y^2 \lesseqgtr px$. Actually, in the cases of the ellipse and hyperbola,

$$y^2 = px \mp px^2/d,$$

where d is the length of the diameter through vertex A. Apollonius derives the bulk of the geometry of the conic sections from the geometrical equivalents of these Cartesian equations. Facts like this cause some to defend the thesis that analytic geometry was really an invention of the Greeks.

Book II of Apollonius' treatise on *Conic Sections* deals with properties of asymptotes and conjugate hyperbolas, and the drawing of tangents. Book III contains an assortment of theorems. Thus there are some area theorems like: *If the tangents at any two points* A *and* B *of a conic intersect in* C *and also intersect the diameters through* B *and* A *in* D *and* E, *then triangles* CDB *and* ACE *are equal in area.* One also finds the harmonic properties of poles and polars (a subject to be considered by us in the later chapter on projective geometry), and theorems concerning the product of the segments of intersecting chords. As an example of the latter there is the theorem (sometimes today referred to as Newton's Theorem): *If two chords* PQ *and* MN, *parallel to two given directions, intersect in* O, *then* (PO)(OQ)/(MO)(ON) *is a constant independent of the position of* O. The well-known focal properties of the central conics occur toward the end of Book III. In the entire extant treatise there is no mention of the focus-directrix property of the conics, nor, for that matter, of the focus of the parabola. This is curious because, according to Pappus, Euclid was aware of these properties. Book IV of the treatise proves the converses of some of those propositions of Book III concerning harmonic properties of poles and polars. There also are some theorems about pairs of intersecting conics. Book V is the most remarkable and original of the extant books. It treats of normals considered as maximum and minimum line segments drawn from a point to the curve. The construction and enumeration of normals from a given point are dealt with. The subject is pushed to the point where one can write down the Cartesian equations of the evolutes (envelopes of normals) of the three conics! Book VI contains theorems and constructions concerning equal and similar conics. Thus it is shown how in a given right cone to find a section equal to a given conic. Book VII contains a number of theorems involving conjugate diameters, such as the one about the constancy of the area of the parallelogram formed by the tangents to a central conic at the extremities of a pair of such diameters.

Conic Sections is a great treatise, but, because of the extent and elaborateness of the exposition and the portentiousness of the statements of many complex propositions, is rather trying to read. Even from the above brief sketch of contents we see that the treatise is far more complete than the usual present-day college course in the subject.

Pappus has given brief indications of the contents of six other works of Apollonius. These are *On Proportional Section* (181 propositions), *On Spatial Section* (124 propositions), *On Determinate Section* (83 propositions), *Tangencies* (124 propositions), *Vergings* (125 propositions), and *Plane Loci* (147

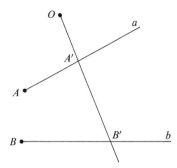

Figure 1.6b

propositions). Only the first of these has survived, and this in Arabic. It deals with the general problem (see Figure 1.6b): Given two lines a and b with the fixed points A on a and B on b, to draw through a given point O a line $OA'B'$ cutting a in A' and b in B' so that $AA'/BB' = k$, a given constant. The exhaustiveness of the treatment is indicated by the fact that Apollonius considers 77 separate cases. The second work dealt with a similar problem, only here we wish to have $(AA')(BB') = k$. The third work concerned itself with the problem: Given four points A, B, C, D on a line, to find a point P on the line such that we have $(AP)(CP)/(BP)(DP) = k$. The work *Tangencies* dealt with the problem of constructing a circle tangent to three given circles, where the given circles are permitted to degenerate into straight lines or points. This problem, now known as the *problem of Apollonius*, has attracted many mathematicians and in the nineteenth century served as a sort of test problem in the competition between synthetic and analytic geometry.* The general problem in *Vergings* was that of inserting a line segment between two given loci such that the line of the segment shall pass through a given point. In the last work, *Plane Loci*, appeared, among many others, the two theorems: (1) *If* A *and* B *are fixed points and* k *a given constant, then the locus of a point* P, *such that* AP/BP = k, *is either a circle (if* k ≠ 1*) or a straight line (if* k = 1*)*, and (2) *If* A, B, ... *are fixed points and* a, b, ..., k *are given constants, then the locus of a point* P, *such that* a(AP)² + b(BP)² + ... = k, *is a circle*. The circle of (1) is known, in modern college geometry texts, as a *circle of Apollonius*. Many attempts have been made to restore, from what little information we know of them, the above lost works of Apollonius.

With the passing of Apollonius the golden age of Greek geometry came to an end, and the lesser geometers who followed did little more than fill in details and perhaps independently develop certain theories the germs of which were already contained in the works of the three great predecessors. In particular, a number of new higher plane curves were discovered and the applications of geometry were exploited. Among these later geometers special mention should be made of Heron, Menelaus, Claudius Ptolemy, and Pappus.

---

* See N. A. Court, "The problem of Apollonius," *The Mathematics Teacher*, vol. 54, no. 6 (Oct. 1961), pp. 444–452.

In geometry Heron largely concerned himself with plane and solid mensuration, and Menelaus and Ptolemy contributed to trigonometry as a handmaiden of astronomy. Pappus, the last of the creative Greek geometers, lived toward the end of the third century A.D., 500 years after Apollonius, and vainly strove with enthusiasm to rekindle fresh life into languishing Greek geometry. His great work, the *Collection*, most of which has come down to us, is a combined commentary and guidebook of the existing geometrical works of his time, sown with numerous original propositions, improvements, extensions, and valuable historical comments. There are many rich geometrical nuggets in the *Collection*, but it proved to be the requiem of Greek geometry, for after Pappus Greek mathematics ceased to be a living study and we find merely its memory perpetuated by minor writers and commentators. Among these were Theon, Proclus, and Eutocius, the first known to us for his edition of Euclid's *Elements*, the second for the *Eudemian Summary* and his *Commentary on Euclid, Book I*, and the third for commentary on Archimedes.

In ancient Greek geometry, both in its form and in its content, we find the fountainhead of the subject. One can scarcely overemphasize the importance to all subsequent geometry of this remarkable bequest of the ancient Greeks.

PROBLEMS

1. In his lost treatise on *Tangencies*, Apollonius considered the problem of drawing a circle tangent to three given circles A, B, C, where each of A, B, C may independently assume either of the degenerate forms of point or straight line. This problem has become known as the *problem of Apollonius*.

   (a) Show that there are ten cases of the problem of Apollonius, depending on whether each of A, B, C is a point, a line, or a circle. What is the number of possible solutions for each case?

   (b) Solve the problem where A, B, C are two points and a line.

   (c) Reduce the problem where A, B, C are two lines and a point to the case of part (b).

   (d) Given the focus and directrix of a parabola p, and a line m. With compass and straightedge find the points of intersection of p and m.

   (e) It is possible to have as many as eight circles tangent to three given circles. Suppose we call a directed circle (that is, a circle with an arrow on it) a *cycle*, and say that two cycles are tangent if and only if their circles are tangent and are directed in the same direction at their point of contact. What is the maximum number of cycles that can be drawn tangent to three given cycles?

2. (a) Solve the following easy verging problem considered by Apollonius in his work *Vergings*: In a given circle to insert a chord of given length and verging to a given point.

   A more difficult verging problem considered by Apollonius is: Given a rhombus with one side produced, to insert a line segment of given length in the exterior angle so that it verges to the opposite vertex.

Let us be given two curves *m* and *n*, and a point *O*. Suppose we permit ourselves to mark, on a given straightedge, a segment *MN*, and then to adjust the straightedge so that it passes through *O* and cuts the curves *m* and *n* with *M* on *m* and *N* on *n*. The line drawn along the straightedge is then said to have been drawn by "the insertion principle." Some problems beyond the Euclidean tools can be solved with these tools if we also permit ourselves to use the insertion principle. Establish the correctness of the following two such constructions.

(b) Let *AB* be a given segment. Draw angle *ABM* = 90° and the angle *ABN* = 120°. Now draw *ACD* cutting *BM* in *C* and *BN* in *D* and such that *CD* = *AB*. Then $(AC)^3 = 2(AB)^3$, and we have a solution, using the insertion principle, of the ancient problem of duplicating a cube. Essentially the above construction was given in publications by Vieta (1646) and Newton (1728).

(c) Let *AOB* be any central angle in a given circle. Through *B* draw a line *BCD* cutting the circle again in *C*, *AO* produced in *D*, and such that *CD* = *OA*, the radius of the circle. Then angle *ADB* = $\frac{1}{3}$ angle *AOB*. This solution of the famous problem of trisecting an angle is implied by a theorem given by Archimedes (*ca.* 240 B.C.).

3. (a) Establish, by analytic geometry, the two theorems (1) and (2) stated in Section 1.6 in connection with Apollonius' work *Plane Loci*.

(b) Establish synthetically the first theorem in part (a) and also the following special case of the second theorem in part (a): *The locus of a point, the sum of the squares of whose distances from two fixed points is constant, is a circle whose center is the midpoint of the segment joining the two points.*

4. In Section 1.6, where the origin of the names *ellipse*, *parabola*, and *hyperbola* is considered, we read: " Actually, in the cases of the ellipse and hyperbola,

$$y^2 = px \mp px^2/d,$$

where *d* is the length of the diameter through vertex *A*." Verify this.

5. Eratosthenes (*ca.* 230 B.C.) made a famous measurement of the earth. He observed at Syene, at noon and at the summer solstice, that a vertical stick had no shadow, while at Alexandria (on the same meridian with Syene) the sun's rays were inclined 1/50 of a complete circle to the vertical. He then calculated the circumference of the earth from the known distance of 5000 stades between Alexandria and Syene. Obtain Eratosthenes' result of 250,000 stades for the circumference of the earth. There is reason to suppose that a stade is about equal to 516.7 feet. Assuming this, calculate from the above result the polar diameter of the earth in miles. (The actual polar diameter of the earth, to the nearest mile, is 7900 miles.)

6. (a) A regular heptagon (seven-sided polygon) cannot be constructed with Euclidean tools. In his work *Metrica*, Heron takes, for an approximate construction, the side of the heptagon equal to the apothem (that is, the radius of the inscribed circle) of a regular hexagon having the same circumcircle. How good an approximation is this?

(b) In *Catoptrica*, Heron proves, on the assumption that light travels by the shortest path, that the angles of incidence and reflection in a mirror are equal. Prove this.

(c) A man wishes to go from his house to the bank of a straight river for a pail of water, which he will then carry to his barn, on the same side of the river as his house. Find the point on the riverbank which will minimize the distance the man must travel.

7. (a) Complete the details of the following indication of Heron's derivation of the formula for the area Δ of a triangle ABC in terms of its sides a, b, c. (1) Let the incircle, with center I and radius r, touch the sides BC, CA, AB in D, E, F, as in Figure 1.6c. On BC produced take G such that $CG = AE$. Draw IH

Figure 1.6c

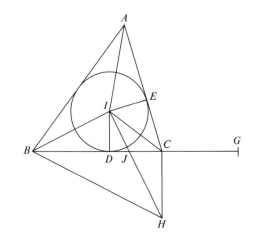

perpendicular to BI to cut BC in J and to meet the perpendicular to BC at C in H. (2) If $s = (a + b + c)/2$, then $\Delta = rs = (BG)(ID)$. (3) B, I, C, H lie on a circle, whence $\angle CHB$ is the supplement of $\angle BIC$ and hence equal to $\angle EIA$. (4) $BC/CG = BC/AE = CH/IE = CJ/JD$. (5) $BG/CG = CD/JD$. (6) $(BG)^2/(CG)(BG) = (CD)(BD)/(JD)(BD) = (CD)(BD)/(ID)^2$. (7) $\Delta = (BG)(ID) = [(BG)(CG)(BD)(CD)]^{1/2} = [s(s - a)(s - b)(s - c)]^{1/2}$.

(b) Derive the formula of part (a) by the following process: Let h be the altitude on side c and let m be the projection of side b on side c. (1) Show that $m = (b^2 + c^2 - a^2)/2c$. (2) Substitute this value for m in $h = (b^2 - m^2)^{1/2}$. (3) Substitute this value for h in $\Delta = (ch)/2$.

8. A *prismatoid* is a polyhedron all of whose vertices lie in two parallel planes. The two faces in these parallel planes are called the *bases* of the prismatoid, the perpendicular distance between the two planes is called the *altitude* of the prismatoid, and the section parallel to the bases and midway between them is called the *midsection* of the prismatoid. Let us denote the volume of the prismatoid by V, the areas of the upper base, lower base, and midsection by U, L, M, and the altitude by h. In books on solid geometry it is shown that

$$V = h(U + L + 4M)/6.$$

(a) In Book II of the *Metrica*, Heron gives, as the volume of a prismatoid having similarly oriented rectangular bases with corresponding pairs of dimensions a, b and a', b',

$$V = h[(a + a')(b + b')/4 + (a - a')(b - b')/12].$$

Show that this result is equivalent to that given by the prismatoid formula above.

(b) Though a sphere is not a prismatoid, show that an application of the prismatoid formula to the sphere yields a correct expression for its volume.

9. Claudius Ptolemy is the author of the famous *Almagest*, the great definitive Greek work on astronomy. The first of the thirteen books of the *Almagest* contains a table giving the lengths of the chords of all central angles, in a circle of radius 60, by half-degree intervals from $\frac{1}{2}°$ to 180°. This is, of course, essentially a table of sines. Accompanying the table is a succinct explanation of its derivation from the fruitful theorem now known as Ptolemy's Theorem: *In a cyclic quadrilateral the product of the diagonals is equal to the sum of the products of the two pairs of opposite sides.*

(a) Designating the length of the chord of a central angle θ, in a circle of radius 60, by crd θ, show that $\sin \theta = $ (crd 2θ)/120.

(b) Prove Ptolemy's Theorem.

Derive, from Ptolemy's Theorem, the following relations:

(c) If a and b are the chords of two arcs of a circle of unit radius, then

$$s = (a/2)(4 - b^2)^{1/2} + (b/2)(4 - a^2)^{1/2}$$

is the chord of the sum of the two arcs.

(d) If a and b, $a \geqslant b$, are chords of two arcs of a circle of unit radius, then

$$d = (a/2)(4 - b^2)^{1/2} - (b/2)(4 - a^2)^{1/2}$$

is the chord of the difference of the two arcs.

(e) If t is the chord of an arc of a circle of unit radius, then

$$s = [2 - (4 - t^2)^{1/2}]^{1/2}$$

is the chord of half the arc.

In a circle of unit radius, crd 60° = 1, and it may be shown that crd 36° = the larger segment of the radius when divided in extreme and mean ratio (that is, when the longer segment is a mean proportional between the shorter segment and the whole radius) = 0.6180. By part (d), crd 24° = crd (60° − 36°) = 0.4158. By part (e) we may calculate the chords of 12°, 6°, 3°, 90', and 45' obtaining crd 90' = 0.0524 and crd 45' = 0.0262. Now it is easily shown that if $0° < b < a < 90°$,

$$\sin a/\sin b < a/b.$$

Therefore crd 60'/crd 45' < 60/45 = $\frac{4}{3}$, or crd 1° < ($\frac{4}{3}$)(0.0262) = 0.0349. Also, crd 90'/crd 60' < 90/60 = $\frac{3}{2}$, or crd 1° > ($\frac{3}{2}$)(0.0542) = 0.0349. Therefore crd 1° = 0.0349. By part (e) we may find crd $\frac{1}{2}°$. Now one can construct a table of chords for $\frac{1}{2}°$ intervals. This is the gist of Ptolemy's method of constructing his table of chords.

(f) From a knowledge of the graphs of the functions $\sin x$ and $\tan x$ show that $(\sin x)/x$ decreases, and $(\tan x)/x$ increases, as x increases from 0° to 90°, and thus establish the inequalities

$$\sin a/\sin b < a/b < \tan a/\tan b,$$

where $0° < b < a < 90°$.

(g) Show that the relations of parts (c), (d), (e) are equivalent to the trigonometrical formulas for sin $(\alpha + \beta)$, sin $(\alpha - \beta)$, and sin $(\theta/2)$.

(h) Establish the following interesting results as consequences of Ptolemy's Theorem: If P lies on the arc AB of the circumcircle of

(1) an equilateral triangle ABC, then $PC = PA + PB$;

(2) a square $ABCD$, then $(PA + PC)PC = (PB + PD)PD$;

(3) a regular pentagon $ABCDE$, then $PC + PE = PA + PB + PD$;

(4) a regular hexagon $ABCDEF$, then $PD + PE = PA + PB + PC + PF$.

10. (a) In Book III of Pappus' *Collection* we find the following interesting geometrical representation of some means. Take B on segment AC, B not being the midpoint O of AC. Erect the perpendicular to AC at B to cut the semicircle on AC in D, and let F be the foot of the perpendicular from B on OD. Show that OD, BD, FD represent the arithmetic mean, the geometric mean, and the harmonic mean of the segments AB and BC, and show that, if $AB \neq BC$,

$$\text{arith. mean} > \text{geom. mean} > \text{harm. mean}.$$

(b) In Book III of the *Collection*, Pappus gives the following neat construction for the harmonic mean of the two given segments OA and OB in Figure 1.6d.

Figure 1.6d

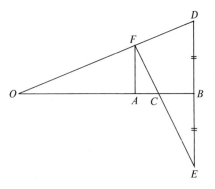

On the perpendicular to OB at B mark off $BD = BE$, and let the perpendicular to OB at A cut OD in F. Draw FE to cut OB in C. Then OC is the sought harmonic mean. Prove this.

(c) Let x be the side of a square inscribed in a triangle such that one side of the square lies along the base a of the triangle. Show that x is half the harmonic mean between a and h, where h is the altitude on the base a.

(d) Let a and b represent the lengths of two vertical poles, and let wires from the tip of each pole to the base of the other intersect at a distance x above ground. Prove that x is half the harmonic mean between a and b.

(e) Show that if a square is inscribed in a right triangle so as to include the right angle of the triangle, then the side of the square is equal to the product of the legs of the right triangle divided by the sum of these legs.

(f) In Figure 1.6e, show that x is half the harmonic mean between a and b. (This is the basis of a nomogram for the lens formula of optics, wherein a, b, x

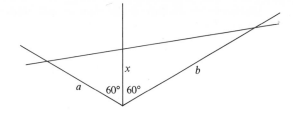

Figure 1.6e

represent the object distance, the image distance, and the focal length for a lens. Given any two of these distances, Figure 1.6e gives us a ready construction of the third.)

(g) Show that the line segment through the intersection of the diagonals of a trapezoid, parallel to the bases of the trapezoid, and intercepted by the sides of the trapezoid is the harmonic mean of the bases of the trapezoid.

11. Prove the following elegant extension of the Pythagorean Theorem given by Pappus in Book IV of his *Collection. Let* ABC *(see Figure 1.6f) be any triangle*

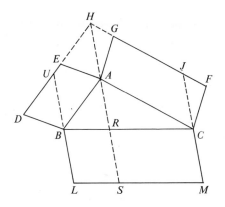

Figure 1.6f

and ABDE, ACFG *any parallelograms described externally on* AB *and* AC. *Let* DE *and* FG *meet in* H *and draw* BL *and* CM *equal and parallel to* HA. *Then*

$$\square \text{ BCML} = \square \text{ ABDE} + \square \text{ ACFG}.$$

12. In Book VII of the *Collection*, Pappus anticipated one of the centroid theorems sometimes credited to P. Guldin (1577–1642). These theorems may be stated as follows: (1) *If a plane arc be revolved about an axis in its plane, but not cutting the arc, the area of the surface of revolution so formed is equal to the product of the length of the arc and the length of the path traced by the centroid of the arc.* (2) *If a plane area be revolved about an axis in its plane, but not intersecting the area, the volume of revolution so formed is equal to the product of the area and the length of the path traced by the centroid of the area.* Using these theorems, find

(a) The volume and surface area of the torus formed by revolving a circle of radius r about an axis, in the plane of this circle, at distance $R > r$ from the center of the circle.

(b) The centroid of a semicircular arc.

(c) The centroid of a semicircular area.

1.7 THE TRANSMISSION OF GREEK GEOMETRY TO THE OCCIDENT

The closing period of ancient times was dominated by Rome. One Greek center after another fell before the power of the Roman armies, and in 146 B.C. Greece became a province of the Roman Empire, though Mesopotamia was not conquered until 65 B.C. and Egypt held out until 30 B.C. Conditions proved more and more stifling to original scientific work, and a gradual decline in creative thinking set in. The arrival of the Barbarians in the west and the eventual collapse of the slave market, with their disastrous effects on Roman economy, found science reduced to a mediocre level. The famous Alexandrian school gradually faded with the breakup of ancient society, becoming completely extinct in 641 A.D., when Alexandria was taken by the Arabs.

The period starting with the fall of the Roman Empire in the middle of the fifth century and extending into the eleventh century is known as Europe's Dark Ages, for during this period civilization in western Europe reached a very low ebb. Schooling became almost nonexistent, Greek learning all but disappeared, and many of the arts and crafts bequeathed by the ancient world were forgotten. Only the monks of the Catholic monasteries and a few cultured laymen preserved a slender thread of Greek and Latin learning. The period was marked by great physical violence and intense religious faith. The old social order gave way, and society became feudal and ecclesiastical.

The Romans had never taken to abstract mathematics but had contented themselves with merely a few practical aspects of the subject that were associated with commerce and civil engineering. With the fall of the Roman Empire and the subsequent closing of much of east-west trade and the abandonment of state engineering projects, even these interests waned, and it is no exaggeration to say that very little in mathematics, beyond the development of the Christian calendar, was accomplished in the West during the whole of the half millennium covered by the Dark Ages.

During this bleak period of learning, the people of the east, especially the Hindus and the Arabs, became the major custodians of mathematics. However, the Greek concept of rigorous thinking—in fact, the very idea of deductive proof—seemed distasteful to the Hindu way of doing things. Although the Hindus excelled in computation, contributed to the devices of algebra, and played an important role in the development of our present positional numeral system, they produced almost nothing of importance in geometry or in basic mathematical methodology. Hindu mathematics of this period is largely empirical and lacks those outstanding Greek characteristics of clarity and logicality.

The spectacular episode of the rise and decline of the Arabian Empire occurred during the period of Europe's Dark Ages. Within a decade following Mohammed's flight from Mecca to Medina in 622 A.D., the scattered and disunited tribes of the Arabian peninsula were consolidated by a strong religious fervor into a powerful nation. Within a century, force of arms had extended the Moslem rule and influence over a territory reaching from India, through Persia, Mesopotamia, and northern Africa, clear into Spain. Of considerable importance for the preservation of much of world culture was the manner in which the Arabs seized upon Greek and Hindu erudition. The Bagdad caliphs not only governed wisely and well but many became patrons of learning and invited distinguished scholars to their courts. Numerous Hindu and Greek works in astronomy, medicine, and mathematics were industriously translated into the Arabic tongue and thus were saved until later European scholars were able to retranslate them into Latin and other languages. But for the work of the Arabian scholars a great part of Greek and Hindu science would have been irretrievably lost over the long period of the Dark Ages.

Not until the latter part of the eleventh century did Greek classics in science and mathematics begin once again to filter into Europe. There followed a period of transmission during which the ancient learning preserved by Moslem culture was passed on to the western Europeans through Latin translations made by Christian scholars travelling to Moslem centers of learning, and through the opening of western European commercial relations with the Levant and the Arabian world. The loss of Toledo by the Moors to the Christians in 1085 was followed by an influx of Christian scholars to that city to acquire Moslem learning. Other Moorish centers in Spain were infiltrated, and the twelfth century became, in the history of mathematics, a century of translators. One of the most industrious translators of the period was Gherardo of Cremona, who translated into Latin more than 90 Arabian works, among which were Ptolemy's *Almagest* and Euclid's *Elements*. At the same time Italian merchants came in close contact with eastern civilization, thereby picking up useful arithmetical and algebraical information. These merchants played an important part in the European dissemination of the Hindu-Arabic numeral system.

The thirteenth century saw the rise of the universities at Paris, Oxford, Cambridge, Padua, and Naples. Universities were to become potent factors in the development of mathematics, since many mathematicians associated themselves with one or more such institutions. During this century Campanus made a Latin translation of Euclid's *Elements* which later, in 1482, became the first printed version of Euclid's great work.

The fourteenth century was a mathematically barren one. It was the century of the Black Deadt, which swept away more than a third of the population of Europe, and during this century the Hundred Years War, with its political and economic upheavals in northern Europe, got well under way.

The fifteenth century witnessed the start of the European Renaissance in art and learning. With the collapse of the Byzantine Empire, culminating in the fall of Constantinople to the Turks in 1453, refugees flowed into Italy, bringing with them treasures of Greek civilization. Many Greek classics, up to that time known only through the often inadequate Arabic translations, could now be studied from original sources. Also, about the middle of the century, occurred the invention of printing, which revolutionized the book trade and enabled knowledge to be disseminated at an unprecedented rate. Mathematical activity in this century was largely centered in the Italian cities and in the central European cities of Nuremberg, Vienna, and Prague, and it concentrated on arithmetic, algebra, and trigonometry, under the practical influence of trade, navigation, astronomy, and surveying.

In the sixteenth century the development of arithmetic and algebra continued, the most spectacular mathematical achievement of the century—and the first really deep mathematical accomplishment beyond the Greeks and Arabs—being the discovery, by Italian mathematicians, of the algebraic solution of cubic and quartic equations. A decided stimulus to the further development of geometry was the translation, in 1533, of Proclus' *Commentary on Euclid, Book I*. The first important translation into Latin of Books I–IV of Apollonius' *Conic Sections* was made by Commandino in 1566; Books V–VII did not appear in Latin translation until 1661. In 1572 Commandino made a very important Latin translation of Euclid's *Elements* from the Greek. This translation served as a basis for many subsequent translations, including a very influential work by Robert Simson, from which, in turn, so many English editions were derived. By this time a number of the works of Archimedes had also been translated into Latin. With so many of the great Greek works in geometry available, it was inevitable that sooner or later some aspects of the subject should once again claim the attention of researchers.

PROBLEMS

1. Hindu arithmetical problems often involved the Pythagorean relation. Solve the following three such problems, the first two of which are adapted from problems given by Brahmagupta (*ca.* 630), and the last from a problem given by Bhāskara (*ca.* 1150).

(a) Two ascetics lived at the top of a cliff of height h, whose base was distant d from a neighboring village. One descended the cliff and walked to the village. The other, being a wizard, flew up a height x and then flew in a straight line to the village. The distance traversed by each was the same. Find x. (In the original problem $h = 100$ and $d = 200$).

(b) A bamboo 18 cubits high was broken by the wind. Its top touched the ground 6 cubits from the root. Tell the length of the segments of the bamboo.

(c) A snake's hole is at the foot of a pillar which is 15 cubits high, and a peacock is perched on its summit. Seeing a snake, at a distance of thrice the pillar's height, gliding toward his hole, he pounces obliquely upon him. Say quickly at how many cubits from the snake's hole do they meet, both proceeding an equal distance?

2. Most remarkable in Hindu geometry, and solitary in its excellence, is a study made by Brahmagupta concerning cyclic quadrilaterals. Establish the following chain of theorems, which embody some of Brahmagupta's findings. It should be known that Brahmagupta knew Ptolemy's Theorem on the cyclic quadrilateral (see Problem 9, Section 1.6.)

(a) The product of two sides of a triangle is equal to the product of the altitude on the third side and the diameter of the circumscribed circle.

(b) Let $ABCD$ be a cyclic quadrilateral of diameter δ. Denote the lengths of the sides AB, BC, CD, DA by a, b, c, d, the diagonals BD and AC by m and n, and the angle between either diagonal and the perpendicular upon the other by θ. Show that

$$m\delta \cos \theta = ab + cd, \qquad n\delta \cos \theta = ad + bc.$$

(c) Show, for the above quadrilateral, that

$$m^2 = (ac + bd)(ab + cd)/(ad + bc),$$

$$n^2 = (ac + bd)(ad + bc)/(ab + cd).$$

(d) If, in the above quadrilateral, the diagonals are perpendicular to each other, then

$$\delta^2 = (ad + bc)(ab + cd)/(ac + bd).$$

3. (a) Brahmagupta gave the formula

$$K^2 = (s - a)(s - b)(s - c)(s - d)$$

for the area K of a cyclic quadrilateral of sides a, b, c, d and semi-perimeter s.* Show that Heron's formula for the area of a triangle is a special case of this formula.

(b) Using Brahmagupta's formula of part (a) show that the area of a quadrilateral possessing both an inscribed and a circumscribed circle is equal to the square root of the product of its four sides.

(c) Show that a quadrilateral has perpendicular diagonals if and only if the sum of the squares of one pair of opposite sides is equal to the sum of the squares of the other pair of opposite sides.

---

* For a derivation of this formula see, for example, E. W. Hobson, *A Treatise on Plane Trigonometry* (New York: Macmillan Company, 4th ed. 1902), p. 204, or R. A. Johnson, *Modern Geometry* (Boston: Houghton Mifflin Company, 1929), p. 81. A formula giving the area of an arbitrary quadrilateral is

$$K^2 = (s - a)(s - b)(s - c)(s - d) - abcd \cos^2 [(A + C)/2],$$

where A and C are a pair of opposite vertex angles of the quadrilateral. There is also the interesting Bretschneider formula,

$$16K^2 = 4m^2n^2 - (a^2 - b^2 + c^2 - d^2)^2,$$

where m and n are the diagonals of the quadrilateral. See Prob. E 1376, *The American Mathematical Monthly*, vol. 67, no. 3 (Mar. 1960), p. 291.

(d) Brahmagupta showed that if $a^2 + b^2 = c^2$ and $A^2 + B^2 = C^2$, then any quadrilateral having aC, cB, bC, cA for consecutive sides has perpendicular diagonals. Prove this.

(e) If (a, b, c), (A, B, C) are two Pythagorean triples (that is, a, b, c, A, B, C are positive integers such that $a^2 + b^2 = c^2$ and $A^2 + B^2 = C^2$), then the cyclic quadrilateral having consecutive sides aC, cB, bC, cA is called a *Brahmagupta trapezium*. Find the sides, diagonals, circumradius, and area of the Brahmagupta trapezium determined by the two Pythagorean triples $(3, 4, 5)$ and $(5, 12, 13)$.

4. Many students of high school geometry have seen Bhāskara's dissection proof of the Pythagorean Theorem, in which the square on the hypotenuse is cut up as indicated in Figure 1.7a, into four triangles, each congruent to the given

Figure 1.7a
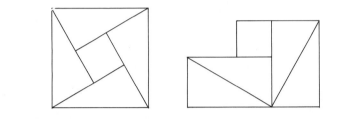

triangle, plus a square with side equal to the difference of the legs of the given triangle. The pieces are easily rearranged to give the sum of the squares on the two legs. Bhāskara drew the figure and offered no further explanation than the word "Behold!" Supply a proof.

5. (a) One of the oldest methods for approximating the real roots of an equation is the rule known as *regula duorum falsorum*, often called the *rule of double false position*. This method seems to have originated in India and was used by the Arabians. In brief the method is this: Let x_1 and x_2 be two numbers lying close to and on opposite sides of a real root x of the equation $f(x) = 0$. Then the intersection with the x axis of the chord joining the points $(x_1, f(x_1))$, $(x_2, f(x_2))$ gives an approximation x_3 to the sought root. Show, by elementary geometry, that

$$x_3 = [x_2 f(x_1) - x_1 f(x_2)]/[f(x_1) - f(x_2)].$$

The process can now be reapplied with the appropriate pair x_1, x_3 or x_3, x_2.

(b) Compute, by double false position, to three decimal places, the root of $x^3 - 36x + 72 = 0$ which lies between 2 and 3.

(c) Compute, by double false position, to three decimal places, the root of $x - \tan x = 0$ which lies between 4.4 and 4.5.

6. As in Problem 5 above, some simple geometry often leads to a procedure for successively approximating the real roots of an equation. Along these lines consider the following:

(a) Show that if x_1 is a sufficiently close approximation to a real root x of the equation $f(x) = 0$, then

$$x_2 = x_1 - f(x_1)/f'(x_1),$$

where $f'(x)$ is the derivative of $f(x)$, is a better approximation. Etc. (This is the so-called Newton-Raphson method of successively approximating the real roots of an equation.)

(b) As a variant of the method of double false position, let a and x_1 be two numbers lying sufficiently close to a real root x of the equation $f(x) = 0$. Then

$$x_2 = [x_1 f(a) - af(x_1)]/[f(a) - f(x_1)]$$

is a better approximation. Etc.

(c) Suppose we wish to solve the equation $F(x) = 0$, and suppose that by transposition we can write the equation in the form $f(x) = g(x)$, where $f(x)$ is such that we can solve $f(x) = c$, where c is a constant, for x. Then, if x_1 is a suitable approximation to a real root x of the equation $F(x) = 0$, show that the solution x_2 of $f(x) = g(x_1)$ is a better approximation. Etc.

(d) Use the method of part (c) to solve the equation $x - \tan x = 0$.

(e) Use the method of part (c) to solve the equation $\cos x - e^x/3 = 0$.

(f) By means of the hyperbola $xy = k$, $k > 0$, show that if x_1 is an approximation to $\sqrt{k}$, then $x_2 = (x_1 + k/x_1)/2$ is a better approximation. Etc. (This procedure was used by Heron of Alexandria.)

(g) Obtain the procedure in part (f) from the Newton-Raphson method applied to $f(x) = x^2 - k$.

(h) By the Newton-Raphson method applied to $f(x) = x^n - k$, n a positive integer, show that if x_1 is an approximation to $\sqrt[n]{k}$, then

$$x_2 = [(n - 1)x_1 + k/x_1^{n-1}]/n$$

is a better approximation. Etc.

The student may wish to look up, in a text on the theory of equations, the so-called *Fourier's Theorem*, which states a guarantee under which the Newton-Raphson method is bound to succeed.

7. (a) Given line segments of lengths a, b, n, construct, with Euclidean tools, a line segment of length $m = a^3/bn$.

(b) Omar Khayyam (1044?–1123?) was the first to handle every type of cubic equation that possesses a positive root. Complete the details of the following sketch of Khayyam's geometrical solution of the cubic

$$x^3 + b^2x + a^3 = cx^2,$$

where a, b, c, x are thought of as lengths of line segments. Khayyam stated this type of cubic rhetorically as "a cube, some sides, and some numbers are equal to some squares."

In Figure 1.7b, construct $AB = a^3/b^2$ (by part (a)) and $BC = c$. Draw a semicircle on AC as diameter and let the perpendicular to AC at B cut it in D. On BD mark off $BE = b$, and through E draw EF parallel to AC. Find G on BC such that $(BG)(ED) = (BE)(AB)$ and complete the rectangle $DBGH$. Through H draw the rectangular hyperbola having EF and ED for asymptotes, and let it cut the semicircle in J. Let the parallel to DE through J cut EF in K and BC in L. Show, successively, that: (1) $(EK)(KJ) = (BG)(ED) = (BE)(AB)$, (2) $(BL)(LJ) = (BE)(AL)$. (3) $(LJ)^2 = (AL)(LC)$, (4) $(BE)^2/(BL)^2 = (LJ)^2/(AL)^2 = LC/AL$, (5) $(BE)^2(AL) = (BL)^2(LC)$, (6) $b^2(BL + a^3/b^2) = (BL)^2(c - BL)$, (7) $(BL)^3 + b^2(BL) + a^3 = c(BL)^2$. Thus BL is a root of the given

Figure 1.7b

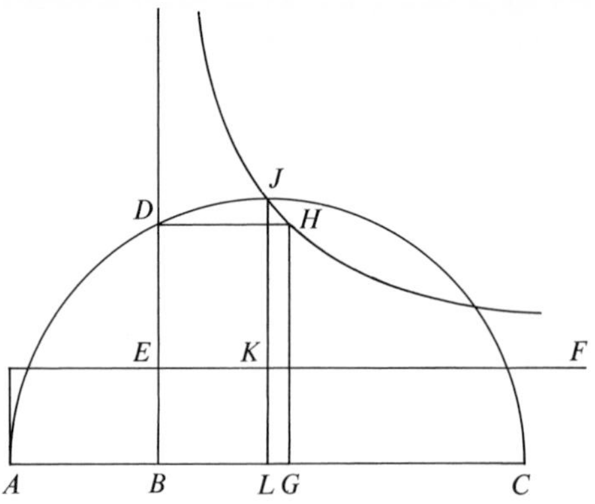

cubic equation. (See J. L. Coolidge, *The Mathematics of Great Amateurs*, New York: Oxford University Press, 1949, Chap. 2, "Omar Khayyam.")

8. The Arabians were interested in constructions on a spherical surface. Consider the following problems, to be solved with Euclidean tools and appropriate plane constructions.
   (a) Given a material sphere, find its diameter.
   (b) On a given material sphere locate the vertices of an inscribed cube.
   (c) On a given material sphere locate the vertices of an inscribed regular tetrahedron.

9. The poverty of geometry in western Europe during the Dark Ages is illustrated by the following two problems considered by the famous French scholar and churchman, Gerbert (*ca.* 950–1003), who became Pope Sylvester II.
   (a) In his *Geometry* Gerbert solved the problem, considered very difficult at the time, of determining the legs of a right triangle whose hypotenuse and area are given. Solve this problem.
   (b) Gerbert expressed the area of an equilateral triangle of side a as $(a/2)$ $(a - a/7)$. Show that this is not correct and is equivalent to taking $\sqrt{3} = 1.714$.

10. A *regular star-polygon* is the figure formed by connecting with straight lines every ath point, starting at some given one, of the n points which divide a circumference into n equal parts, where a and n are relatively prime and $n > 2$. Such a star-polygon is represented by the symbol $\{n/a\}$, and is sometimes called a regular *n-gram*. When $a = 1$ we have a regular polygon. Star-polygons made their appearance in the ancient Pythagorean school, where the $\{\frac{5}{2}\}$ star-polygon, or pentagram, was used as a badge of recognition. Star-polygons also occur in the geometry of Boethius (*ca.* 475–524) and the translations of Euclid from the Arabic by Adelard of Bath (*ca.* 1120) and Johannes Campanus (*ca.* 1260). Thomas Bradwardine (1290–1349), who died as Archbishop of Canterbury, developed some of their properties. They were considered also by Regiomontanus (1436–1476), Charles de Bouelles (1470–1533), and Johann Kepler (1571–1630).
    (a) Construct, with the aid of a protractor, the star-polygons $\{\frac{5}{2}\}$, $\{\frac{7}{2}\}$, $\{\frac{7}{3}\}$, $\{\frac{8}{3}\}$, $\{\frac{9}{2}\}$, $\{\frac{9}{4}\}$, $\{\frac{10}{3}\}$.
    (b) Let $\phi(n)$, called the *Euler ϕ function*, denote the number of numbers less than n and prime to it. Show that there are $[\phi(n)]/2$ regular *n*-grams.

(c) Show that if n is prime there are $(n - 1)/2$ regular n-grams.

(d) Show that the sum of the angles at the "points" of the regular $\{n/a\}$ star-polygon is given by $|n - 2a|180°$. (This result was given by Bradwardine.) For the extension to polyhedra, see H. S. M. Coxeter, *Regular Polytopes*, New York: Pitman Publishing Corporation, 1947, Chap. VI, "Star-Polyhedra."

11. Johann Müller, more generally known as Regiomontanus, from the Latinized form of his birthplace of Königsberg ("king's mountain"), was perhaps the ablest and most influential mathematician of the fifteenth century. His treatise *De triangulis omnimodis*, which was written about 1464 but posthumously published in 1533, was the first systematic European exposition of plane and spherical trigonometry considered independently of astronomy. Solve the following three problems the first two of which are found in this work.

(a) Determine a triangle given the difference of two sides, the altitude on the third side, and the difference of the segments into which the altitude divides the third side.

(b) Determine a triangle given a side, the altitude on this side, and the ratio of the other two sides.

(c) Construct a cyclic quadrilateral given the four sides.

12. Solve the following geometrical problem found in the *Sūma* of Luca Pacioli (*ca*. 1445–*ca*. 1509). The radius of the inscribed circle of a triangle is 4 and the segments into which one side is divided by the point of contact are 6 and 8. Determine the other two sides.

1.8 THE CASE FOR EMPIRICAL, OR EXPERIMENTAL, GEOMETRY

In Section 1.2 the empirical nature of pre-Hellenic geometry was considered, and it was pointed out that conclusions reached in this way may be incorrect and therefore cannot be regarded as established. To be assured that a conclusion incontestably follows from preliminary premises, deductive reasoning is necessary, and no amount of supporting empirical evidence will, of itself, suffice. For this reason the Greeks found in deductive reasoning the vital element of mathematical method, and they insisted that geometry—indeed, all mathematics—be developed via this process.

Since deductive reasoning has the advantage that its conclusions are unquestionable if the premises are accepted, whereas empirical procedure always leaves some room for doubt, one might overhastily outlaw all empiricism and experimentation from geometry. This would be a grave mistake, for though today all *recorded* geometry is wholly deductive and completely bereft of any inductive element, it is probably quite true that few, if any, significant geometrical facts were ever found without some preliminary empirical work of one form or another. Before a geometrical statement can be proved or disproved by deduction, it must first be thought of, or conjectured, and a conjecture is nothing but a guess made more or less plausible by intuition, observation, analogy, experimentation, or some other form of empirical procedure. Deduction is a convincing formal mode of exposition, but it is hardly a means of discovery. It is a set of complicated machinery that needs material to work upon, and the material is usually furnished by empirical considerations. Even the steps of a deductive proof or disproof are not dictated to us by the deductive apparatus itself, but must be arrived at by trial and error, experience, and shrewd guessing. Indeed, skill in the art of good guessing is one of the prime ingredients in the make-up of a worthy geometer.

To succeed in geometry, either as a creator or simply as a problem solver, one must be willing to experiment, to draw and test innumerable figures, to try this and to try that. Galileo (1564–1642), in 1599, attempted to ascertain the area under one arch of the cycloid curve* by balancing a cycloidal template against circular templates of the size of the generating circle. He incorrectly conjectured that the area under an arch is very nearly, but not exactly, three times the area of the circle. The first published mathematical demonstration that the area is exactly three times that of the generating circle was furnished, in 1644, by his pupil, Evangelista Torricelli (1608–1647), with the use of early integration methods.

Blaise Pascal (1623–1662), when a very young boy, "discovered" that the sum of the angles of a triangle is a straight angle by a simple experiment involving the folding of a paper triangle.

By actually constructing a right circular cone, three times filling it with sand and then emptying the contents into a right circular cylinder of the same radius and height, one would conjecture that the volume of a right circular cone is one third the product of its altitude and the area of its circular base.

Suppose one should take a horizontal circular disc and drive an upright nail into its center, and then coil a thick cord on the disc, in spiral fashion about the nail, until the disc is covered. This would require a piece of the cord of a certain length. Now take a material hemisphere of the same radius as the disc and drive a nail into its pole. As before, coil a thick cord, now on the hemisphere, in spiral fashion about the nail, until the hemisphere is covered. It would be found that approximately twice as much cord is needed in the second experiment as was needed in the first one. From this it can be conjectured that the area of a sphere is four times the area of one of its great circles.

Many first-rate conjectures concerning maxima and minima problems in the calculus of variations were first obtained by soap film experiments.

Archimedes, in his treatise on *Method*, has described how he first came to realize, by mechanical considerations, that the volume of a sphere is given by $4\pi r^3/3$, where r is the radius of the sphere. Here, briefly, is his process. Place the sphere with its polar diameter along a horizontal x axis with the north pole N at the origin (see Figure 1.8a). Construct the cylinder and the cone of revolution obtained by rotating the rectangle $NABS$ and the triangle

* A cycloid is the curve traced by a fixed point on the circumference of a circle which rolls, without slipping, along a straight line.

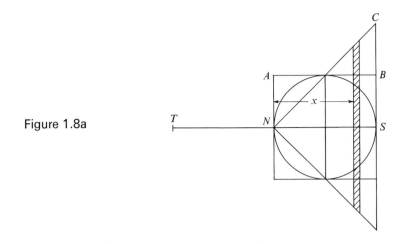

Figure 1.8a

NCS about the x axis. Now cut from the three solids thin vertical slices (assuming that they are flat cylinders) at distance x from N and of thickness Δx. The volumes of these slices are, approximately

$$\text{sphere:}\quad \pi x(2r - x)\Delta x,$$
$$\text{cylinder:}\quad \pi r^2 \Delta x.$$
$$\text{cone:}\quad \pi x^2 \Delta x.$$

Let us hang at T the slices from the sphere and the cone, where $TN = 2r$. Their combined moment* about N is

$$[\pi x(2r - x)\Delta x + \pi x^2 \Delta x]2r = 4\pi r^2 x \Delta x.$$

This, we observe, is four times the moment of the slice cut from the cylinder when that slice is left where it is. Adding a large number of these slices together we find

$$2r[\text{vol. of sphere} + \text{vol. of cone}] = 4r[\text{vol. of cylinder}],$$

or

$$2r[\text{vol. of sphere} + 8\pi r^3/3] = 8\pi r^4,$$

or

$$\text{vol. of sphere} = 4\pi r^3/3.$$

But Archimedes' mathematical conscience would not permit him to accept the above mechanical argument as a proof, and he accordingly supplied a rigorous demonstration employing the so-called Greek method of exhaustion.

One should not deprecate experiments and approaches of the above kind, for there is no doubt that much geometry has been "discovered" by such means. Of course, once a geometrical conjecture has been formulated, we

---

* By the *moment* of a volume about a point is here meant the product of the volume and the perpendicular distance from the point to the vertical line passing through the centroid of the volume.

must, like Archimedes, establish or disestablish it by deductive reasoning, and thus completely settle the matter one way or the other. Many a geometrical conjecture has been discarded by the outcome of just one carefully drawn figure, or by the examination of some extreme case.

A very fruitful way of making geometrical conjectures is by the employment of analogy, though it must be confessed that many conjectures so made are ultimately proved to be incorrect. An astonishing amount of space geometry has been discovered via analogy from similar situations in the plane, and in the geometry of higher dimensional spaces analogy has played a very successful role.

There is much to be said, at the elementary level of instruction, for empirical, or experimental, geometry, and many teachers feel it wise to precede a first course in demonstrative geometry with a few weeks of experimental geometry. The work of these weeks acquaints the student with many geometrical concepts, and can be designed to emphasize both the values and the shortcomings of empirical geometry. Such instructional procedure follows the thesis that, in general, the learning program should parallel the historical development.

PROBLEMS

1. Does the process called "mathematical induction" have any induction in it? If so, where?

2. Show empirically, by a simple experiment involving the folding of a paper triangle, that the sum of the angles of a triangle is a straight angle.

3. If we cut off the top of a triangle by a line parallel to the base of the triangle, a trapezoid will remain, and the area of a trapezoid is given by the product of its altitude and the arithmetic average of its two bases. Now, if we cut off the top of a pyramid by a plane parallel to the base of the pyramid, a frustum will remain. Reasoning by analogy, obtain the incorrect Babylonian formula (see Section 1.2) for the volume of a frustum of a pyramid. (This illustrates that conclusions reached by analogy cannot be regarded as established.)

4. Two ladders, 60 feet long and 40 feet long, lean from opposite sides across an alley lying between two buildings, the feet of the ladders resting against the bases of the buildings. If the ladders cross each other at a distance of 10 feet above the alley, how wide is the alley? Find an approximate solution from drawings. An algebraic treatment of this problem requires the solution of a quartic equation. If a and b represent the lengths of the ladders, c the height at which they cross, and x the width of the alley, one can show that

$$(a^2 - x^2)^{-1/2} + (b^2 - x^2)^{-1/2} = c^{-1}.$$

5. To trisect a central angle AOB of a circle, someone suggests that we trisect the chord AB and then join these points of trisection with O. While this construction may look somewhat reasonable for small angles, show, by taking an angle almost equal to 180°, that the construction is patently false.

6. How good are the following empirical straightedge and compass trisections of an angle of 30°?

(a) Let *AOB* be the given angle, with *OA* = *OB*. On *AB* as diameter draw a semicircle lying on the same side of *AB* as is the point *O*. Take *D* and *E* on the semicircle such that *AD* = *DE* = *EB*. Take *F* on *DE* such that *DF* = *DE*/4. Then *OF* is a sought trisector.

(b) Take the given angle *AOB* as a central angle in a circle. Let *D* be diametrically opposite *B*. Take *C* as the midpoint of *DO* and *M* as the midpoint of arc *AB*. Then angle *MCB* is (approximately) one-third the given angle *AOB*. (This very simple method was devised by M. d'Ocagne in 1934 as an approximate trisection of an arbitrary angle *AOB*; it is surprisingly accurate for small angles.)

7. The three altitudes of a triangle are concurrent. Are the four altitudes of a tetrahedron concurrent?

8. By considering a triangle as a thin homogeneous plate made up of a family of fibers (thin strips) parallel to a certain side of the triangle, devise a mechanical argument showing that the medians of a triangle are concurrent.

9. A line segment has 2 zero-dimensional bounding elements (2 end-points), and its interior is one-dimensional.

A triangle has 3 zero-dimensional and 3 one-dimensional bounding elements (3 vertices and 3 sides), and its interior is two-dimensional.

A tetrahedron has 4 zero-dimensional, 6 one-dimensional, and 4 two-dimensional bounding elements (4 vertices, 6 edges, 4 faces), and its interior is three-dimensional.

From this information, how many zero-, one-, two-, and three-dimensional bounding elements might you expect a pentatope (the four-dimensional analog of a tetrahedron) to have?

10. Does Figure 1.8b suggest a proof of the relation

$$1 + 2 + 3 + \cdots + n = n(n + 1)/2?$$

Figure 1.8b

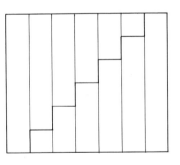

11. List the three-space analogs of the planar concepts *parallelogram, rectangle, circle* if (a) a tetrahedron, (b) a pyramid, is considered as the three-space analog of a triangle.

12. State theorems in three-space that are analogs of the following theorems in the plane.

(a) The bisectors of the angles of a triangle are concurrent at the center of the inscribed circle of the triangle.

(b) The area of a circle is equal to the area of a triangle the base of which has the same length as the circumference of the circle and the altitude of which is equal to the radius of the circle.

(c) The foot of the altitude of an isosceles triangle is the midpoint of the base.

13. Two lines through the vertex of an angle and symmetrical with respect to the bisector of the angle are called a pair of *isogonal conjugate lines* of the angle. There is an attractive theorem about triangles which states that if three lines through the vertices of a triangle are concurrent, then the three isogonal conjugate lines through the vertices of the triangle are also concurrent. Try to construct an analogous definition and theorem for the tetrahedron.

14. Let F, V, E denote the number of faces, vertices, and edges of a polyhedron. For the tetrahedron, cube, triangular prism, pentagonal prism, square pyramid, pentagonal pyramid, cube with one corner cut off, cube with a square pyramid erected on one face, we find that $V - E + F = 2$. Do you feel that this formula holds for *all* polyhedra? Consider the one pictured in Figure 1.8c.

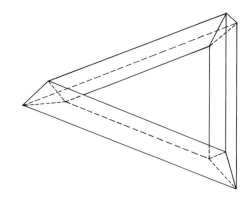

Figure 1.8c

15. There are convex polyhedra all faces of which are triangles (for instance a tetrahedron), all faces of which are quadrilaterals (for instance a cube), all faces of which are pentagons (for instance a regular dodecahedron). Do you think the list can be continued?

16. Consider an ellipse with semiaxes a and b. If $a = b$ the ellipse becomes a circle and the two expressions

$$P = \pi(a + b), \qquad P' = 2\pi(ab)^{1/2}$$

each become $2\pi a$, which gives the perimeter of the circle. This suggests that P or P' may give the perimeter E of any ellipse. Discuss.

17. (a) Consider a convex polyhedron P and let C be any point in its interior. We can imagine a suitable heterogeneous distribution of mass within P such that the center of gravity of P will coincide with C. If the polyhedron is thrown upon a horizontal floor, it will come to rest on one of its faces. Show that this yields a mechanical argument for the geometrical proposition: "Given a

convex polyhedron P and a point C in its interior, then there exists a face F of P such that the foot of the perpendicular from C to the plane of F lies in the interior of F."

(b) Give a geometrical proof of the proposition of part (a).

18. Construct a mechanical device for finding the point within an acute-angled triangle the sum of whose distances from the vertices of the triangle is a minimum.

19. Given n line segments $a_1, a_2, \ldots, a_n$, it is not obvious that there exists a convex cyclic n-gon having these segments for consecutive sides, nor is it apparent how we might compute the radius of the circumcircle. Show the existence of such an n-gon and find its circumradius by a simple paperfolding experiment.

20. How can one convince a freshman class that if the inside of a race track is a noncircular ellipse, and the track is of constant width, then the outside is not an ellipse?

BIBLIOGRAPHY

AABOE, ASGER, *Episodes from the Early History of Mathematics* (New Mathematical Library, No. 13). New York: Random House and The L. W. Singer Company, 1964.

ALLMAN, G. J., *Greek Geometry from Thales to Euclid*. Dublin: University Press, 1889.

APOLLONIUS OF PERGA, *Conics*, 3 vols., tr. by R. Catesby Taliaferro. Classics of the St. John's program. Annapolis, Md.: R. C. Taliaferro, 1939.

BOYER, C. B., *The History of the Calculus and Its Conceptual Development*. New York: Dover Publications, Inc., 1949.

———, *A History of Mathematics*. New York: John Wiley & Sons, Inc., 1968.

CHACE, A. B., L. S. BULL, H. P. MANNING, and R. C. ARCHIBALD, eds., *The Rhind Mathematical Papyrus*, 2 vols. Buffalo, N.Y.: Mathematical Association of America, 1927–1929.

COOLIDGE, J. L., *A History of Geometrical Methods*. New York: Oxford University Press, 1940.

———, *A History of the Conic Sections and Quadric Surfaces*. New York: Oxford University Press, 1945.

DANTZIG, TOBIAS, *The Bequest of the Greeks*. New York: Charles Scribner's Sons, 1955.

DATTA, B., *The Science of the Sulba: A Study in Early Hindu Geometry*. Calcutta: University of Calcutta, 1932.

———, and A. N. SINGH, *History of Hindu Mathematics*, 2 vols. Lahore: Motilah Banarshi Das, 1935–1938.

EVES, HOWARD, *An Introduction to the History of Mathematics*. New York: Holt, Rinehart and Winston, 3rd ed., 1969.

GOW, JAMES, *A Short History of Greek Mathematics*. New York: Hafner Publishing Company, 1923.

HEATH, T. L., *A Manual of Greek Mathematics*. New York: Oxford University Press, 1931.

——, *History of Greek Mathematics*, 2 vols. New York: Oxford University Press, 1921.

——, *The Thirteen Books of Euclid's Elements*, 2d ed., 3 vols. New York: Cambridge University Press, 1926. Reprinted by Dover Publications, Inc., 1956.

——, *Apollonius of Perga, Treatise on Conic Sections, edited in Modern Notation with Introductions Including an Essay on the Earlier History of the Subject*. New York: Cambridge University Press, 1896.

——, *The Works of Archimedes, with the Method of Archimedes*. New York: Cambridge University Press, 1897 and 1912. Reprinted by Dover Publications, Inc. (no date).

HUGHES, BARNABAS, *Regiomontanus on Triangles*. Madison, Wis.: The University of Wisconsin Press, 1964.

LOOMIS, E. S., *The Pythagorean Proposition*, 2d ed. Ann Arbor, Mich.: priv. ptd., Edwards Brothers, 1940. Reprinted by National Council of Teachers of Mathematics, Washington, D.C.

MIKAMI, YOSHIO, *The Development of Mathematics in China and Japan*. New York: Hafner Publishing Company, 1913. Reprinted by Chelsea Publishing Co., 1961.

NEUGEBAUER, OTTO, *The Exact Sciences in Antiquity*. Princeton, N.J.: Princeton University Press, 1952.

——, and A. J. SACHS, eds., *Mathematical Cuneiform Texts*, American Oriental Series, vol. 29. New Haven, Conn.: American Oriental Society, 1946.

POLYA, G., *How to Solve It, a New Aspect of Mathematical Method*. Princeton N.J.: Princeton University Press, 1945.

——, *Mathematics and Plausible Reasoning*, 2 vols. Princeton, N.J.: Princeton University Press, 1954.

——, *Mathematical Discovery*, vol. 1, New York: John Wiley and Sons, Inc., 1962.

SMITH, D. E., and YOSHIO MIKAMI, *A History of Japanese Mathematics*. Chicago: The Open Court Publishing Company, 1914.

USPENSKII, V. A., *Some Applications of Mechanics to Mathematics*, tr. by Halina Moss. New York: Blaisdell Publishing Company, 1961. Original Russian volume published in 1958.

VAN DER WAERDEN, B. L., *Science Awakening*, tr. by Arnold Dresden. Groningen, Holland: P. Noordhoff, Ltd., 1954.

2

Modern Elementary
Geometry

The period following the European Renaissance, and running into present times, is known in the history of mathematics as *the modern era*. One of the ways in which mathematicians of the modern era have extended geometry beyond that inherited from the Greeks has been by the discovery of a host of further propositions about circles and rectilinear figures deduced from those listed in Euclid's *Elements*. This material is referred to as *modern elementary geometry*, and it constitutes a sequel to, or expansion of, Euclid's *Elements*. The nineteenth century witnessed an astonishing growth in this area of geometry, and the number of papers in the subject which have since appeared is almost unbelievable. In 1906, Maximilian Simon (in his *Über die Entwickelung der Elementargeometrie im XIX Jahrhundert*) attempted the construction of a catalogue of contributions to elementary geometry made during the nineteenth century; it has been estimated that this catalogue contains upward of 10,000 references! Research in the field has not abated, and it would seem that the geometry of the triangle and its associated points, lines, and circles must be inexhaustible. Much of the material has been extended to the tetrahedron and its associated points, lines, planes, and spheres, resulting in an enormous and beautiful expansion of elementary solid geometry. Many of the special points, lines, circles, planes, and spheres have been named after original or subsequent investigators. Among these names are Gergonne, Nagel, Feuerbach, Hart, Casey, Brocard, Lemoine, Tucker, Neuberg, Simson, McCay, Euler, Gauss, Bodenmiller, Fuhrmann, Schoute, Spieker, Taylor, Droz-Farny, Morley, Miquel, Hagge, Peaucellier, Steiner, Tarry, and many others.

Here is a singularly fascinating, highly intricate, extensive, and challenging field of mathematics that can be studied and pursued with very little mathematical prerequisite and with a fair chance of making an original discovery. The subject, though elementary, is often far from easy. Large portions of the material have been summarized and organized into textbooks bearing the title of modern, or college, geometry. On the grounds that "you cannot lead anyone farther than you have gone yourself," a course following one of these texts is highly desirable for anyone preparing to teach geometry in high school.

In this chapter we shall briefly consider a few selected topics from modern elementary geometry. Though we cannot hope in such a short space to do justice to the subject, the topics we choose will illustrate some of the flavor and attractiveness of this area of geometry, and all of them will be useful to us in later parts of our work. Much of modern elementary geometry has, in one way or another, entered into the main stream of mathematics.

In Chapters 3, 4, and 5 certain additional topics belonging to modern elementary geometry will be encountered.

2.1 SENSED MAGNITUDES

One of the innovations of modern elementary geometry is the employment, when it proves useful, of sensed, or signed, magnitudes. It was the extension of the number system so as to include both positive and negative numbers that led to this forward step in geometry, and though Albert Girard, René Descartes, and others introduced negative segments into geometry during the seventeenth century, the idea of sensed magnitudes was first systematically exploited in the early nineteenth century by L. N. M. Carnot (in his *Géométrie de position* of 1803) and especially by A. F. Möbius (in his *Der barycentrische Calcul* of 1827). By means of the concept of sensed magnitudes, several separate statements or relations can often be combined into a single embracive statement or relation, and a single proof can frequently be formulated for a theorem that would otherwise require the treatment of a number of different cases.

We start a study of sensed magnitudes with some definitions and a notation.

2.1.1 DEFINITIONS AND NOTATION. Sometimes we shall choose one direction along a given straight line as the positive direction, and the other direction as the negative direction. A segment *AB* on the line will then be considered *positive* or *negative* according as the direction from *A* to *B* is the positive or negative direction of the line, and the symbol $\overline{AB}$ (in contrast to *AB*) will be used to denote the resulting signed distance from the point

A to the point *B*. Such a segment $\overline{AB}$ is called a *sensed*, or *directed*, *segment*; point *A* is called the *initial point* of the segment and point *B* is called the *terminal point* of the segment. The fact that $\overline{AB}$ and $\overline{BA}$ are equal in magnitude but opposite in direction is indicated by the equation

$$\overline{AB} = -\overline{BA},$$

or by the equivalent equation

$$\overline{AB} + \overline{BA} = 0.$$

Of course $\overline{AA} = 0$.

2.1.2 DEFINITIONS. Points which lie on the same straight line are said to be *collinear*. A set of collinear points is said to constitute a *range* of points, and the straight line on which they lie is called the *base* of the range. A range which consists of all the points of its base is called a *complete range*.

We are now in a position to establish a few basic theorems about sensed line segments.

2.1.3 THEOREM. *If* A, B, C *are any three collinear points, then*

$$\overline{AB} + \overline{BC} + \overline{CA} = 0.$$

If the points *A*, *B*, *C* are distinct, then *C* must lie between *A* and *B*, or on the prolongation of $\overline{AB}$, or on the prolongation of $\overline{BA}$. We consider these three cases in turn.

If *C* lies between *A* and *B*, then $\overline{AB} = \overline{AC} + \overline{CB}$, or $\overline{AB} - \overline{CB} - \overline{AC} = 0$, or $\overline{AB} + \overline{BC} + \overline{CA} = 0$.

If *C* lies on the prolongation of $\overline{AB}$, then $\overline{AB} + \overline{BC} = \overline{AC}$, or $\overline{AB} + \overline{BC} - \overline{AC} = 0$, or $\overline{AB} + \overline{BC} + \overline{CA} = 0$.

If *C* lies on the prolongation of $\overline{BA}$, then $\overline{CA} + \overline{AB} = \overline{CB}$, or $\overline{AB} - \overline{CB} + \overline{CA} = 0$, or $\overline{AB} + \overline{BC} + \overline{CA} = 0$.

The situations where one or more of the points *A*, *B*, *C* coincide are easily disposed of.

This theorem illustrates one of the economy features of sensed magnitudes. Without the concept of directed line segments, three separate equations would have to be given to describe the possible relations connecting the three unsigned distances *AB*, *BC*, *CA* between pairs of the three distinct collinear points *A*, *B*, *C*.

2.1.4 THEOREM. *Let* O *be any point on the line of segment* AB. *Then* $\overline{AB} = \overline{OB} - \overline{OA}$.

This is an immediate consequence of Theorem 2.1.3, for by that theorem we have $\overline{AB} + \overline{BO} + \overline{OA} = 0$, whence $\overline{AB} = -\overline{BO} - \overline{OA} = \overline{OB} - \overline{OA}$.

2.1.5 EULER'S THEOREM (1747). *If* A, B, C, D *are any four collinear points, then*

$$\overline{AD} \cdot \overline{BC} + \overline{BD} \cdot \overline{CA} + \overline{CD} \cdot \overline{AB} = 0.$$

The theorem follows by noting that, by Theorem 2.1.4, the left member of the above equation may be put in the form

$$\overline{AD}(\overline{DC} - \overline{DB}) + \overline{BD}(\overline{DA} - \overline{DC}) + \overline{CD}(\overline{DB} - \overline{DA}),$$

which, upon expansion and reduction, is found to vanish identically.

The notion of directed segment leads to the following very useful definition of the ratio in which a point on a line divides a segment on that line.

2.1.6 DEFINITIONS. If A, B, P are distinct collinear points, we define *the ratio in which* P *divides the segment* $\overline{AB}$ to be the ratio $\overline{AP}/\overline{PB}$. It is to be noticed that the value of this ratio is independent of any direction assigned to line AB. If P lies between A and B, the division is said to be *internal*; otherwise the division is said to be *external*. Denoting the ratio $\overline{AP}/\overline{PB}$ by r we note that if P lies on the prolongation of $\overline{BA}$, then $-1 < r < 0$; if P lies between A and B, then $0 < r < \infty$; if P lies on the prolongation of $\overline{AB}$, then $-\infty < r < -1$.

If A and B are distinct and P coincides with A, we set $\overline{AP}/\overline{PB} = 0$. If A and B are distinct and P coincides with B, the ratio $\overline{AP}/\overline{PB}$ is undefined and we indicate this by writing $\overline{AP}/\overline{PB} = \infty$.

Workers in modern elementary geometry have devised several ways of assigning a sense to angles lying in a common plane, and each way has its own uses. The way we are about to describe is particularly useful in relations involving trigonometric functions.

2.1.7 DEFINITIONS AND NOTATION. We may consider an $\measuredangle AOB$ as generated by the rotation of side OA about point O until it coincides with side OB, the rotation not exceeding $180°$. If the rotation is counterclockwise the angle is said to be *positive*; if the rotation is clockwise the angle is said to be *negative*, and the symbol $\measuredangle \overline{AOB}$ (in contrast to $\measuredangle AOB$) will be used to denote the resulting signed rotation. Such an $\measuredangle \overline{AOB}$ is called a *sensed*, or *directed, angle*; point O is called the *vertex* of the angle; side OA is called the *initial side* of the angle; side OB is called the *terminal side* of the angle. If $\measuredangle AOB$ is not a straight angle, then the fact that $\measuredangle \overline{AOB}$ and $\measuredangle \overline{BOA}$ are equal in magnitude but opposite in direction is indicated by the equation

$$\measuredangle \overline{AOB} = - \measuredangle \overline{BOA},$$

or by the equivalent equation

$$\measuredangle \overline{AOB} + \measuredangle \overline{BOA} = 0.$$

It is sometimes convenient to assign a sense to the areas of triangles lying in a common plane.

2.1.8 DEFINITIONS AND NOTATION. A triangle ABC will be considered as *positive* or *negative* according as the tracing of the perimeter from A to B to C to A is counterclockwise or clockwise. Such a signed triangular area is called a *sensed*, or *directed*, *area*, and will be denoted by $\triangle \overline{ABC}$ (in contrast to $\triangle ABC$).

Some subsequent developments in this chapter will make use of the following two important theorems.

2.1.9 THEOREM. *If vertex* A *of triangle* ABC *is joined to any point* L *on line* BC, *then*

$$\frac{\overline{BL}}{\overline{LC}} = \frac{AB \sin \overline{BAL}}{AC \sin \overline{LAC}}.$$

Let h denote the length of the perpendicular from A to line BC. The reader may then check that for all possible figures,

$$\frac{\overline{BL}}{\overline{LC}} = \frac{h\overline{BL}}{h\overline{LC}} = \frac{2\triangle \overline{ABL}}{2\triangle \overline{ALC}} = \frac{(AB)(AL) \sin \overline{BAL}}{(AL)(AC) \sin \overline{LAC}} = \frac{AB \sin \overline{BAL}}{AC \sin \overline{LAC}}.$$

2.1.10 THEOREM. *If* a, b, c, d *are four distinct lines passing through a point* V, *then*

$$(\sin \overline{AVC}/\sin \overline{CVB})/(\sin \overline{AVD}/\sin \overline{DVB})$$

is independent of the positions of A, B, C, D *on the lines* a, b, c, d, *respectively, so long as they are all distinct from* V.

The value of the expression certainly will not change if any one of the points is taken at a different position of its line on the same side of V. The reader can easily show that the value of the expression also will not change if any one of the points is taken at a position of its line on the opposite side of V.

We now close the present section with the following convenient definitions.

2.1.11 DEFINITIONS. Straight lines which lie in a plane and pass through a common point are said to be *concurrent*. A set of concurrent coplanar lines is said to constitute a *pencil* of lines, and the point through which they all pass is called the *vertex* of the pencil. A pencil which consists of all the lines through its vertex is called a *complete pencil*. A line in the plane of a pencil and not passing through the vertex of the pencil is called a *transversal* of the pencil.

PROBLEMS

1. If $A_1, A_2, \ldots, A_n$ are n collinear points, show that

$$\overline{A_1 A_2} + \overline{A_2 A_3} + \cdots + \overline{A_{n-1} A_n} + \overline{A_n A_1} = 0.$$

2. If A, B, P are collinear and M is the midpoint of AB, show that $\overline{PM} = (\overline{PA} + \overline{PB})/2$.

3. If O, A, B are collinear, show that $\overline{OA}^2 + \overline{OB}^2 = \overline{AB}^2 + 2(\overline{OA})(\overline{OB})$.

4. If O, A, B, C are collinear and $\overline{OA} + \overline{OB} + \overline{OC} = 0$ and if P is any point on the line AB, show that $\overline{PA} + \overline{PB} + \overline{PC} = 3\,\overline{PO}$.

5. If on the same line we have $\overline{OA} + \overline{OB} + \overline{OC} = 0$ and $\overline{O'A'} + \overline{O'B'} + \overline{O'C'} = 0$, show that $\overline{AA'} + \overline{BB'} + \overline{CC'} = 3\,\overline{OO'}$.

6. If A, B, C are collinear and P, Q, R are the midpoints of BC, CA, AB respectively, show that the midpoints of CR and PQ coincide.

7. Let a and b be two given (positive) segments. Construct points P and Q on AB such that $\overline{AP}/\overline{PB} = a/b$ and $\overline{AQ}/\overline{QB} = -a/b$.

8. Show that if two points divide a line segment AB in equal ratios, then the two points coincide.

9. If r denotes the ratio $\overline{OA}/\overline{OB}$ and r' the ratio $\overline{OA'}/\overline{OB'}$, where O, A, B, A', B' are collinear, show that

$$rr'\overline{BB'} + r\overline{A'B} + r'\overline{B'A} + \overline{AA'} = 0.$$

10. If M is the midpoint of side BC of triangle ABC, and if $AB < AC$, prove that $\angle MAC < \angle BAM$.

11. If AL is the bisector of angle A in triangle ABC, show that $\overline{BL}/\overline{LC} = AB/AC$.

12. If AL is the bisector of exterior angle A of triangle ABC, where $AB \neq AC$, show that $\overline{BL}/\overline{LC} = -AB/AC$.

13. Prove Stewart's Theorem: If A, B, C are any three points on a line and P any point, then $\overline{PA}^2 \cdot \overline{BC} + \overline{PB}^2 \cdot \overline{CA} + \overline{PC}^2 \cdot \overline{AB} + \overline{BC} \cdot \overline{CA} \cdot \overline{AB} = 0$. (This theorem was stated, without proof, by Matthew Stewart (1717–1785) in 1746; it was rediscovered and proved by Thomas Simpson (1710–1761) in 1751, by L. Euler in 1780, and by L. N. M. Carnot in 1803. The case where P lies on the line ABC is found in Pappus' *Collection*.)

14. Find the lengths of the medians of a triangle having sides a, b, c.

15. Find the lengths of the angle bisectors of a triangle having sides a, b, c.

16. Give a direct proof of the Steiner-Lehmus Theorem: If the bisectors of the base angles of a triangle are equal, the triangle is isosceles. (This problem was proposed in 1840 by D. C. Lehmus (1780–1863) to Jacob Steiner (1796–1863).)

17. Show that the sum of the squares of the distances of the vertex of the right angle of a right triangle from the two points of trisection of the hypotenuse is equal to $\frac{5}{9}$ the square of the hypotenuse.

18. If A, B, C are three collinear points and a, b, c are the tangents from A, B, C to a given circle, then

$$a^2\overline{BC} + b^2\overline{CA} + c^2\overline{AB} + \overline{BC} \cdot \overline{CA} \cdot \overline{AB} = 0.$$

19. If A, B, C, D, O are any five coplanar points, show that

(a) $\sin \overline{AOD} \sin \overline{BOC} + \sin \overline{BOD} \sin \overline{COA} + \sin \overline{COD} \sin \overline{AOB} = 0$.
(b) $\triangle \overline{AOD} \triangle \overline{BOC} + \triangle \overline{BOD} \triangle \overline{COA} + \triangle \overline{COD} \triangle \overline{AOB} = 0$.

20. Using 19 (a), prove Ptolemy's Theorem: If $ABCD$ is a cyclic quadrilateral, then $AD \cdot BC + AB \cdot CD = AC \cdot BD$.

21. If O is any point in the plane of triangle ABC, show that

$$\triangle \overline{OBC} + \triangle \overline{OCA} + \triangle \overline{OAB} = \triangle \overline{ABC}.$$

22. If P is any point in the plane of the parallelogram $ABCD$, show that $\triangle \overline{PAB} + \triangle \overline{PCD} = \triangle \overline{ABC}$.

23. If A, B, C, D, P, Q are any six distinct collinear points, show that

$$(\overline{AP} \cdot \overline{AQ})/(\overline{AB} \cdot \overline{AC} \cdot \overline{AD}) + (\overline{BP} \cdot \overline{BQ})/(\overline{BC} \cdot \overline{BD} \cdot \overline{BA})$$
$$+ (\overline{CP} \cdot \overline{CQ})/(\overline{CD} \cdot \overline{CA} \cdot \overline{CB}) + (\overline{DP} \cdot \overline{DQ})/(\overline{DA} \cdot \overline{DB} \cdot \overline{DC}) = 0.$$

24. Generalize Problem 23 for n points A, B, ... and $n - 2$ points P, Q, ...

25. Complete the proof of Theorem 2.1.10.

2.2 INFINITE ELEMENTS

Another innovation of modern elementary geometry is the creation of some ideal elements called "points at infinity," "the line at infinity" in a plane, and "the plane at infinity" in space. The purpose of introducing these ideal elements is to eliminate certain bothersome case distinctions in plane and solid geometry which arise from the possibility of lines and planes being either parallel or intersecting. It follows that with these ideal elements many theorems can be given a single universal statement, whereas without these ideal elements the statements have to be qualified to take care of various exceptional situations. Subterfuges of this sort are common in mathematics. Consider, for example, the discussion of quadratic equations in elementary algebra. The equation $x^2 - 2x + 1 = 0$ actually has only the one root, $x = 1$, but for the sake of uniformity it is agreed to say that the equation has *two equal roots*, each equal to 1. Again, in order that the equation $x^2 + x + 1 = 0$ have any root at all, it is agreed to extend the number system so as to include imaginary numbers. With these two conventions—that a repeated root is to count as two roots and that imaginary roots are to be accepted equally with real roots—we can assert as a universal statement that "every quadratic equation with real coefficients has exactly two roots."

The introduction into geometry of the notion of points at infinity is usually credited to Johann Kepler (1571–1630), but it was Gérard Desargues (1593–1662) who, in a treatment of the conic sections (his *Brouillon projet*) published in 1639, first used the idea systematically. This work of Desargues marks the first essential advance in synthetic geometry since the time of the ancient Greeks.

Restricting ourselves for the time being to plane geometry, consider two lines l_1 and l_2, where l_1 is held fast while l_2 rotates about a fixed point O in the plane but not on line l_1. As l_2 approaches the position of parallelism with l_1, the point P of intersection of l_1 and l_2 recedes farther and farther along line l_1, and in the limiting position of parallelism the point P ceases to exist. To accommodate this exceptional situation, we agree to augment the set of points on l_1 by an ideal point, called *the point at infinity* on l_1, and we say that when l_1 and l_2 are parallel they intersect in this ideal point on l_1.

If our introduction of an ideal point at infinity on a line is not to create more exceptions than it removes, it must be done in such a way that two distinct points, ordinary or ideal, determine one and only one line, and such that two distinct lines intersect in one and only one point. As a first consequence of this we see that two parallel lines must intersect in the same ideal point, no matter in which direction the lines are traversed.

Suppose l_1 and l_2 are two parallel lines intersecting in the ideal point I, and let O be any ordinary point not on l_1 or l_2. Since O and I are to determine a line l_3, and since l_3 cannot intersect l_1 or l_2 a second time, we see that l_3 must be the parallel to l_1 and l_2 through point O. That is, the ideal point I lies on all three of the parallel lines l_1, l_2, l_3, and, by the same argument, on all lines parallel to l_1. Thus the members of a family of parallel lines must share a common ideal point at infinity.

It is easy to see that a different ideal point must be assigned for a different family of parallel lines. For let line m_1 cut line l_1 in an ordinary point P, and let m_2 be a line parallel to m_1. Then m_1 and m_2 intersect in an ideal point J which must be distinct from I, since otherwise the distinct lines m_1 and l_1 would intersect in the two points P and I.

Now consider two distinct ideal points, I and J. The line l which they are to determine cannot pass through any ordinary point P of the plane. For if it did, then line l_1 determined by P and I and line l_2 determined by P and J would be a pair of distinct ordinary lines each of which would be contained in line l because of the collinearity of P, I, J. It follows that the line l determined by a pair of ideal points can contain only ideal points and must therefore be an ideal line, which we call a *line at infinity*.

Finally, we see that in the plane there can be only one line at infinity. For if l_1 and l_2 should be two distinct lines at infinity, these lines would have to intersect in an ideal point I. A line l_3 passing through an ordinary point O and not passing through I would have to intersect l_1 and l_2 in distinct ideal points J and K respectively. The line through J and K would then contain the ordinary point O, which we have seen is impossible.

The above discussion leads to the following convention and theorem.

2.2.1 CONVENTION AND DEFINITIONS. We agree to add to the points of the plane a collection of ideal points, called *points at infinity*, such that

1. each ordinary line of the plane contains exactly one ideal point,

2. the members of a family of parallel lines in the plane share a common ideal point, distinct families having distinct ideal points.

The collection of added ideal points is regarded as an ideal line, called the *line at infinity*, which contains no ordinary points.

The plane, augmented by the above ideal points, will be referred to as the *extended plane*.

2.2.2 THEOREM. *In the extended plane, any two distinct points determine one and only one line and any two distinct lines intersect in one and only one point.*

If a point P recedes indefinitely along the line determined by two ordinary points A and B, then $\overline{AP}/\overline{PB}$ approaches the limiting value -1. This motivates the following definition.

2.2.3 DEFINITION. *If A and B are any two ordinary points, and I the ideal point, on a given line, then we define $\overline{AI}/\overline{IB}$ to be -1.*

The reader may care to supply an analysis leading to the following convention, definitions, and theorem for three-dimensional space.

2.2.4 CONVENTION AND DEFINITIONS. We agree to add to the points of space a collection of ideal points, called *points at infinity*, and ideal lines, called *lines at infinity*, such that

1. each ordinary line of space contains exactly one ideal point,
2. the members of a family of parallel lines in space share a common ideal point, distinct families having distinct ideal points,
3. each ordinary plane of space contains exactly one ideal line,
4. the members of a family of parallel planes in space share a common ideal line, distinct families having distinct ideal lines,
5. the ideal line of an ordinary plane in space consists of the ideal points of the ordinary lines of that plane.

The collection of added ideal points and ideal lines is regarded as an ideal plane, called the *plane at infinity*, which contains no ordinary points or lines.

Three-dimensional space augmented by the above ideal elements, will be referred to as *extended three-dimensional* space.

2.2.5 THEOREM. *In extended three-dimensional space, any two distinct coplanar lines intersect in one and only one point, any non-incident line and plane intersect in one and only one point, any three non-coaxial planes (that is, planes not sharing a common line) intersect in one and only one point, any two distinct planes intersect in one and only one line, any two distinct points determine one and only one line, any two distinct intersecting lines* *determine one and only one plane, any non-incident point and line determine one and only one plane, any three non-collinear points determine one and only one plane.*

There still remains the important matter of whether the above conventions can ever lead to a contradiction. One is reminded of efforts to extend the concept of quotient so as to include fractions having zero denominators. We shall later show that our conventions about infinite elements cannot lead to any contradiction.

PROBLEMS

1. Develop an analysis leading to Convention 2.2.4 and Theorem 2.2.5.

2. Which of the following statements are true for the ordinary plane and which are true for the extended plane?
    (a) There is a one-to-one correspondence between the lines through a fixed point O and the points of a fixed line l, not passing through O, such that corresponding lines and points are in incidence.
    (b) The bisector of an exterior angle of an ordinary triangle divides the opposite side externally in the ratio of the adjacent sides.
    (c) Every straight line possesses one and only one ideal point.
    (d) Every straight line possesses infinitely many ordinary points.
    (e) If a triangle is the figure determined by any three non-concurrent straight lines, then every triangle encloses a finite area.
    (f) If parallel lines are lines lying in the same plane and having no ordinary point in common, then through a given point O there passes one and only one line parallel to a given line l not containing O.
    (g) If A and B are distinct ordinary points and r is any real number, there is a unique point P on line AB such that $\overline{AP}/\overline{PB} = r$.

3. Translate the following theorems of ordinary three-dimensional space into the language of infinite elements, and then supply simple proofs.
    (a) Through a given point there is one and only one plane parallel to a given plane not containing the given point.
    (b) Two lines which are parallel to a third line are parallel to each other.
    (c) If a line is parallel to each of two intersecting planes it is parallel to their line of intersection.
    (d) If a line is parallel to the line of intersection of two intersecting planes, then it is parallel to each of the two planes.
    (e) If a line l is parallel to a plane p, then any plane containing l and intersecting p, cuts p in a line parallel to l.
    (f) Through a given line one and only one plane can be passed parallel to a given skew line.
    (g) Through a given point one and only one plane can be passed parallel to each of two skew lines, neither of which contains the given point.
    (h) All the lines through a point and parallel to a given plane lie in a plane parallel to the first plane.

(i) If a plane contains one of two parallel lines but not the other, then it is parallel to the other.

(j) The intersection of a plane with two parallel planes are parallel lines.

4. Define prism and cylinder in terms of pyramid and cone respectively.

5. If A, B, C, P, Q are any five collinear points, show that

$$(\overline{AP} \cdot \overline{AQ})/(\overline{AB} \cdot \overline{AC}) + (\overline{BP} \cdot \overline{BQ})/(\overline{BC} \cdot \overline{BA}) + (\overline{CP} \cdot \overline{CQ})/(\overline{CA} \cdot \overline{CB}) = 1.$$

6. If A, B, C, D, P are any five collinear points, show that

$$\overline{AP}/(\overline{AB} \cdot \overline{AC} \cdot \overline{AD}) + \overline{BP}/(\overline{BC} \cdot \overline{BD} \cdot \overline{BA}) + \overline{CP}/(\overline{CD} \cdot \overline{CA} \cdot \overline{CB})$$
$$+ \overline{DP}/(\overline{DA} \cdot \overline{DB} \cdot \overline{DC}) = 0.$$

7. Generalize Problem 5 for n points A, B, ... and $n - 1$ points P, Q, ...

8. Generalize Problem 6 for n points A, B, ... and $n - k$ points P, Q, ..., where $1 < k < n - 1$.

2.3 THE THEOREMS OF MENELAUS AND CEVA

The theorems of Menelaus and Ceva, in their original versions, are quite old, for the one dates back to ancient Greece and the other to 1678. It is when they are stated in terms of sensed magnitudes that they assume a particularly modern appearance.

Menelaus of Alexandria was a Greek astronomer who lived in the first century A.D. Though his works in their original Greek are all lost to us, we know of some of them from remarks made by later commentators, and his three-book treatise *Sphaerica* has been preserved for us in the Arabic. This work throws considerable light on the Greek development of trigonometry. Book I is devoted to establishing for spherical triangles many of the propositions of Euclid's *Elements* that hold for plane triangles, such as the familiar congruence theorems, theorems about isosceles triangles, and so on. In addition, Menelaus establishes the congruence of two spherical triangles having the angles of one equal to the angles of the other (for which there is no analogue in the plane) and the fact that the sum of the angles of a spherical triangle is greater than two right angles. Book II contains theorems of interest in astronomy. In Book III is developed the spherical trigonometry of the time, largely deduced from the spherical analogue of the plane proposition now commonly referred to as *Menelaus' Theorem*. Actually, the plane case is assumed by Menelaus as well known and is used by him to establish the spherical case. A good deal of spherical trigonometry can be deduced from the spherical version of the theorem by taking special triangles and special transversals. L. N. M. Carnot made the theorem of Menelaus basic in his *Essai sur la théorie des transversales* of 1806.

Though the theorem of Ceva is a close companion theorem to that of Menelaus, it seems to have eluded discovery until 1678, when the Italian Giovanni Ceva (*ca.* 1647–1736) published a work containing both it and the then apparently long forgotten theorem of Menelaus.

The theorems of Menelaus and Ceva, in their modern dress, are powerful theorems, and they deal elegantly with many problems involving collinearity of points and concurrency of lines. We now turn to a study of these two remarkable theorems. The reader should observe how the convention as to points at infinity eliminates the separate consideration of a number of otherwise exceptional situations.

2.3.1 DEFINITION. A point lying on a side line of a triangle, but not coinciding with a vertex of the triangle, will be called a *menelaus point* of the triangle for this side.

2.3.2 MENELAUS' THEOREM. *A necessary and sufficient condition for three menelaus points* D, E, F *for the sides* BC, CA, AB *of an ordinary triangle* ABC *to be collinear is that*

$$(\overline{BD}/\overline{DC})(\overline{CE}/\overline{EA})(\overline{AF}/\overline{FB}) = -1.$$

Necessity. Suppose (see Figure 2.3a) *D, E, F* are collinear on a line *l*

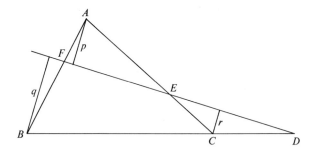

Figure 2.3a

which is not the line at infinity. Drop perpendiculars *p, q, r* on *l* from *A, B, C.* Then, disregarding signs,

$$BD/DC = q/r, \ CE/EA = r/p, \ AF/FB = p/q.$$

It follows that

$$(\overline{BD}/\overline{DC})(\overline{CE}/\overline{EA})(\overline{AF}/\overline{FB}) = \pm 1.$$

Since, however, *l* must cut one or all three sides externally, we see that we can have only the − sign. If *l* is the line at infinity, the proof is simple.

Sufficiency. Suppose

$$(\overline{BD}/\overline{DC})(\overline{CE}/\overline{EA})(\overline{AF}/\overline{FB}) = -1$$

and let *EF* cut *BC* in *D′*. Then *D′* is a menelaus point and, by the above,

$$(\overline{BD'}/\overline{D'C})(\overline{CE}/\overline{EA})(\overline{AF}/\overline{FB}) = -1.$$

It follows that $\overline{BD}/\overline{DC} = \overline{BD'}/\overline{D'C}$, or that $D \equiv D'$. That is, *D, E, F* are collinear.

2.3.3 TRIGONOMETRIC FORM OF MENELAUS' THEOREM. *A necessary and sufficient condition for three menelaus points* D, E, F *for the sides* BC, CA, AB *of an ordinary triangle* ABC *to be collinear is that*

$$(\sin \overline{BAD}/\sin \overline{DAC})(\sin \overline{CBE}/\sin \overline{EBA})(\sin \overline{ACF}/\sin \overline{FCB}) = -1.$$

For we have, by Theorem 2.1.9,

$$\overline{BD}/\overline{DC} = (AB \sin \overline{BAD})/(AC \sin \overline{DAC}),$$
$$\overline{CE}\ \overline{EA} = (BC \sin \overline{CBE})/(BA \sin \overline{EBA}),$$
$$\overline{AF}/\overline{FB} = (CA \sin \overline{ACF})/(CB \sin \overline{FCB}).$$

It follows that

$$(\sin \overline{BAD}/\sin \overline{DAC})(\sin \overline{CBE}/\sin \overline{EBA})(\sin \overline{ACF}/\sin \overline{FCB}) = -1$$

if and only if

$$(\overline{BD}/\overline{DC})(\overline{CE}/\overline{EA})(\overline{AF}/\overline{FB}) = -1.$$

Hence the theorem.

2.3.4 DEFINITION. A line passing through a vertex of a triangle, but not coinciding with a side of the triangle, will be called a *cevian line* of the triangle for this vertex. A cevian line will be identified by the vertex to which it belongs and the point in which it cuts the opposite side, as cevian line *AD* through vertex *A* of triangle *ABC* and cutting the opposite side *BC* in the point *D*.

2.3.5 CEVA'S THEOREM. *A necessary and sufficient condition for three cevian lines* AD, BE, CF *of an ordinary triangle* ABC *to be concurrent is that*

$$(\overline{BD}/\overline{DC})(\overline{CE}/\overline{EA})(\overline{AF}/\overline{FB}) = +1.$$

Necessity. Suppose (see Figure 2.3b) *AD*, *BE*, *CF* are concurrent in *P*.

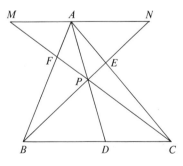

Figure 2.3b

Without loss of generality we may assume that *P* does not lie on the parallel through *A* to *BC*. Let *BE*, *CF* intersect this parallel in *N* and *M*. Then, disregarding signs,

$$BD/DC = AN/MA, \quad CE/EA = BC/AN, \quad AF/FB = MA/BC,$$

whence

$$(\overline{BD}/\overline{DC})(\overline{CE}/\overline{EA})(\overline{AF}/\overline{FB}) = \pm 1.$$

That the sign must be $+$ follows from the fact that either none or two of the points D, E, F divide their corresponding sides externally.

Sufficiency. Suppose

$$(\overline{BD}/\overline{DC})(\overline{CE}/\overline{EA})(\overline{AF}/\overline{FB}) = +1$$

and let BE, CF intersect in P and draw AP to cut BC in D'. Then AD' is a cevian line. Hence, by the above, we have

$$(\overline{BD'}/\overline{D'C})(\overline{CE}/\overline{EA})(\overline{AF}/\overline{FB}) = +1.$$

It follows that $\overline{BD'}/\overline{D'C} = \overline{BD}/\overline{DC}$, or that $D \equiv D'$. That is, AD, BE, CF are concurrent.

2.3.6 TRIGONOMETRIC FORM OF CEVA'S THEOREM. *A necessary and sufficient condition for three cevian lines* AD, BE, CF *of an ordinary triangle* ABC *to be concurrent is that*

$$(\sin \overline{BAD}/\sin \overline{DAC})(\sin \overline{CBE}/\sin \overline{EBA})(\sin \overline{ACF}/\sin \overline{FCB}) = +1.$$

The reader can easily supply a proof similar to that given for Theorem 2.3.3.

PROBLEMS

1. Supply a proof of Theorem 2.3.6.

2. Prove the "necessary" part of Menelaus' Theorem by drawing a line (see Figure 2.3a) through C parallel to DEF to cut AB in L.

3. Prove the "necessary" part of Menelaus' Theorem using the feet of the perpendiculars from the vertices of the triangle on any line perpendicular to the transversal.

4. Derive the "necessary" part of Ceva's Theorem by applying Menelaus' Theorem (see Figure 2.3b) to triangle ABD with transversal CPF and to triangle ADC with transversal BPE.

5. Points E and F are taken on the sides CA, AB of a triangle ABC such that $\overline{CE}/\overline{EA} = \overline{AF}/\overline{FB} = k$. If EF cuts BC in D, show that $\overline{CD} = k^2 \overline{BD}$.

6. In Figure 2.3b, prove that $PD/AD + PE/BE + PF/CF = 1$.

7. In Figure 2.3b, prove that $AP/AD + BP/BE + CP/CF = 2$.

8. In Figure 2.3b, prove that $\overline{AF}/\overline{FB} + \overline{AE}/\overline{EC} = \overline{AP}/\overline{PD}$.

9. If the sides AB, BC, CD, DA of a quadrilateral $ABCD$ are cut by a transversal in the points A', B', C', D' respectively, show that

$$(\overline{AA'}/\overline{A'B})(\overline{BB'}/\overline{B'C})(\overline{CC'}/\overline{C'D})(\overline{DD'}/\overline{D'A}) = 1.$$

10. If a transversal cuts the sides AB, BC, CD, DE, . . . of an n-gon $ABCDE \cdots$ in the points A', B', C', D', . . . , show that

$$(\overline{AA'}/\overline{A'B})(\overline{BB'}/\overline{B'C})(\overline{CC'}/\overline{C'D})(\overline{DD'}/\overline{D'E}) \cdots = (-1)^n.$$

(This is a generalization of Menelaus' Theorem.)

11. If on the sides AB, BC, CD, DA of a quadrilateral $ABCD$ points A', B', C', D' are taken such that

$$\overline{AA'} \cdot \overline{BB'} \cdot \overline{CC'} \cdot \overline{DD'} = \overline{A'B} \cdot \overline{B'C} \cdot \overline{C'D} \cdot \overline{D'A},$$

show that $A'B'$ and $C'D'$ intersect on AC, and $A'D'$ and $B'C'$ intersect on BD.

12. Let the sides AB, BC, CD, DA of a nonplanar quadrilateral $ABCD$ be cut by a plane in the points A', B', C', D'. Show that

$$\overline{AA'} \cdot \overline{BB'} \cdot \overline{CC'} \cdot \overline{DD'} = \overline{A'B} \cdot \overline{B'C} \cdot \overline{C'D} \cdot \overline{D'A}.$$

13. Let D', E', F' be menelaus points on the sides $B'C'$, $C'A'$, $A'B'$ of a triangle $A'B'C'$, and let O be a point in space not in the plane of triangle $A'B'C'$. Show that points D', E', F' are collinear if and only if

$$(\sin \overline{B'OD'}/\sin \overline{D'OC'})(\sin \overline{C'OE'}/\sin \overline{E'OA'})(\sin \overline{A'OF'}/\sin \overline{F'OB'}) = -1.$$

14. Let D, E, F be three menelaus points on the sides BC, CA, AB of a spherical triangle ABC. Show that D, E, F lie on a great circle of the sphere if and only if

$$(\sin \widehat{BD}/\sin \widehat{DC})(\sin \widehat{CE}/\sin \widehat{EA})(\sin \widehat{AF}/\sin \widehat{FB}) = -1.$$

(This is the theorem that Menelaus used in Book III of his *Sphaerica*.)

15. If the lines joining a point O to the vertices of a polygon $ABCD \cdots$ of an odd number of sides meet the opposite sides AB, BC, CD, DE, . . . in the points A', B', C', D', . . . , show that

$$(\overline{AA'}/\overline{A'B})(\overline{BB'}/\overline{B'C})(\overline{CC'}/\overline{C'D})(\overline{DD'}/\overline{D'E}) \cdots = 1.$$

(This is a generalization of Ceva's Theorem.)

2.4 APPLICATIONS OF THE THEOREMS OF MENELAUS AND CEVA

We illustrate the power of the theorems of Menelaus and Ceva by now using them to establish three useful and highly attractive theorems. Many further illustrations will be found among the problems at the end of this section.

2.4.1 THEOREM. *If* AD, BE, CF *are any three concurrent cevian lines of an ordinary triangle* ABC, *and if* D' *denotes the point of intersection of* BC *and* FE, *then* D *and* D' *divide* BC, *one internally and one externally, in the same numerical ratio.*

Since *AD*, *BE*, *CF* are concurrent cevian lines (see Figure 2.4a) we have, by Ceva's Theorem,

$$(\overline{BD}/\overline{DC})(\overline{CE}/\overline{EA})(\overline{AF}/\overline{FB}) = +1.$$

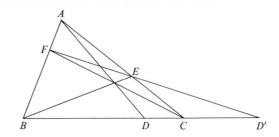

Figure 2.4a

Since D', E, F are collinear menelaus points we have, by Menelaus' Theorem,

$$(\overline{BD'}/\overline{D'C})(\overline{CE}/\overline{EA})(\overline{AF}/\overline{FB}) = -1.$$

It follows that

$$\overline{BD}/\overline{DC} = -\overline{BD'}/\overline{D'C},$$

whence D and D' divide BC, one internally and one externally, in the same numerical ratio.

2.4.2 DEFINITIONS. Two triangles ABC and $A'B'C'$ are said to be *copolar* if AA', BB', CC' are concurrent; they are said to be *coaxial* if the points of intersection of BC and $B'C'$, CA and $C'A'$, AB and $A'B'$ are collinear.

2.4.3 DESARGUES' TWO-TRIANGLE THEOREM. *Copolar triangles are coaxial, and conversely.*

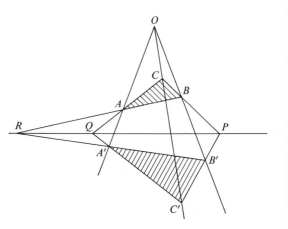

Figure 2.4b

Let the two triangles (see Figure 2.4b) be ABC and $A'B'C'$. Suppose AA', BB', CC' are concurrent in a point O. Let P, Q, R be the points of intersection of BC and $B'C'$, CA and $C'A'$, AB and $A'B'$. Considering the triangles BCO, CAO, ABO in turn, with the respective transversals $B'C'P$, $C'A'Q$, $A'B'R$, we find, by Menelaus' Theorem,

$$(\overline{BP}/\overline{PC})(\overline{CC'}/\overline{C'O})(\overline{OB'}/\overline{B'B}) = -1,$$
$$(\overline{CQ}/\overline{QA})(\overline{AA'}/\overline{A'O})(\overline{OC'}/\overline{C'C}) = -1,$$
$$(\overline{AR}/\overline{RB})(\overline{BB'}/\overline{B'O})(\overline{OA'}/\overline{A'A}) = -1.$$

Setting the product of the three left members of the above equations equal to the product of the three right members, we obtain

$$(\overline{BP}/\overline{PC})(\overline{CQ}/\overline{QA})(\overline{AR}/\overline{RB}) = -1,$$

whence P, Q, R are collinear. Thus copolar triangles are coaxial.

Conversely, suppose P, Q, R are collinear and let O be the point of intersection of AA' and BB'. Now triangles AQA' and BPB' are copolar, and therefore coaxial. That is, O, C, C' are collinear. Thus coaxial triangles are copolar.

2.4.4 PASCAL'S "MYSTIC HEXAGRAM" THEOREM FOR A CIRCLE. *The points L, M, N of intersection of the three pairs of opposite sides* AB *and* DE, BC *and* EF, FA *and* CD *of a* (*not necessarily convex*) *hexagon* ABCDEF *inscribed in a circle lie on a line, called the* Pascal line *of the hexagon.*

Let X, Y, Z (see Figure 2.4c) be the points of intersection of AB and CD, CD and EF, EF and AB, and consider DE, FA, BC as transversals cutting the sides of triangle XYZ. By the Theorem of Menelaus we have

$$(\overline{XL}/\overline{LZ})(\overline{ZE}/\overline{EY})(\overline{YD}/\overline{DX}) = -1,$$
$$(\overline{XA}/\overline{AZ})(\overline{ZF}/\overline{FY})(\overline{YN}/\overline{NX}) = -1,$$
$$(\overline{XB}/\overline{BZ})(\overline{ZM}/\overline{MY})(\overline{YC}/\overline{CX}) = -1.$$

Setting the product of the three left members of the above equations equal to the product of the three right members and rearranging the ratios, we obtain

(1) $$\left(\frac{\overline{XL}}{\overline{LZ}} \cdot \frac{\overline{ZM}}{\overline{MY}} \cdot \frac{\overline{YN}}{\overline{NX}}\right) \left(\frac{\overline{XB} \cdot \overline{XA}}{\overline{XC} \cdot \overline{XD}}\right) \left(\frac{\overline{YC} \cdot \overline{YD}}{\overline{YE} \cdot \overline{YF}}\right) \left(\frac{\overline{ZE} \cdot \overline{ZF}}{\overline{ZB} \cdot \overline{ZA}}\right) = -1.$$

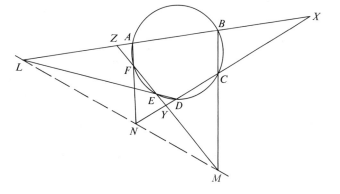

Figure 2.4c

But

$$\overline{XB} \cdot \overline{XA} = \overline{XC} \cdot \overline{XD},$$

$$\overline{YC} \cdot \overline{YD} = \overline{YE} \cdot \overline{YF},$$

$$\overline{ZE} \cdot \overline{ZF} = \overline{ZB} \cdot \overline{ZA},$$

whence each of the last three factors in parentheses in (1) has the value 1. It follows that

$$(\overline{XL}/\overline{LZ})(\overline{ZM}/\overline{MY})(\overline{YN}/\overline{NX}) = -1,$$

or L, M, N are collinear.

Should vertex X, say, of triangle XYZ be an ideal point, so that CD is parallel to AB, the above proof fails. But in this case we may choose a point C' on the circle and near to C, and consider the hexagon $ABC'DEF$ as C' approaches C along the circle. Since the Pascal line exists for all positions of C' as it approaches C, the Pascal line also exists in the limit.

Desargues' two-triangle theorem appears to have been given by Desargues in a work on perspective in 1636, three years before his *Brouillon projet* was published. This theorem has become basic in the present-day theory of projective geometry, and we shall meet it again in later chapters. There we shall also encounter so-called *non-Desarguesian geometries*, or plane geometries in which the two-triangle theorem fails to hold. The great French geometer, Jean-Victor Poncelet (1788–1867), made Desargues' two-triangle theorem the foundation of his theory of homologic figures.

Blaise Pascal (1623–1662) was inspired by the work of Desargues and was in possession of his "mystic hexagram" theorem for a general conic when he was only 16 years old. The consequences of the "mystic hexagram" theorem are very numerous and attractive, and an almost unbelievable amount of research has been expended on the configuration. There are $5!/2 = 60$ possible ways of forming a hexagon from 6 points on a circle, and, by Pascal's theorem, to each hexagon corresponds a *Pascal line*. These 60 Pascal lines pass three by three through 20 points, called *Steiner points*, which in turn lie four by four on 15 lines, called *Plücker lines*. The Pascal lines also concur three by three in another set of points, called *Kirkman points*, of which there are 60. Corresponding to each Steiner point, there are three Kirkman points such that all four lie upon a line, called a *Cayley line*. There are 20 of these Cayley lines, and they pass four by four through 15 points, called *Salmon points*. There are many further extensions and properties of the configuration, and the number of different proofs that have been supplied for the "mystic hexagram" theorem itself is now legion. Some of these alternative proofs will be met in later parts of the book, as will some of the numerous corollaries of the theorem.

PROBLEMS

1. Using Ceva's Theorem prove that: (a) The medians of a triangle are concurrent. (b) The internal angle bisectors of a triangle are concurrent. (c) The altitudes of a triangle are concurrent.

2. If D, E, F are the points of contact of the inscribed circle of triangle ABC with the sides BC, CA, AB respectively, show that AD, BE, CF are concurrent. (This point of concurrency is called the *Gergonne point* of the triangle, after J. D. Gergonne (1771–1859), founder-editor of the mathematics journal *Annales de mathématiques*. Just why the point was named after Gergonne seems not to be known.)

3. Let D, E, F be the points on the sides BC, CA, AB of triangle ABC such that D is half way around the perimeter from A, E half way around from B, and F half way around from C. Show that AD, BE, CF are concurrent. (This point of concurrency is called the *Nagel point* of the triangle, after C. H. Nagel (1803–1882), who considered it in a work of 1836.)

4. Let X and X' be points on a line segment MN symmetric with respect to the midpoint of MN. Then X and X' are called a pair of *isotomic points* for the segment MN. Show that if D and D', E and E', F and F' are isotomic points for the sides BC, CA, AB of triangle ABC, and if AD, BE, CF are concurrent, then AD', BE', CF' are also concurrent. (Two such related points of concurrency are called a pair of *isotomic conjugate points* for the triangle, a term introduced by John Casey in 1889.)

5. Show that the Gergonne and Nagel points of a triangle are a pair of isotomic conjugate points for the triangle. (See Problems 2, 3, 4 for the required definitions.)

6. If, in Problem 4, D, E, F are collinear, show that D', E', F' are also collinear. (Two such related lines as DEF and $D'E'F'$ are sometimes called a pair of *reciprocal transversals* of the triangle ABC, a name used by G. de Longchamps in 1890.)

7. Let OX and OX' be rays through vertex O of angle MON symmetric with respect to the bisector of angle MON. Then OX and OX' are called a pair of *isogonal lines* for the angle MON. Show that if AD and AD', BE and BE', CF and CF' are isogonal cevian lines for the angles A, B, C of a triangle ABC, and if AD, BE, CF are concurrent, then AD', BE', CF' are also concurrent. (This theorem was given by J. J. A. Mathieu in 1865, and was extended to three-space by J. Neuberg in 1884. Two such related points of concurrency are called a pair of *isogonal conjugate points* for the triangle. The orthocenter and circumcenter of a triangle are a pair of isogonal conjugate points. The incenter is its own isogonal conjugate. The isogonal conjugate of the centroid is called the *symmedian point* of the triangle; it enjoys some very attractive properties.)

8. If, in Problem 7, D, E, F are collinear, show that D', E', F' are also collinear.

9. Let AD, BE, CF be three concurrent cevian lines of triangle ABC, and let the circle through D, E, F intersect the sides BC, CA, AB again in D', E', F'. Show that AD', BE', CF' are concurrent.

10. Show that the tangents to the circumcircle of a triangle at the vertices of the triangle intersect the opposite sides of the triangle in three collinear points.

11. If *AD*, *BE*, *CF* are three cevian lines of an ordinary triangle *ABC*, concurrent in a point *P*, and if *EF*, *FD*, *DE* intersect the sides *BC*, *CA*, *AB* of triangle *ABC* in the points *D'*, *E'*, *F'*, show that *D'*, *E'*, *F'* are collinear on a line *p*. (J. J. A. Mathieu, in 1865, called the point *P* the *trilinear pole* of the line *p*, and the line *p* the *trilinear polar* of the point *P*, for the triangle *ABC*. The trilinear polar of the orthocenter of a triangle is called the *orthic axis* of the triangle.)

12. Prove that the external bisectors of the angles of a triangle intersect the opposite sides in three collinear points.

13. Prove that two internal angle bisectors and the external bisector of the third angle of a triangle intersect the opposite sides in three collinear points.

14. Two parallelograms *ACBD* and *A'CB'D'* have a common angle at *C*. Prove that *DD'*, *A'B*, *AB'* are concurrent.

15. Let *ABCD* be a parallelogram and *P* any point. Through *P* draw lines parallel to *BC* and *AB* to cut *BA* and *CD* in *G* and *H* and *AD* and *BC* in *E* and *F*. Prove that the diagonal lines *EG*, *HF*, *DB* are concurrent.

16. If equilateral triangles *BCA'*, *CAB'*, *ABC'* are described externally upon the sides *BC*, *CA*, *AB* of triangle *ABC*, show that *AA'*, *BB'*, *CC'* are concurrent in a point *P*. (The point *P* is the first notable point of the triangle discovered after Greek times. If the angles of triangle *ABC* are each less than 120°, then *P* is the point the sum of whose distances from *A*, *B*, *C* is a minimum. The minimization problem was proposed to Torricelli by Fermat. Torricelli solved the problem and his solution was published in 1659 by his student Viviani.)

17. Show that, in Problem 16, *AA'*, *BB'*, *CC'* are still concurrent if the equilateral triangles are described internally upon the sides of the given triangle *ABC*. (The two points of concurrency of Problems 16 and 17 are known as the *isogonic centers* of triangle *ABC*. The isogonal conjugates of the isogonic centers are called the *isodynamic points* of the triangle, a term given by J. Neuberg in 1885.)

18. Let *ABC* be a triangle right-angled at *B*, and let *BCDD'* and *BAEE'* be squares drawn on *BC* and *BA* externally to the triangle *ABC*. Prove that *CE* and *AD* intersect on the altitude of triangle *ABC* through *B*.

19. If the sides of a triangle *ABC* are cut by a transversal in *D*, *E*, *F*, all exterior to the circumcircle of the triangle, show that the product of the tangent lengths from *D*, *E*, *F* to the circumcircle is equal to *AF · BD · CE*.

20. A transversal cuts the sides *BC*, *CA*, *AB* of a triangle *ABC* in *D*, *E*, *F*. *P*, *Q*, *R* are the midpoints of *EF*, *FD*, *DE*, and *AP*, *BQ*, *CR* intersect *BC*, *CA*, *AB* in *X*, *Y*, *Z*. Show that *X*, *Y*, *Z* are collinear.

21. Let *O* and *U* be two points in the plane of triangle *ABC*. Let *AO*, *BO*, *CO* intersect the opposite sides *BC*, *CA*, *AB* in *P*, *Q*, *R*. Let *PU*, *QU*, *RU* intersect *QR*, *RP*, *PQ* respectively in *X*, *Y*, *Z*. Show that *AX*, *BY*, *CZ* are concurrent.

22. Let *AA'*, *BB'*, *CC'* be three concurrent cevian lines for triangle *ABC*. Let 1 be a point on *BC*, 2 the intersection of 1*B'* and *BA*, 3 of 2*A'* and *AC*, 4 of 3*C'* and *CB*, 5 of 4*B'* and *BA*, 6 of 5*A'* and *AC*, 7 of 6*C'* and *CB*. Show that point 7 coincides with point 1. (This interesting closure theorem is due to O. Nehring, 1942.)

2.5 CROSS RATIO

Another topic which originated with the ancient Greeks, and of which certain aspects were very fully investigated by them, is that now called "cross ratio." In the nineteenth century this concept was revived and considerably improved with the aid of sensed magnitudes and with a highly convenient notation rendered possible by sensed magnitudes. The modern development of the subject is due, independently of each other, to Möbius (in his *Der barycentrische Calcul* of 1827) and Michel Chasles (in his *Aperçu historique sur l'origine et le développement des méthodes en géométrie* of 1829–1837, his *Traité de géométrie supérieure* of 1852, and his *Traité des sections coniques* of 1865). A treatment of the cross-ratio concept freed of metrical considerations was made by Carl George von Staudt (in his *Beiträge zur Geometrie der Lage* of 1847). The cross-ratio concept has become basic in projective geometry, where its power and applicability are of prime importance.

2.5.1 Definition and notation. If A, B, C, D are four distinct points on an ordinary line, we designate the ratio of ratios

$$(\overline{AC}/\overline{CB})/(\overline{AD}/\overline{DB})$$

by the symbol (AB,CD), and call it the *cross ratio* (or *anharmonic ratio*, or *double ratio*) of the range of points A, B, C, D, taken in this order.

Essentially the notation (AB,CD) was introduced by Möbius in 1827. He employed the term *Doppelschnitt-Verhältniss*, and this was later abbreviated by Jacob Steiner to *Doppelverhältniss*, the English equivalent of which is *double ratio*. Chasles used the expression *rapport anharmonique (anharmonic ratio)* in 1837, and William Kingdon Clifford coined the term *cross ratio* in 1878. Staudt used the term *Wurf (throw)*.

The cross ratio of four collinear points depends upon the order in which the points are selected. Since there are twenty-four permutations of four distinct objects, there are twenty-four ways in which a cross ratio of four distinct collinear points may be written. These cross ratios, however, are not all different in value. In fact, we proceed to show that the twenty-four cross ratios may be arranged into six sets of four each, such that the cross ratios in each set have the same value. Indeed, if one of these values be denoted by r, the others are $1/r$, $1 - r$, $1/(1 - r)$, $(r - 1)/r$, and $r/(r - 1)$.

2.5.2 Theorem. *If, in the symbol* $(AB,CD) = r$ *for the cross ratio of four distinct points*, (1) *we interchange any two of the points and at the same time interchange the other two points, the cross ratio is unaltered*, (2) *we interchange only the first pair of points, the resulting cross ratio is* $1/r$, (3) *we interchange only the middle pair of points, the resulting cross ratio is* $1 - r$.

For the first part we must show that

$$(BA,DC) = (CD,AB) = (DC,BA) = (AB,CD) = r,$$

which is easily accomplished by expanding each of the involved cross ratios.

For the second part we must show that

$$(BA,CD) = 1/r,$$

and this too is easily accomplished by expansion.

For the third part we must show that

$$(AC,BD) = 1 - r.$$

This may be accomplished neatly by dividing the Euler identity (Theorem 2.1.5),

$$\overline{AD} \cdot \overline{BC} + \overline{BD} \cdot \overline{CA} + \overline{CD} \cdot \overline{AB} = 0,$$

by $\overline{AD} \cdot \overline{BC}$, obtaining

$$1 + (\overline{BD} \cdot \overline{CA})/(\overline{AD} \cdot \overline{BC}) + (\overline{CD} \cdot \overline{AB})/(\overline{AD} \cdot \overline{BC}) = 0,$$

or, rearranging,

$$(\overline{AB}/\overline{BC})/(\overline{AD}/\overline{DC}) = 1 - (\overline{AC}/\overline{CB})/(\overline{AD}/\overline{DB}),$$

whence $(AC,BD) = 1 - (AB,CD)$.

2.5.3 Theorem. *If* (AB,CD) = r, *then*

(1) (AB,CD) = (BA,DC) = (CD,AB) = (DC,BA) = r,
(2) (BA,CD) = (AB,DC) = (DC,AB) = (CD,BA) = 1/r,
(3) (AC,BD) = (BD,AC) = (CA,DB) = (DB,CA) = 1 − r,
(4) (CA,BD) = (DB,AC) = (AC,DB) = (BD,CA) = 1/(1 − r),
(5) (BC,AD) = (AD,BC) = (DA,CB) = (CB,DA) = (r − 1)/r,
(6) (CB,AD) = (DA,BC) = (AD,CB) = (BC,DA) = r/(r − 1).

The equalities (1) are guaranteed by the first part of Theorem 2.5.2. The equalities (2) are obtained from those in (1) by applying the operation of the second part of Theorem 2.5.2; the equalities (3) are obtained from those in (1) by applying the operation of the third part of Theorem 2.5.2; the equalities (4) are obtained from those in (3) by applying the operation of the second part of Theorem 2.5.2; the equalities (5) are obtained from those in (2) by applying the operation of the third part of Theorem 2.5.2; the equalities (6) are obtained from those in (5) by applying the operation of the second part of Theorem 2.5.2.

2.5.4 Definition and notation. If VA, VB, VC, VD are four distinct coplanar lines passing through an ordinary point V, we designate the ratio of ratios

$$(\sin \overline{AVC}/\sin \overline{CVB})/(\sin \overline{AVD}/\sin \overline{DVB})$$

by the symbol $V(AB,CD)$, and call it the *cross ratio* of the pencil of lines VA, VB, VC, VD, taken in this order. It is to be observed (see Theorem

2.1.10) that this definition is independent of the positions of the points A, B, C, D on their respective lines, so long as they are distinct from V.

2.5.5 THEOREM. *If four distinct parallel lines* a, b, c, d *are cut by two transversals in the points* A, B, C, D *and* A′, B′, C′, D′ *respectively, then* (AB,CD) = (A′B′,C′D′).

This follows immediately from the fact that the segments cut off by the parallel lines on one transversal are proportional to the corresponding segments cut off on the other transversal.

2.5.6 DEFINITION. The *cross ratio* of a pencil of four distinct parallel lines, *a*, *b*, *c*, *d* is taken to be the cross ratio of the range *A*, *B*, *C*, *D* cut off by the parallel lines on any transversal to these lines.

We now state and prove the theorem which gives to cross ratio its singular power in projective geometry.

2.5.7 THEOREM. *The cross ratio of any pencil of four distinct lines is equal to the cross ratio of the corresponding four points in which any ordinary transversal cuts the pencil.*

If the vertex of the pencil is a point at infinity, the theorem follows from Definition 2.5.6.

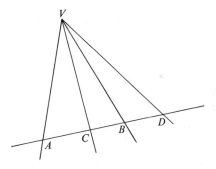

Figure 2.5a

Suppose the vertex *V* of the pencil is not at infinity, and let *A*, *B*, *C*, *D* be the points in which the pencil is cut by an ordinary transversal (see Figure 2.5a). Then, by Theorem 2.1.9,

$$\overline{AC}/\overline{CB} = (VA \sin \overline{AVC})/(VB \sin \overline{CVB}),$$
$$\overline{AD}/\overline{DB} = (VA \sin \overline{AVD})/(VB \sin \overline{DVB}),$$

whence

$$(\overline{AC}/\overline{CB})/(\overline{AD}/\overline{DB}) = (\sin \overline{AVC}/\sin \overline{CVB})/(\sin \overline{AVD}/\sin \overline{DVB}).$$

It follows that (AB,CD) = V(AB,CD), and the theorem is established.

PROBLEMS

1. If we extend the definition of the cross ratio of four collinear points so that two, but no more than two, of the four points may coincide, show that given three distinct collinear points A, B, C there exists a unique point D collinear with them such that $(AB,CD) = r$, where r has any real value or is infinite. Also show that two points coincide if and only if $r = 0$, 1, or ∞.

2. If A, B, C, D are four distinct collinear points, show that the pairs A, B and C, D do or do not separate each other according as (AB,CD) is negative or positive.

3. Given three distinct points A, B, C on a line l, construct a fourth point D collinear with them such that (AB,CD) shall have a given value r.

4. If P, Q, R, S, T are collinear points, show that:
   (a) $(PQ,RT)(PQ,TS) = (PQ,RS)$.
   (b) $(PT,RS)(TQ,RS) = (PQ,RS)$.

5. If O, A, B, C, A', B', C' are collinear and if $\overline{OU} \cdot \overline{OA'} = \overline{OB} \cdot \overline{OB'} = \overline{OC} \cdot \overline{OC'}$, show that $(AB',BC) = (A'B,B'C')$.

6. If $(AB,CP) = m$ and $(AB,CQ) = n$, show that $(AC,PQ) = (n - 1)/(m - 1)$.

7. Investigate the cases where two of the six values of the cross ratios of four collinear points are equal.

8. Show that the six cross ratios of four collinear points can be represented by $\cos^2 \theta$, $\csc^2 \theta$, $-\tan^2 \theta$, $\sec^2 \theta$, $\sin^2 \theta$, $-\cot^2 \theta$. Show that 2θ is the angle of intersection of the circles described on AB and CD as diameters, it being supposed that the points are in the order A, C, B, D. (This result is due to J. Casey.)

2.6 APPLICATIONS OF CROSS RATIO

We first state a number of useful corollaries to Theorem 2.5.7. The proofs can easily be supplied by the reader.

2.6.1 COROLLARY. *If* A, B, C, D *and* A', B', C', D' *are two coplanar ranges on distinct bases such that* (AB,CD) = (A'B',C'D'), *and if* AA', BB', CC' *are concurrent, then* DD' *passes through the point of concurrence.*

2.6.2 COROLLARY. *If* A, B, C, D *and* A', B', C', D' *are two coplanar ranges on distinct bases such that* (AB,CD) = (A'B',C'D'), *and if* A *and* A' *coincide, then* BB', CC', DD' *are concurrent.*

2.6.3 COROLLARY. *If* VA, VB, VC, VD *and* V'A, V'B, V'C, V'D *are two coplanar pencils on distinct vertices* V *and* V' *such that* V(AB,CD) = V' (AB,CD), *and if* A, B, C *are collinear, then* D *lies on the line of collinearity.*

2.6.4 COROLLARY. *If* VA, VB, VC, VD *and* V'A, V'B, V'C, V'D *are two coplanar pencils on distinct vertices* V *and* V' *such that* V(AB,CD) = V'(AB,CD), *and if* A *lies on* VV', *then* B, C, D *are collinear.*

The next theorem, which is an immediate consequence of the elementary angle relations in a circle, gives an important cross-ratio property of the circle.

2.6.5 THEOREM. *If* A, B, C, D *are any four distinct points on a circle, and if* V *and* V' *are any two points on the circle, then* V(AB,CD) = V'(AB,CD), *where* VA, *say, is taken as the tangent to the circle at* A *if* V *should coincide with* A.

2.6.6 DEFINITION AND NOTATION. If *A, B, C, D* are four distinct points on a circle, and *V* is any fifth point on the circle, we shall designate the cross ratio *V(AB,CD)*, which, by Theorem 2.6.5, is independent of the position of *V*, by the symbol (*AB,CD*), and we shall call it the *cross ratio* of the *cyclic range* of points *A, B, C, D*, taken in this order.

2.6.7 THEOREM. *If* A, B, C, D *are four distinct points on a circle, then* (AB,CD) = e(AC/CB)/(AD/DB), *where* AC, CB, AD, DB *are chord lengths, and where* e *is* −1 *or* +1 *according as the pairs* A, B *and* C, D *do or do not separate each other.*

For, denoting the center of the circle by O and its radius by r, we have

$$AC = 2r \sin(AOC/2) = 2r \sin AVC, \text{ etc.},$$

whence

$$(AB,CD) = V(AB,CD) = (\sin \overline{AVC}/\sin \overline{CVB})/(\sin \overline{AVD}/\sin \overline{DVB})$$
$$= \pm(AC/CB)/(AD/DB).$$

The reader can easily show that we must take the minus sign if the pairs of rays VA, VB and VC, VD separate each other, and that otherwise we must take the plus sign.

We now illustrate the power of cross ratio by proving anew Desargues' two-triangle theorem (Theorem 2.4.3) and Pascal's "mystic hexagram" theorem for a circle (Theorem 2.4.4). Further illustrations will be found among the problems at the end of this section and in various other parts of the book.

2.6.8 DESARGUES' TWO-TRIANGLE THEOREM. *Copolar triangles are coaxial, and conversely.*

Let the two triangles (see Figure 2.6a) be ABC and $A'B'C'$. Suppose AA', BB', CC' are concurrent in a point O. Let P, Q, R be the points of intersection of BC and $B'C'$, CA and $C'A'$, AB and $A'B'$. Let CC' cut AB in X, $A'B'$ in X', and PQ in Y. Then, by successive applications of Theorem 2.5.7,

$$C(YP,QR) = (XB,AR) = O(XB,AR) = (X'B',A'R)$$
$$= C'(X'B',A'R) = C'(YP,QR).$$

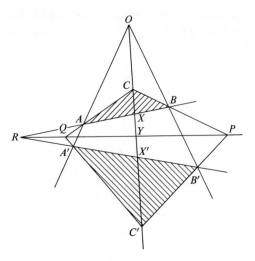

Figure 2.6a

Since Y lies on CC', it follows (by Corollary 2.6.4) that P, Q, R are collinear.

Conversely, suppose P, Q, R are collinear. Then, again by successive applications of Theorem 2.5.7,

$$(RA,XB) = C(RA,XB) = C(RQ,YP) = C'(RQ,YP) = (RA',X'B').$$

It now follows (by Corollary 2.6.2) that AA', XX' (or CC'), BB' are concurrent.

2.6.9 PASCAL'S "MYSTIC HEXAGRAM" THEOREM FOR A CIRCLE. *The points* L, M, N *of intersection of the three pairs of opposite sides* AB *and* DE, BC *and* EF, FA *and* CD *of a* (*not necessarily convex*) *hexagon* ABCDEF *inscribed in a circle lie on a line.*

Let AF and ED (see Figure 2.6b) intersect in H, and EF and CD in K. Then (by Theorem 2.6.5) $A(EB,DF) = C(EB,DF)$, whence (by Theorem 2.5.7) $(EL,DH) = (EM,KF)$. It now follows (by Corollary 2.6.2) that LM, DK, HF are concurrent, or that L, M, N are collinear.

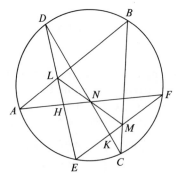

Figure 2.6b

PROBLEMS

1. Supply proofs for Corollaries 2.6.1, 2.6.2, 2.6.3, and 2.6.4.

2. Prove Theorem 2.6.5.

3. Let a, b, c, d be four distinct fixed tangents to a given circle and let p be a variable fifth tangent. If p cuts a, b, c, d in A, B, C, D, show that (AB,CD) is a constant independent of the position of p.

4. Prove Brianchon's Theorem for a circle: If $ABCDEF$ is a (not necessarily convex) hexagon circumscribed about a circle, then AD, BE, CF are concurrent.

5. If two transversals cut the sides BC, CA, AB of a triangle ABC in points P, Q, R and P', Q', R', show that

$$(BC,PP')(CA,QQ')(AB,RR') = 1.$$

6. If A, B, C, D, A', B', C', D', M, N are collinear, and if $(AA',MN) = (BB',MN) = (CC',MN) = (DD',MN)$, show that $(AB,CD) = (A'B',C'D')$.

7. Prove Pappus' Theorem: If A, C, E and B, D, F are two sets of three points on distinct lines, then the points of intersection of AB and DE, BC and EF, FA and CD are collinear.

8. In a hexagon $AC'BA'CB'$, BB', $C'A$, $A'C$ are concurrent and CC', $A'B$, $B'A$ are concurrent. Prove that AA', $B'C$, $C'B$ are also concurrent.

2.7 HOMOGRAPHIC RANGES AND PENCILS

This section is devoted to a brief consideration of homographic ranges and pencils.

2.7.1 DEFINITIONS. If the points of two complete ranges are paired in one-to-one correspondence such that the cross ratio of each four points of one range is equal to the cross ratio of the corresponding four points of the other range, then the two ranges are said to be *homographic* (to each other).

If the lines of two complete pencils are paired in one-to-one correspondence such that the cross ratio of each four lines of one pencil is equal to the cross ratio of the corresponding four lines of the other pencil, then the two pencils are said to be *homographic* (to each other).

2.7.2 THEOREM. (1) *If the joins of three distinct pairs of corresponding points of two homographic ranges on distinct bases are concurrent in a point* O, *then the joins of all pairs of corresponding points of the two ranges pass through* O. (2) *If the intersections of three distinct pairs of corresponding lines of two homographic pencils on distinct vertices are collinear on a line* 1, *then the intersections of all pairs of corresponding lines of the two pencils lie on* 1.

This theorem is an immediate consequence of Corollaries 2.6.1 and 2.6.3.

2.7.3 THEOREM. (1) *If two homographic ranges on distinct bases have a pair of corresponding points coincident, then the joins of all other pairs of* *corresponding points are concurrent.* (2) *If two homographic pencils on distinct vertices have a pair of corresponding lines coincident, then the intersections of all other pairs of corresponding lines are collinear.*

This theorem is an immediate consequence of Corollaries 2.6.2 and 2.6.4.

2.7.4 THEOREM. *A necessary and sufficient condition that two complete ranges be homographic is that the points of the two ranges be paired in one-to-one correspondence such that the signed distances* x *and* x' *of a variable pair of corresponding points* X *and* X' *measured from fixed origins on the bases of the two ranges be connected by a relation of the form*

$$rxx' + sx + tx' + u = 0,$$

where r, s, t, u *are real numbers and* ru − st ≠ 0.

Necessity. Suppose the ranges are homographic. Let A and A', B and B', C and C' be three distinct fixed pairs of corresponding points and X and X' an arbitrary pair of corresponding points of the two ranges. Then $(AB,CX) = (A'B',C'X')$, or

$$(\overline{AC}/\overline{CB})/(\overline{AX}/\overline{XB}) = (\overline{A'C'}/\overline{C'B'})/(\overline{A'X'}/\overline{X'B'}),$$

or, letting corresponding lower case letters represent the signed distances of the points from the fixed origins,

$$\frac{(c - a)(b - x)}{(b - c)(x - a)} = \frac{(c' - a')(b' - x')}{(b' - c')(x' - a')} \,,$$

which reduces to

$$rxx' + sx + tx' + u = 0$$

where

$$r = (c' - a')(b - c) - (c - a)(b' - c'),$$
$$s = a'(c - a)(b' - c') - b'(c' - a')(b - c),$$
$$t = b(c - a)(b' - c') - a(c' - a')(b - c),$$
$$u = b'a(c' - a')(b - c) - ba'(c - a)(b' - c').$$

A little algebra shows that

$$ru - st = (c - a)(c' - a')(b - c)(b' - c')(a - b)(a' - b'),$$

which is different from zero since A, B, C and A', B', C' are distinct points on their bases.

Sufficiency. Suppose the ranges are in one-to-one correspondence with

$$rxx' + sx + tx' + u = 0, \qquad ru - st \neq 0.$$

Then

$$x' = -(sx + u)/(rx + t).$$

Let A and A', B and B', C and C', D and D' be any four distinct pairs of corresponding points. Now

$$(A'B',C'D') = \frac{\overline{A'C'}/\overline{C'B'}}{\overline{A'D'}/\overline{D'B'}} = \frac{(c'-a')(b'-d')}{(b'-c')(d'-a')}.$$

But

$$c' - a' = -\frac{sc+u}{rc+t} + \frac{sa+u}{ra+t} = \frac{(ru-st)(c-a)}{(rc+t)(ra+t)},$$

with similar expressions for $b' - d'$, $b' - c'$, $d' - a'$. Therefore

$$\frac{(c'-a')(b'-d')}{(b'-c')(d'-a')} = \frac{(c-a)(b-d)}{(b-c)(d-a)},$$

since all other factors cancel. That is, $(A'B',C'D') = (AB,CD)$.

There are two classes of problems in the solution of which Theorem 2.7.3 is often applicable, namely: (1) those in which it is required to prove that the locus of a variable point is a straight line, (2) those in which it is required to prove that a variable straight line passes through a fixed point. In (1) we obtain two homographic pencils having a common line, namely the line joining the vertices of the pencils, and having the different positions of the variable point as the intersections of pairs of corresponding lines of the pencils. In (2) we obtain two homographic ranges having a common point, namely the point of intersection of the bases of the ranges, and having the different positions of the variable line as the joins of pairs of corresponding points of the ranges. Some problems of this sort are found among those appearing at the end of the section. Theorem 2.7.4 often furnishes an easy way of showing that two complete ranges are homographic.

PROBLEMS

1. Prove that two ranges (pencils) which are homographic to the same range (pencil) are homographic to each other.

2. A and B are two fixed points and X and X' are two variable points on two given lines intersecting in a point O. If $\overline{OX} + \overline{OX'} = \overline{OA} + \overline{OB}$, show that the locus of the point of intersection of AX' and BX is a straight line.

3. Given a variable triangle ABC whose sides BC, CA, AB pass through fixed points P, Q, R respectively. If the vertices B and C move along given lines through a point O collinear with Q and R, find the locus of vertex A.

4. Given a variable triangle ABC whose vertices A, B, C move along fixed lines p, q, r respectively. If the sides CA and AB pass through given points on a line l concurrent with q and r, show that side BC passes through a fixed point.

5. Three points F, G, H are taken on side BC of a triangle ABC. Through G a variable line is drawn cutting AB and AC in L and M respectively. Prove that the intersection K of FL and HM is a straight line passing through A.

6. A point P, which moves along a given straight line, is joined to two fixed points B and C, and the lines PB, PC cut another line in X and Y. Find the locus of the point of intersection of BY and CX.

7. A, D, C are three fixed collinear points, E is a fixed point not on line ADC, B is a variable point on line CE. The lines AE and BD intersect in Q, CQ and DE intersect in R, BR and AC intersect in P. Show that P is a fixed point.

8. Let (A) and (A') be two homographic ranges. If I corresponds to the point at infinity on (A') and J' to the point at infinity on (A), show that $(\overline{IX})(\overline{J'X'})$ is constant for all pairs of corresponding points X and X'.

9. If I and J' are fixed points, and X and X' are variable points, on two straight lines, and if $(\overline{IX})(\overline{J'X'}) = $ constant, then X and X' generate homographic ranges in which I and J' are the points corresponding to the points at infinity on the two straight lines.

10. (a) Let a variable circle through a fixed point V and cutting a fixed line l not through V in a given angle θ cut l in points P and P', P lying to the left of P' when l is considered as horizontal. Show that (P) and (P') are homographic ranges.
(b) Let $J'VI$ be an isosceles triangle with base lying on l and having base angles θ, J' lying to the left of I. Show that $(\overline{IP})(\overline{J'P'}) = $ constant.

11. If the vertices of a polygon move on fixed concurrent lines, and all but one of the sides pass through fixed points, show that this side and each diagonal will pass through fixed points.

12. If each side of a polygon passes through one of a set of collinear points while all but one of its vertices slide on fixed lines, show that the remaining vertex and each intersection of two sides will describe lines.

13. Let four lines through a point V cut a circle in A, A'; B, B'; C, C'; D, D' respectively. Show that $(AB,CD) = (A'B',C'D')$.

2.8 HARMONIC DIVISION

Very important and particularly useful is the special cross ratio having the value -1. Since, as we shall see, there is an intimate connection between such a cross ratio and three numbers in harmonic progression, such a cross ratio is referred to as a "harmonic division."

2.8.1 DEFINITIONS. If A, B, C, D are four collinear points such that $(AB,CD) = -1$ (so that C and D divide AB one internally and the other externally in the same numerical ratio), the segment AB is said to be *divided harmonically* by C and D, the points C and D are called *harmonic conjugates* of each other with respect to A and B, and the four points A, B, C, D are said to constitute a *harmonic range*. If $V(AB,CD) = -1$, then VA, VB, VC, VD are said to constitute a *harmonic pencil*.

2.8.2 THEOREM. *If* C *and* D *divide* AB *harmonically, then* A *and* B *divide* CD *harmonically.*

For if $(AB,CD) = -1$, then (by Theorem 2.5.2(1)) so also does $(CD,AB) = -1$.

2.8.3 THEOREM. *The harmonic conjugate with respect to A and B of the midpoint of* AB *is the point at infinity on* AB.

Let M be the midpoint of AB. Then $\overline{AM}/\overline{MB} = 1$. It follows that if D is the harmonic conjugate of M with respect to A and B we must have $\overline{AD}/\overline{DB} = -1$. Thus (see Definition 2.2.3) D is the point at infinity on line AB.

The next two theorems furnish useful criteria for four collinear points to form a harmonic range.

2.8.4 THEOREM. $(AB,CD) = -1$ *if and only if* $2/\overline{AB} = 1/\overline{AC} + 1/\overline{AD}$.

Suppose $(AB,CD) = -1$. Then $\overline{AC}/\overline{CB} = -\overline{AD}/\overline{DB}$, whence

$$\overline{CB}/(\overline{AB} \cdot \overline{AC}) = \overline{BD}/(\overline{AB} \cdot \overline{AD}),$$

or

$$(\overline{AB} - \overline{AC})/(\overline{AB} \cdot \overline{AC}) = (\overline{AD} - \overline{AB})/(\overline{AB} \cdot \overline{AD}).$$

That is,

$$1/\overline{AC} - 1/\overline{AB} = 1/\overline{AB} - 1/\overline{AD},$$

or

$$2/\overline{AB} = 1/\overline{AC} + 1/\overline{AD}.$$

The converse may be established by reversing the above steps.

2.8.5 THEOREM. $(AB,CD) = -1$ *if and only if* $\overline{OB}^2 = \overline{OC} \cdot \overline{OD}$, *where* O *is the midpoint of* AB.

Suppose $(AB,CD) = -1$. Then $\overline{AC}/\overline{CB} = -\overline{AD}/\overline{DB}$, whence

$$(\overline{OC} - \overline{OA})/(\overline{OB} - \overline{OC}) = -(\overline{OD} - \overline{OA})/(\overline{OB} - \overline{OD}),$$

or, since $\overline{OA} = -\overline{OB}$,

$$(\overline{OC} + \overline{OB})/(\overline{OB} - \overline{OC}) = (\overline{OD} + \overline{OB})/(\overline{OD} - \overline{OB}).$$

It now follows that

$$(\overline{OC} + \overline{OB})(\overline{OD} - \overline{OB}) = (\overline{OD} + \overline{OB})(\overline{OB} - \overline{OC}),$$

or, upon multiplying out and simplifying,

$$\overline{OB}^2 = \overline{OC} \cdot \overline{OD}.$$

The converse may be established by reversing the above steps.

We now look at the connection between "harmonic division" and "harmonic progression."

2.8.6 DEFINITION. The sequence of numbers $\{a_1, a_2, \ldots, a_n\}$ is said to be a *harmonic progression* if the sequence of numbers $\{1/a_1, 1/a_2, \ldots, 1/a_n\}$ is an arithmetic progression.

2.8.7 THEOREM. *The sequence of numbers* $\{a_1, a_2, a_3\}$ *is a harmonic progression if and only if* $2/a_2 = 1/a_1 + 1/a_3$.

The proof follows readily from Definition 2.8.6.

2.8.8 THEOREM. *If* $(AB,CD) = -1$, *then* $\{\overline{AC}, \overline{AB}, \overline{AD}\}$ *is a harmonic progression.*

This is a consequence of Theorems 2.8.4 and 2.8.7.

We conclude this section with a brief consideration of complete quadrilaterals and complete quadrangles, and their useful harmonic properties.

2.8.9 DEFINITIONS. A *complete quadrilateral* (see Figure 2.8a) is the figure

Figure 2.8a

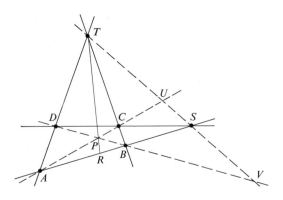

formed by four coplanar lines no three of which are concurrent. The four lines are called the *sides* of the complete quadrilateral, and the six points of intersection of pairs of the sides are called the *vertices* of the complete quadrilateral. Pairs of vertices not lying on any common side are called *opposite vertices* of the complete quadrilateral. The lines through the three pairs of opposite vertices are called the *diagonal lines* of the complete quadrilateral, and the triangle determined by the three diagonal lines is called the *diagonal 3-line* of the complete quadrilateral.

A *complete quadrangle* (see Figure 2.8b) is the figure formed by four coplanar points no three of which are collinear. The four points are called the *vertices* of the complete quadrangle, and the six lines determined by pairs of the vertices are called the *sides* of the complete quadrangle. Pairs of sides not passing through any common vertex are called *opposite sides* of the complete quadrangle. The points of intersection of the three pairs

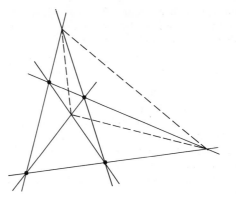

Figure 2.8b

of opposite sides are called the *diagonal points* of the complete quadrangle, and the triangle determined by the three diagonal points is called the *diagonal 3-point* of the complete quadrangle.

2.8.10 THEOREM. *On each diagonal line of a complete quadrilateral there is a harmonic range consisting of the two vertices of the quadrilateral and the two vertices of the diagonal 3-line lying on it.*

Let us show (see Figure 2.8a) that $(UV,TS) = -1$. We have

$$(UV,TS) = P(UV,TS) = P(AB,RS) = (AB,RS).$$

But, by Theorem 2.4.1, $(AB,RS) = -1$. We may now easily show that $(PU,AC) = (PV,DB) = (UV,TS) = -1$.

2.8.11 THEOREM. *At each diagonal point of a complete quadrangle there is a harmonic pencil consisting of the two sides of the quadrangle and the two sides of the diagonal 3-point passing through it.*

This follows immediately from Theorem 2.4.1.

PROBLEMS

1. Justify each of the following methods of constructing the harmonic conjugate D of a given point C with respect to a given segment AB.

(a) Take any point P not on line AB and connect P to A, B, C. Through B draw the parallel to AP, cutting line PC in M, and on this parallel mark off $\overline{BN} = \overline{MB}$. Then PN cuts line AB in the sought point D.

(b) Draw the circle on AB as diameter. If C lies between A and B, draw CT perpendicular to AB to cut the circle in T. Then the tangent to the circle at T cuts line AB in the sought point D. If C is not between A and B, draw one of the tangents CT to the circle, T being the point of contact of the tangent. Then the sought point D is the foot of the perpendicular dropped from T on AB.

(c) Connect any point P not on line AB with A, B, C. Through A draw any line (other than *AB* or *AP*) to cut *PC* and *PB* in *M* and *N* respectively. Draw *BM* to cut *PA* in *G*. Now draw *GN* to cut line *AB* in the sought point *D*. (Note that this construction uses only a straightedge.)

2. If $(AB,CD) = -1$ and *O* and *O'* are the midpoints of *AB* and *CD* respectively, show that $(OB)^2 + (O'C)^2 = (OO')^2$.

3. (a) Show that the lines joining any point on a circle to the vertices of an inscribed square form a harmonic pencil.
   (b) Show, more generally, that the lines joining any point on a circle to the extremities of a given diameter and to the extremities of a given chord perpendicular to the diameter form a harmonic pencil.
   (c) Triangle *ABC* is inscribed in a circle of which *DE* is the diameter perpendicular to side *AC*. If lines *DB* and *EB* intersect *AC* in *L* and *M*, show that $(AC,LM) = -1$.
   (d) Show that the diameter of a circle perpendicular to one of the sides of an inscribed triangle is divided harmonically by the other two sides.

4. (a) If *L*, *M*, *N* are the midpoints of the sides *BC*, *CA*, *AB* of a triangle *ABC*, show that $L(MN,AB) = -1$.
   (b) If *P*, *Q*, *R* are the feet of the altitudes on sides *BC*, *CA*, *AB* of a triangle *ABC*, show that $P(QR,AB) = -1$.
   (c) The bisector of angle *A* of triangle *ABC* intersects the opposite side in *T*. *U* and *V* are the feet of the perpendiculars from *B* and *C* upon line *AT*. Show that $(AT,UV) = -1$.

5. Let *BC* be a diameter of a given circle, let *A* be a point on *BC* produced, and let *P* and *Q* be the points of contact of the tangent to the circle from point *A*. Show that $P(AQ,CB) = -1$.

6. If $P(AB,CD) = -1$ and if *PC* is perpendicular to *PD*, show that *PC* and *PD* are the bisectors of angle *APB*.

7. If *O* is any point on the altitude *AP* of triangle *ABC*, and *BO* and *CO* intersect *AC* and *AB* in *E* and *F* respectively, show that *PA* bisects angle *EPF*.

8. Given four collinear points *A*, *B*, *C*, *D*, find points *P* and *Q* such that $(AB,PQ) = (CD,PQ) = -1$.

9. In triangle *ABC* we have $(BC,PP') = (CA,QQ') = (AB,RR') = -1$. Show that *AP'*, *BQ'*, *CR'* are concurrent if and only if *P*, *Q*, *R* are collinear.

10. Prove, in Figure 2.8a, that
$$TA \cdot TC \cdot SB \cdot SD = SA \cdot SC \cdot TB \cdot TD.$$

11. If $(AB,CD) = -1$ and *O* is the midpoint of *CD*, show that $\overline{AC} \cdot \overline{AD} = \overline{AB} \cdot \overline{AO}$.

12. If *P*, *P'* divide one diameter of a circle harmonically and *Q*, *Q'* divide another diameter harmonically, prove that *P*, *Q*, *P'*, *Q'* are concyclic.

13. Two circles intersect in points *A* and *B*. A common tangent touches the circles at *P* and *Q* and cuts a third circle through *A* and *B* in *L* and *M*. Prove that $(PQ,LM) = -1$.

14. A circle Σ inscribed in a semicircle touches the diameter *AB* of the semicircle at a point *C*. Prove that the diameter of Σ is the harmonic mean between *AC* and *CB*.

15. In Figure 2.8a, prove that UD, VA, PS are concurrent.

16. If $(AB,CD) = -1$ and O is the midpoint of AB, prove that $\overline{OC}/\overline{OD} = (AC)^2/(AD)^2$.

17. In Figure 2.8a, prove that the midpoints X, Y, Z of DB, AC, ST are collinear.

18. A secant from an external point A cuts a circle in C and D, and cuts the chord of contact of the tangents to the circle from A in B. Prove that $(AB,CD) = -1$.

19. If A, B, C are collinear, P the harmonic conjugate of A with respect to B and C, Q the harmonic conjugate of B with respect to C and A, and R the harmonic conjugate of C with respect to A and B, show that A is the harmonic conjugate of P with respect to Q and R.

20. If P', Q' are the harmonic conjugates of P and Q with respect to A and B, show that the segments PQ, $P'Q'$ subtend equal or supplementary angles at any point on the circle described on AB as diameter.

21. Prove that the geometric mean of two positive numbers is the geometric mean of the arithmetic mean and harmonic mean of the two numbers.

22. If $(AB,PQ) = -1$ and O is collinear with A, B, P, Q, show that $2(\overline{OB}/\overline{AB}) = \overline{OP}/\overline{AP} + \overline{OQ}/\overline{AQ}$.

23. In triangle ABC, D and D', E and E', F and F' are harmonic conjugates with respect to B and C, C and A, A and B respectively. Prove that corresponding sides of triangles DEF, $D'E'F'$ intersect on the sides of triangle ABC.

24. Through a given point O a variable line is drawn cutting two fixed lines in P and Q, and on OPQ point X is taken such that $1/\overline{OX} = 1/\overline{OP} + 1/\overline{OQ}$. Find the locus of X.

25. Through a given point O a variable line is drawn cutting n fixed lines in P_1, $P_2, \ldots, P_n$, and on the variable line a point X is taken such that $1/\overline{OX} = 1/\overline{OP_1} + 1/\overline{OP_2} + \cdots + 1/\overline{OP_n}$. Find the locus of X.

2.9 ORTHOGONAL CIRCLES

For later purposes we shall need certain parts of the elementary geometry of circles that did not make their appearance until the nineteenth century. The material we are concerned about is that centered around the concepts of orthogonal circles, the power of a point with respect to a circle, the radical axis of a pair of circles, the radical center and radical circle of a trio of circles, and coaxial pencils of circles. Though one can see the notion of power of a point with respect to a circle foreshadowed in Propositions 35 and 36 of Book III of Euclid's *Elements*, the concept was first crystallized and developed by Louis Gaultier in a paper published in 1813 in the *Journal de l'École Polytechnique*. Here we find, for the first time, the terms *radical axis* and *radical center*; the term *power* was introduced somewhat later by Jacob Steiner. The initial studies of orthogonal circles and coaxial pencils of circles were made in the early nineteenth century by Gaultier, Poncelet, Steiner, J. B. Durrende, and others. This rather recent elementary geometry of the circle has found valuable application in various parts of mathematics and physics.

We start with a definition of orthogonal curves.

2.9.1 DEFINITIONS. By the *angles of intersection* of two coplanar curves at a point which they have in common is meant the angles between the tangents to the curves at the common point. If the angles of intersection are right angles, the two curves are said to be *orthogonal*.

The facts stated about circles in the following theorem are quite obvious.

2.9.2 THEOREM. (1) *The angles of intersection at one of the common points of two intersecting circles are equal to those at the other common point.* (2) *If two circles are orthogonal, a radius of either, drawn to a point of intersection, is tangent to the other; conversely, if the radius of one of two intersecting circles, drawn to a point of intersection, is tangent to the other, the circles are orthogonal.* (3) *Two circles are orthogonal if and only if the square of the distance between their centers is equal to the sum of the squares of their radii.* (4) *If two circles are orthogonal, the center of each lies outside the other.*

We now establish a few deeper facts about orthogonal circles that will be useful to us in a later chapter.

2.9.3 THEOREM. *If two circles are orthogonal, then any diameter of one which intersects the other is cut harmonically by the other; conversely, if a diameter of one circle is cut harmonically by a second circle, then the two circles are orthogonal.*

Let O (see Figure 2.9a) be the center of one of a pair of orthogonal circles

Figure 2.9a

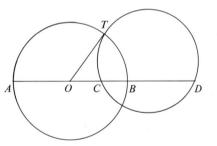

and let a diameter AOB of this circle cut the other in points C and D. Let T be a point of intersection of the two circles. Then $(OB)^2 = (OT)^2 = (\overline{OC})(\overline{OD})$, since (by Theorem 2.9.2 (2)) OT is tangent to the second circle. It now follows (by Theorem 2.8.5) that $(AB,CD) = -1$.

Conversely, if $(AB,CD) = -1$, then (by Theorem 2.8.5) $(OT)^2 = (OB)^2 = (\overline{OC})(\overline{OD})$, and OT is tangent to the second circle, whence (by Theorem 2.9.2 (2)) the two circles are orthogonal.

2.9.4 DEFINITION AND CONVENTION. We shall call a circle or a straight line a "*circle*" (with quotation marks), and we shall adopt the convention that two straight lines are *tangent* if and only if they either coincide or are parallel. With this convention, it is perfectly clear in all cases what is meant by two "circles" being tangent to one another. Some authors refer to a "circle" as a *stircle*.

2.9.5 THEOREM. *There is one and only one "circle" orthogonal to a given circle* Σ *and passing through two given interior points* A *and* B *of* Σ.

Let O (see Figure 2.9b) be the center of Σ. If A, O, B are collinear, the

Figure 2.9b

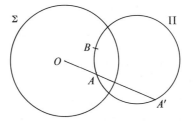

diameter AOB is a straight line through A and B and orthogonal to Σ. If A, O, B are not collinear, let A' be the harmonic conjugate of A with respect to the endpoints of the diameter of Σ passing through A. Then the circle BAA' passes through A and B and (by Theorem 2.9.3) is orthogonal to Σ. Thus in any event, there is at least one "circle" through the points A and B and orthogonal to Σ. To show that there is only one such "circle," let Π represent any "circle" through A and B and orthogonal to Σ. If Π is a straight line then it must be a diameter of Σ. That is, A, O, B are collinear and Π coincides with the straight line considered earlier. If Π is a circle, then (by Theorem 2.9.3) it must also pass through the point A', and thus coincide with the circle BAA' considered earlier. This proves the theorem.

2.9.6 THEOREM. *There is a unique "circle" orthogonal to a given circle and tangent to a given line* l *at an ordinary point* A *of* l *not on* Σ.

Let O (see Figure 2.9c) be the center of Σ. It is easy to show that if O lies on l, then l is the unique "circle" satisfying the given conditions. If O does not lie on l, it is easy to show that the circle passing through A and having its center at the point of intersection of the perpendicular to l at A and the perpendicular bisector of AA', where A' is the harmonic conjugate of A with respect to the endpoints of the diameter of Σ through A, is the unique "circle" satisfying the given conditions.

As mentioned earlier, Propositions III 35 and III 36 of Euclid's *Elements* contain the germs of the notion of power of a point with respect to a circle. With the aid of sensed magnitudes, these two propositions can be combined into the following single statement.

Figure 2.9c

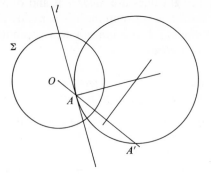

2.9.7 THEOREM. *If* P *is a fixed point in the plane of a given circle* Σ, *and if a variable line* 1 *through* P *intersects* Σ *in points* A *and* B, *then the product* $\overline{PA} \cdot \overline{PB}$ *is independent of the position of* 1.

Let O be the center of Σ. If P coincides with O, or lies on Σ, or is an ideal point, the theorem is obvious. Otherwise (see Figures 2.9d$_1$ and 2.9d$_2$),

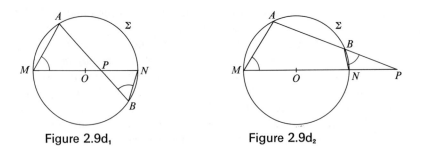

Figure 2.9d₁   Figure 2.9d₂

draw the diameter MN through the point P and connect A with M and B with N. The two triangles PMA and PBN are equiangular, and therefore similar, whence

$$\overline{PA}/\overline{PM} = \overline{PN}/\overline{PB} \quad \text{or} \quad \overline{PA} \cdot \overline{PB} = \overline{PM} \cdot \overline{PN},$$

and the theorem follows since the right-hand side of the last equality is independent of the position of *l*.

The preceding theorem justifies the following definition.

2.9.8 DEFINITION. The *power of a point with respect to a circle* is the product of the signed distances of the point from any two points on the circle and collinear with it.

It follows that the power of a point with respect to a circle is positive, zero, or negative according as the point lies outside, on, or inside the circle.

If the point lies outside the circle, its power with respect to the circle is equal to the square of the tangent from the point to the circle; if the point lies inside the circle, its power with respect to the circle is the negative of the square of half the chord perpendicular to the diameter passing through the given point. We thus have:

2.9.9 THEOREM. *Let* P *be a point in the plane of a circle* Σ *of center* O *and radius* r. *Then the power of* P *with respect to* Σ *is equal to* $(OP)^2 - r^2$.

We leave it to the reader to show that Theorem 2.9.3 can be rephrased as follows:

2.9.10 THEOREM. *A necessary and sufficient condition for two circles to be orthogonal is that the power of the center of either with respect to the other be equal to the square of the corresponding radius.*

PROBLEMS

1. Establish Theorem 2.9.2.

2. Establish Theorem 2.9.10.

3. Show that if d is the distance between the centers of two intersecting circles, c is the length of their common chord, r and r' their radii, then the circles are orthogonal if and only if $cd = 2rr'$.

4. If a line drawn through a point of intersection of two circles meets the circles again in P and Q respectively, show that the circles with centers P and Q, each orthogonal to the other circle, are orthogonal to each other.

5. If AB is a diameter of a circle and if any two lines AC and BC meet the circle again at P and Q respectively, show that circle CPQ is orthogonal to the given circle.

6. Two orthogonal circles intersect in points P and Q. If C is a point on one of the circles, and if CP and CQ cut the other circle in A and B, prove that AB is a diameter of this other circle.

7. Let H be the orthocenter of a triangle ABC. Show that the circles on AH and BC as diameters are orthogonal.

8. If the quadrilateral whose vertices are the centers and the points of intersection of two circles is cyclic, prove that the circles are orthogonal.

9. If AB is a diameter and M any point of a circle of center O, show that the two circles AMO and BMO are orthogonal.

10. Let ABC be a triangle having altitudes AD, BE, CF and orthocenter H. Circles are drawn having A, B, C for centers and $(AH)(AD)$, $(BH)(BE)$, $(CH)(CF)$ for the squares of their respective radii. Prove that each circle is orthogonal to the other two.

2.10 THE RADICAL AXIS OF A PAIR OF CIRCLES

In this section we continue our study of some of the modern elementary geometry of the circle by considering the locus of a point that has equal powers with respect to two given circles. This leads to the important concept of a coaxial pencil of circles.

2.10.1   DEFINITION.   The locus of a point whose powers with respect to two given circles are equal is called the *radical axis* of the two given circles.

2.10.2   THEOREM.   *The radical axis of two nonconcentric circles is a straight line perpendicular to the line of centers of the two circles.*

Consider two nonconcentric circles with centers O and O' and radii r and r' (see Figure 2.10a), and let P be any point on the radical axis of the

Figure 2.10a

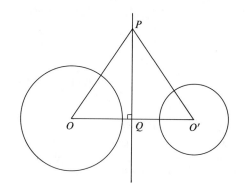

two circles. Let Q be the foot of the perpendicular from P on OO'. Then (by Theorem 2.9.9)

$$(PO)^2 - r^2 = (PO')^2 - r'^2.$$

Subtracting $(PQ)^2$ from each side we get

$$(OQ)^2 - r^2 = (QO')^2 - r'^2,$$

or

$$(\overline{OQ} + \overline{QO'})(\overline{OQ} - \overline{QO'}) = r^2 - r'^2,$$

whence

(1) $$\overline{OQ} - \overline{QO'} = (r^2 - r'^2)/\overline{OO'}.$$

Now there is only one point Q on OO' satisfying relation (1). For if R is any such point we have

$$\overline{OQ} - \overline{QO'} = \overline{OR} - \overline{RO'},$$

or

$$(\overline{OR} + \overline{RQ}) - \overline{QO'} = \overline{OR} - (\overline{RQ} + \overline{QO'}),$$

or

$$\overline{OR} - \overline{QR} - \overline{QO'} = \overline{OR} + \overline{QR} - \overline{QO'},$$

and $\overline{QR} = 0$, or R coincides with Q. It follows that if a point is on the radical axis of the two circles it lies on the perpendicular to the line of centers at the point Q. Conversely, by reversing the above steps, it can be shown that any point on the perpendicular to OO' at Q lies on the radical axis of the two circles. Therefore the radical axis of the two circles is the perpendicular to OO' at the point Q.

2.10.3 REMARK. If, in equation (1) of the proof of Theorem 2.10.2, $r' \neq r$ and O' approaches O, Q approaches an ideal point. The radical axis of two unequal concentric circles is therefore frequently defined to be the line at infinity in the plane of the circles. The radical axis of two equal concentric circles is left undefined, and it is to be understood that any statement about radical axes is not intended to include this situation.

2.10.4 THEOREM. *The radical axes of three circles with noncollinear centers, taken in pairs, are concurrent.*

Let P be the intersection of the radical axis of the first and second circles with that of the second and third circles. Then P has equal powers with respect to all three circles, and thus must also lie on the radical axis of the first and third circles.

2.10.5 DEFINITION. The point of concurrence of the radical axes of three circles with noncollinear centers, taken in pairs, is called the *radical center* of the three circles.

2.10.6 THEOREM. (1) *The center of a circle which cuts each of two circles orthogonally lies on the radical axis of the two circles.* (2) *If a circle whose center lies on the radical axis of two circles is orthogonal to one of them, it is also orthogonal to the other.*

This is an immediate consequence of Theorem 2.9.10.

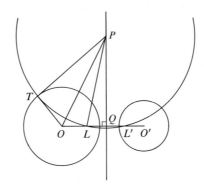

Figure 2.10b₁

2.10.7 Theorem. (1) *All the circles which cut each of two given noninter-secting circles orthogonally intersect the line of centers of the two given circles in the same two points.* (2) *A circle which cuts each of two given intersecting circles orthogonally does not intersect the line of centers of the two given circles.*

(1) Let a circle with center P cut two given circles with centers O and O' orthogonally. Then (by Theorem 2.10.6) P lies on the radical axis of the two given circles. Referring to Figure $2.10b_1$, we then have $OQ > OT$, whence $PT > PQ$, and the common orthogonal circle intersects OO' in points L and L'. Now

$$(PL)^2 = (LQ)^2 + (QP)^2$$

and also

$$(PL)^2 = (PT)^2 = (PO)^2 - (OT)^2 = (OQ)^2 + (QP)^2 - (OT)^2,$$

whence

$$(LQ)^2 = (OQ)^2 - (OT)^2.$$

This last equation shows that the position of L is independent of that of P. Hence every circle orthogonal to the two given circles passes through point L. Similarly, every such circle passes through L'.

Figure $2.10b_2$

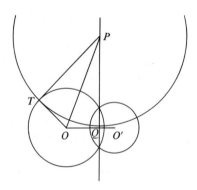

(2) Referring to Figure $2.10b_2$ we have $OQ < OT$, whence $PT < PQ$, and the common orthogonal circle fails to intersect OO'.

2.10.8 Definitions. A set of circles is said to form a *coaxial pencil* if the same straight line is the radical axis of any two circles of the set; the straight line is called the *radical axis* of the coaxial pencil.

Coaxial pencils of circles are very useful in certain mathematical and physical investigations. We leave to the reader the easy task of establishing the two following important theorems about such sets of circles.

2.10.9 Theorem. (1) *The centers of the circles of a coaxial pencil are* collinear. (2) *If two circles of a coaxial pencil intersect, every circle of the pencil passes through the same two points; if two circles of a coaxial pencil are tangent at a point, all circles of the pencil are tangent to one another at the same point; if two circles of a coaxial pencil do not intersect, no two circles of the pencil intersect.* (3) *The radical axis of a coaxial pencil of circles is the locus of a point whose powers with respect to all the circles of the pencil are equal.*

2.10.10 DEFINITION. By Theorem 2.10.9 (2) there are three types of co-axial pencils of circles, and these are called an *intersecting coaxial pencil,* a *tangent coaxial pencil,* and a *nonintersecting coaxial pencil.*

2.10.11 THEOREM. (1) *All the circles orthogonal to two given nonintersecting circles belong to an intersecting coaxial pencil whose line of centers is the radical axis of the two given circles.* (2) *All the circles orthogonal to two given tangent circles belong to a tangent coaxial pencil whose line of centers is the common tangent to the two given circles.* (3) *All the circles orthogonal to two given intersecting circles belong to a nonintersecting coaxial pencil whose line of centers is the line of the common chord of the two given circles.*

We close the section with a particularly pretty application of some of the foregoing theory to the complete quadrilateral.

2.10.12 THEOREM. *The three circles on the diagonals of a complete quadrilateral as diameters are coaxial; the orthocenters of the four triangles determined by the four sides of the quadrilateral taken three at a time are collinear; the midpoints of the three diagonals are collinear on a line perpendicular to the line of collinearity of the four orthocenters.*

Referring to Figure 2.10c, let A, B, C, D, E, F be the six vertices of a

Figure 2.10c

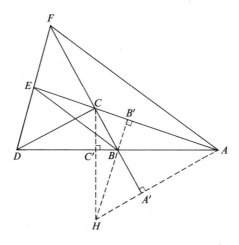

complete quadrilateral, H the orthocenter of triangle ABC, and A', B', C' the feet of the altitudes of triangle ABC. Since A, C, C', A' and B', C, C', B are sets of concyclic points,

$$(\overline{HA})(\overline{HA'}) = (\overline{HB})(\overline{HB'}) = (\overline{HC})(\overline{HC'}).$$

But AA', BB' CC' are chords of the circles having the diagonals AF, BE, CD of the complete quadrilateral as diameters. If follows that H has the same power with respect to all three of these circles. Similarly it can be shown that the orthocenters of triangles ADE, BDF, CEF each have equal powers with respect to the three circles. It follows that the three circles are coaxial, the four orthocenters are collinear on their radical axis, and the centers of the circles (that is, the midpoints of the three diagonals) are collinear on a line perpendicular to the line of collinearity of the four orthocenters.

PROBLEMS

1. Show that if the radical center of three circles with noncollinear centers is exterior to each of the three circles, it is the center of a circle orthogonal to all three circles. (This circle is called the *radical circle* of the three circles.)

2. Establish Theorem 2.10.9.

3. Establish Theorem 2.10.11.

4. Prove that the radical axis of two circles having a common tangent bisects the segment on the common tangent determined by the points of contact.

5. Justify the following construction of the radical axis of two nonconcentric nonintersecting circles. Draw any circle cutting the given circles in A, A' and B, B' respectively. Through P, the intersection of AA' and BB', draw the perpendicular to the line of centers of the given circles. This perpendicular is the required radical axis.

6. (a) Prove that the radical center of the three circles constructed on the sides of a triangle as diameters is the orthocenter of the triangle.
   (b) Let AD, BE, CF be cevian lines of triangle ABC. Prove that the radical center of circles constructed on AD, BE, CF as diameters is the orthocenter of the triangle.

7. If the common chord of two intersecting circles C_1 and C_2 is a diameter of C_2, circle C_2 is said to be *bisected* by circle C_1, and circle C_1 is said to *bisect* circle C_2. Prove the following theorems concerning bisected circles.
   (a) If circle C_2 is bisected by circle C_1, the square of the radius of C_2 is equal to the negative of the power of the center of C_2 with respect to C_1.
   (b) If point P lies inside a circle C_1, then P is the center of a circle C_2 which is bisected by C_1.
   (c) If the radical center of three circles with noncollinear centers lies inside the three circles, then it is the center of a circle which is bisected by each of the three circles.
   (d) The locus of the center of a circle which bisects two given nonconcentric circles is a straight line parallel to the radical axis of the two given circles. (This line is called the *antiradical axis* of the two given circles.)

(e) The circles having their centers on a fixed line and bisecting a given circle form a coaxial pencil of circles.

8. Prove that if each of a pair of circles cuts each of a second pair orthogonally, then the radical axis of either pair is the line of centers of the other.

9. (a) Through a given point draw a circle that is orthogonal to two given circles.
(b) Through a given point draw a circle that is coaxial with two given circles.

10. Construct a circle such that tangents to it from three given points shall have given lengths.

11. Prove Casey's Power Theorem: The difference of the powers of a point with respect to two circles is equal to twice the product of the distance of the point from the radical axis and the distance between the centers of the circles.

12. (a) What is the locus of a point whose power with respect to a given circle is constant?
(b) What is the locus of a point the sum of whose powers with respect to two circles is constant?
(c) What is the locus of a point the difference of whose powers with respect to two circles is constant?
(d) What is the locus of a point the ratio of whose powers with respect to two circles is constant?

13. (a) Prove that if a circle has its center on the radical axis of a coaxial pencil of circles and is orthogonal to one of the circles of the pencil, it is orthogonal to all the circles of the pencil.
(b) Prove that if a circle is orthogonal to two circles of a coaxial pencil of circles, it is orthogonal to all the circles of the pencil.

14. Show that the radical axes of the circles of a coaxial pencil with a circle not belonging to the pencil are concurrent.

15. Given three lines, and on each a pair of points such that a circle passes through each two pairs, show that either the three lines are concurrent or the six points lie on one circle.

BIBLIOGRAPHY

ALTSHILLER-COURT, NATHAN, *College Geometry, an Introduction to the Modern Geometry of the Triangle and the Circle.* New York: Barnes and Noble Inc., 1952.

———, *Modern Pure Solid Geometry.* New York: The Macmillan Company, 1935.

COOLIDGE, J. L., *A Treatise on the Circle and the Sphere.* New York: Oxford University Press, 1916.

COXETER, H. S. M., and S. L. GREITZER, *Geometry Revisited* (New Mathematical Library, No. 19). New York: Random House and The L. W. Singer Company, 1967.

DAUS, P. H., *College Geometry.* Englewood Cliffs, N.J.: Prentice-Hall, Inc., 1941.

DAVIS, D. R., *Modern College Geometry.* Reading, Mass.: Addison-Wesley Publishing Company, Inc., 1949.

Dupuis, N. F., *Elementary Synthetic Geometry of the Point, Line and Circle in the Plane*. New York: The Macmillan Company, 1914.

Durell, C. V., *Modern Geometry*. New York: The Macmillan Company, 1928.

Forder, H. G., *Higher Course Geometry*. New York: Cambridge University Press, 1949.

Johnson, R. A., *Modern Geometry, an Elementary Treatise on the Geometry of the Triangle and the Circle*. Boston: Houghton Mifflin Company, 1929.

Kay, D. C., *College Geometry*. New York: Holt, Rinehart and Winston, Inc., 1969.

Lachlan, R., *An Elementary Treatise on Modern Pure Geometry*. New York: The Macmillan Company, 1893.

M'Clelland, W. J., *A Treatise on the Geometry of the Circle*. New York: The Macmillan Company, 1891.

Maxwell, E. A., *Geometry for Advanced Pupils*. New York: Oxford University Press, 1949.

Miller, L. H., *College Geometry*. New York: Appleton-Century-Crofts, Inc., 1957.

Narayan, Shanti, and M. L. Kochhar, *A Text Book of Modern Pure Geometry*. Delhi: S. Chand and Company (no date).

Pedoe, D., *Circles*. New York: Pergamon Press, 1957.

Perfect, Helen, *Topics in Geometry*. New York: The Macmillan Company, 1963.

Richardson, G., and A. S. Ramsey, *Modern Plane Geometry*. New York: The Macmillan Company, 1928.

Russell, J. W., *A Sequel to Elementary Geometry*. New York: Oxford University Press, 1907.

Shively, L. S., *An Introduction to Modern Geometry*. New York: John Wiley and Sons, Inc., 1939.

Taylor, E. H., and G. C. Bartoo, *An Introduction to College Geometry*. New York: The Macmillan Company, 1949.

Winsor, A. S., *Modern Higher Plane Geometry*. Boston: The Christopher Publishing House, 1941.

3

Elementary
Transformations

One of the most useful methods exploited by geometers of the modern era is that of cleverly transforming a figure into another which is better suited to a geometrical investigation. The gist of the idea is this. We wish to solve a difficult problem connected with a given figure. We *transform* the given figure into another which is related to it in a definite way and such that under the transformation the difficult problem concerning the original figure becomes a simpler problem concerning the new figure. We *solve* the simpler problem related to the new figure, and then *invert* the transformation to obtain the solution of the more difficult problem related to the original figure.

The idea of solving a difficult problem by means of an appropriate transformation is not peculiar to geometry but is found throughout mathematics.* For example, if one were asked to find the Roman numeral representing the product of the two given Roman numerals LXIII and XXIV, one would *transform* the two given Roman numerals into the corresponding Hindu-Arabic numerals, 63 and 24, *solve* the related problem in the Hindu-Arabic notation by means of the familiar multiplication algorithm to obtain the product 1512, then *invert* this result back into Roman notation, finally

---

* See M. S. Klamkin and D. J. Newman, "The philosophy and applications of transform theory," *SIAM Review*, vol. 3, no. 1, Jan. 1961, pp. 10–36.

obtaining MDXII as the answer to the original problem. By an appropriate transformation, a difficult problem has been converted into an easy problem.

Again, suppose we wish to show that the equation

$$x^7 - 2x^5 + 10x^2 - 1 = 0$$

has no root greater than 1. By the substitution $x = y + 1$ we *transform* the given equation into

$$y^7 + 7y^6 + 19y^5 + 25y^4 + 15y^3 + 11y^2 + 17y + 8 = 0.$$

Since the roots of this new equation are equal to the roots of the original equation diminished by 1 ($y = x - 1$), we must show that the new equation has no root greater than 0. We *solve* this problem simply by noting that all the coefficients in the new equation are positive, whence y cannot also be positive and yet yield a zero sum. Now if we *invert* the transformation we obtain the desired result.

Geometrical transformations, as indeed transformations in other areas of mathematics, are useful not only in solving problems, but also in discovering new facts. We *transform* a given figure into a new figure; by studying the new figure we *discover* some property in it; then we *invert* to obtain a property of the original figure. In this chapter we shall examine some elementary geometrical transformations that can frequently be used to simplify the solution of geometrical problems or to discover new geometrical facts. Applications of the *transform-solve-invert* and *transform-discover-invert* procedures will appear both in this chapter and, along with further transformations, in other parts of the book.

In a later chapter we shall look more deeply into the idea of geometrical transformation, for besides serving as a tool for solution and discovery, the concept leads to a great unifying and codifying principle in geometry.

3.1 TRANSFORMATION THEORY

As an illustration of the principal concept to be introduced in this section, consider the set B of all books in some specific library and the set P of all positive integers. Let us associate with each book of the library the number of pages in the book. In this way we make correspond to each element of set B a unique element of set P, and we say that "the set B has been mapped into the set P." As another illustration, let N be the set of all names listed in some given telephone directory and let A be the set of twenty-six letters of the alphabet. Let us associate with each name in the directory the last letter of the surname, thus making correspond to each element of set N a unique element of set A. This correspondence defines "a mapping of set N into set A." These are examples of the following formal definition.

3.1.1 DEFINITIONS AND NOTATION. If A and B are two (not necessarily distinct) sets, then a *mapping* of set A into set B is a correspondence that associates with each element a of A a unique element b of B. We write $a \to b$, and call b the *image* (or *map*) of a under the mapping, and we say that element a has been *carried into* (or *mapped into*) element b by the mapping. If every element of B is the image of some element of A, then we say that set A has been mapped *onto* set B.

Thus if A is the set $\{1,2,3,4\}$ and B the set $\{a,b,c\}$, the associations

$$1 \to a, \ 2 \to b, \ 3 \to b, \ 4 \to a$$

define a mapping of set A into set B. This, however, is not a mapping of set A onto set B, since element c of B is not the image under the mapping of any element of set A. On the other hand, the mapping induced by the associations

$$1 \to a, \ 2 \to b, \ 3 \to b, \ 4 \to c$$

is a mapping of set A onto set B, for now every element of B is the unique image of some element of A.

A very important kind of mapping of a set A onto a set B is one in which distinct elements of set A have distinct images in set B. We assign a special name to such mappings.

3.1.2 DEFINITION. A mapping of a set A onto a set B in which distinct elements of A have distinct images in B is called a *transformation* (or *one-to-one mapping*) of A onto B.

3.1.3 DEFINITIONS AND NOTATION. If, in Definition 3.1.2, A and B are the same set, then the mapping is a transformation of a set A onto itself. In this case there may be an element of A which corresponds to itself. Such an element is called an *invariant element* (or *double element*) of the transformation. A transformation of a set A onto itself in which every element is an invariant element is called the *identity transformation* on A, and will, when no ambiguity is involved, be denoted by I.

3.1.4 DEFINITION AND NOTATION. It is clear that a transformation of set A onto set B defines a second transformation, of set B onto set A, wherein an element of B is carried into the element of A of which it was the image under the first transformation. This second transformation is called the *inverse* of the first transformation. If T represents a transformation of a set A onto a set B, then the inverse transformation will be denoted by T^{-1}.

Thus if, among the married couples of a certain city, we let A be the set of husbands and B the set of wives, then the mapping which associates with each man of set A his wife in set B is a transformation of set A onto set B. The inverse of this transformation is the mapping of B onto A in which each woman in set B is associated with her husband in set A.

We now introduce the notion of product of two transformations, and examine some properties of products.

3.1.5 DEFINITIONS AND NOTATION. Let T_1 be a transformation of set A onto set B and T_2 a transformation of set B onto set C. The performance of transformation T_1 followed by transformation T_2 induces a transformation T of set A onto set C, wherein an element a of A is associated with the element c of C which is the image under T_2 of the element b of B which is the image under T_1 of element a of A. Transformation T is called the *product*, T_2T_1, of transformations T_1 and T_2, taken in this order. If the product transformation T_2T_1 exists, we say that T_2 is *compatible* with T_1.

Note that in the product T_2T_1, transformation T_1 is to be performed first, then transformation T_2. That is, we perform the component transformations from *right to left*. This is purely a convention and we could, as some writers do, have agreed to write the product the other way about. We adopt the present convention because it better fits into the algebraic treatment of transformations to be given in a later chapter.

3.1.6 THEOREM. *If* T_2 *is compatible with* T_1, *it does not follow that* T_1 *is compatible with* T_2. *If, however, both* T_1 *and* T_2 *are transformations of a set* A *onto itself, then necessarily both* T_2 *is compatible with* T_1 *and* T_1 *is compatible with* T_2.

The reader can easily construct an example where T_2T_1 exists but T_1T_2 does not exist. The second part of the theorem is quite obvious.

3.1.7 DEFINITION. A transformation T of a set A onto itself is said to be *involutoric* if $T^2 \equiv TT = I$.

3.1.8 THEOREM. *A product of two compatible transformations, even if each is a transformation of a set* A *onto itself, is not necessarily commutative; that is, if* T_1T_2 *and* T_2T_1 *both exist, we do not necessarily have* $T_1T_2 = T_2T_1$.

Let A be the set of all points of a plane on which a rectangular coordinate framework has been superimposed. Let T_1 be the transformation of A onto itself which carries each point of A into a point one unit in the direction of the positive x axis, and let T_2 be the transformation of A onto itself which rotates each point of A counterclockwise about the origin through $90°$. Under T_2T_1 the point $(1,0)$ is carried into the point $(0,2)$, whereas under T_1T_2 it is carried into the point $(1,1)$. It follows that $T_1T_2 \neq T_2T_1$.

3.1.9 THEOREM. *Multiplication of compatible transformations is associative; that is, if* T_1, T_2, T_3 *are transformations such that* T_2 *is compatible with* T_1 *and* T_3 *with* T_2, *then* $T_3(T_2T_1) = (T_3T_2)T_1$.

For both $T_3(T_2T_1)$ and $(T_3T_2)T_1$ denote the resultant transformation obtained by first performing T_1, then T_2, then T_3.

We leave to the reader the establishment of the following three theorems.

3.1.10   THEOREM.   *If* T *is a transformation of set* A *onto itself, then* (1) $TI = IT = T$, (2) $TT^{-1} = T^{-1}T = I$.

3.1.11   THEOREM.   *If* T *and* S *are transformations of a set* A *onto itself, and if* $TS = I$, *then* $S = T^{-1}$.

3.1.12   THEOREM.   *If transformation* T_2 *is compatible with transformation* T_1, *then* $(T_2T_1)^{-1} = T_1^{-1}T_2^{-1}$.

We conclude with a definition which will be basic in a later chapter.

3.1.13   DEFINITIONS.   A nonempty set of transformations of a set A onto itself is said to constitute a *transformation group* if the inverse of every transformation of the set is in the set and if the product of any two transformations of the set is in the set. If, in addition, the product of every two transformations of the set is commutative, then the transformation group is said to be *abelian* (or *commutative*).

PROBLEMS

1. If A represents the set of all integers, which of the following mappings of A into itself are mappings of A onto itself? Which are transformations of A onto itself?

   (a) $a \rightarrow a + 5$ (b) $a \rightarrow a + a^2$
   (c) $a \rightarrow a^5$ (d) $a \rightarrow 2a - 1$
   (e) $a \rightarrow 5 - a$ (f) $a \rightarrow a - 5$

2. If R represents the set of all real numbers, which of the following mappings of R into itself are mappings of R onto itself? Which are transformations of R onto itself?

   (a) $r \rightarrow 2r - 1$ (b) $r \rightarrow r^2$
   (c) $r \rightarrow r^3$ (d) $r \rightarrow 1 - r$
   (e) $r \rightarrow r + r^2$ (f) $r \rightarrow 5r$

3. If R represents the set of all real numbers, is the mapping indicated by the association $r \rightarrow r^3 - r$ a mapping of R onto itself? Is it a transformation of R onto itself?

4. (a) Generalize Definition 3.1.5 for the situation where T_1 is a mapping of set A *into* set B and T_2 is a mapping of set B *into* set C.
   (b) Let T_1 and T_2 be the mappings of the set N of natural numbers into itself indicated by the associations $n \rightarrow n^2$ and $n \rightarrow 2n + 3$ respectively. Find the associations for the mappings T_1T_2, T_2T_1, T_1^2, T_2^2, $(T_1T_2)T_1$, $T_1(T_2T_1)$.

5. Supply a proof for Theorem 3.1.6.

6. Establish Theorem 3.1.10.

7. Establish Theorem 3.1.11.

8. Establish Theorem 3.1.12.

9. If T is a transformation of set A onto set B, show that $(T^{-1})^{-1} = T$.

10. If T is an involutoric transformation, show that $T = T^{-1}$.

11. If T_1, T_2, T_3 are transformations of a set A onto itself, show that $(T_3 T_2 T_1)^{-1} = T_1^{-1} T_2^{-1} T_3^{-1}$.

12. DEFINITION. If T and S are two transformations of set A onto itself, the transformation STS^{-1} is called the *transform of* T *by* S.
    (a) Show that the transform of the inverse of T by S is the inverse of the transform of T by S.
    (b) If $TS = ST$, show that each transformation is its own transform by the other.
    (c) If T_1, T_2, S are all transformations of set A onto itself, show that the product of the transforms of T_1 and T_2 by S is the transform of $T_2 T_1$ by S.

13. In abstract algebra a *group* is defined to be a set G of elements in which a binary operation $*$ is defined satisfying the following four postulates:
    G1: *For all* a, b *in* G, a $*$ b *is in* G.
    G2: *For all* a, b, c *in* G, (a $*$ b) $*$ c = a $*$ (b $*$ c).
    G3: *There exists an element* i *of* G *such that, for all* a *in* G, a $*$ i = a.
    G4: *For each element* a *of* G *there exists an element* a^{-1} *of* G *such that* a $*$ a^{-1} = i.

    Show that a transformation group is a group in the sense of abstract algebra, where transformation multiplication plays the role of the binary operation.

3.2 FUNDAMENTAL POINT TRANSFORMATIONS OF THE PLANE

Let S be the set of all points of an ordinary plane. In this section we consider some fundamental transformations of the set S onto itself.

3.2.1 DEFINITIONS AND NOTATION. Let $\overline{AB}$ be a directed line segment in the plane. By the *translation* $T(AB)$ we mean the transformation of S onto itself which carries each point P of the plane into the point P' of the plane such that $\overline{PP'}$ is equal and parallel to $\overline{AB}$. The directed segment $\overline{AB}$ is called the *vector* of the translation.

3.2.2 DEFINITIONS AND NOTATION. Let O be a fixed point of the plane and θ a given sensed angle. By the *rotation* $R(O,\theta)$ we mean the transformation of S onto itself which carries each point P of the plane into the point P' of the plane such that $OP' = OP$ and $\not\angle \overline{POP'} = \theta$. Point O is called the *center* of the rotation, and θ is called the *angle* of the rotation.

3.2.3 DEFINITIONS AND NOTATION. Let l be a fixed line of the plane. By the *reflection* $R(l)$ *in line* l we mean the transformation of S onto itself which carries each point P of the plane into the point P' of the plane such that l is the perpendicular bisector of PP'. The line l is called the *axis* of the reflection.

3.2.4 DEFINITIONS AND NOTATION. Let O be a fixed point of the plane. By the *reflection* (or *half-turn*) $R(O)$ *in* (*about*) *point* O we mean the transformation of S onto itself which carries each point P of the plane into the point P' of the plane such that O is the midpoint of PP'. Point O is called the *center* of the reflection.

3.2.5 DEFINITIONS AND NOTATION. Let O be a fixed point of the plane and k a given nonzero real number. By the *homothety* (or *expansion*, or *dilatation*, or *stretch*) $H(O,k)$ we mean the transformation of S onto itself which carries each point P of the plane into the point P' of the plane such that $\overline{OP'} = k\ \overline{OP}$. The point O is called the *center* of the homothety, and k is called the *ratio* of the homothety.

There are certain products of the above transformations which also are of fundamental importance.

3.2.6 DEFINITIONS AND NOTATION. Let l be a fixed line of the plane and $\overline{AB}$ a given directed segment on l. By the *glide-reflection* $G(l,AB)$ we mean the product $R(l)T(AB)$. The line l is called the *axis* of the glide-reflection, and the directed segment $\overline{AB}$ on l is called the *vector* of the glide-reflection.

3.2.7 DEFINITIONS AND NOTATION. Let l be a fixed line of the plane and O a fixed point on l, and let k be a given nonzero real number. By the *stretch-reflection* $S(O,k,l)$ we mean the product $R(l)H(O,k)$. The line l is called the *axis* of the stretch-reflection, the point O is called the *center* of the stretch-reflection, and k is called the *ratio* of the stretch-reflection.

3.2.8 DEFINITIONS AND NOTATION. Let O be a fixed point of the plane, k a given nonzero real number, and θ a given sensed angle. By the *homology* (or *stretch-rotation*, or *spiral rotation*) $H(O,k,\theta)$ we mean the product $R(O,\theta)H(O,k)$. Point O is called the *center* of the homology, k the *ratio* of the homology, and θ the *angle* of the homology.

The following theorems are easy consequences of the above definitions.

3.2.9 THEOREM. *If* n *is an integer, then* $R(O,(2n + 1)180°) = R(O) = H(O, -1)$.

3.2.10 THEOREM. *If* n *is an integer, then* (1) $H(O,k,n360°) = H(O,k)$, (2) $H(O,k,(2n + 1)180°\ H) = (O, -k)$.

3.2.11 THEOREM. $T(BC)T(AB) = T(AB)T(BC) = T(AC)$.

3.2.12 THEOREM. $R(O,\theta_1)R(O,\theta_2) = R(O,\theta_2)R(O,\theta_1) = R(O,\theta_1 + \theta_2)$.

3.2.13 THEOREM. $R(O,\theta)H(O,k) = H(O,k)R(O,\theta) = H(O,k,\theta)$.

3.2.14 THEOREM. *If $\overline{AB}$ is on* l, *then* $R(l)T(AB) = T(AB)R(l) = G(l,AB)$.

3.2.15 THEOREM. *If O is on* l, *then* $R(l)H(O,k) = H(O,k)R(l) = S(O,k,l)$.

3.2.16 THEOREM. (1) $[T(AB)]^{-1} = T(BA)$, (2) $[R(O,\theta)]^{-1} = R(O, -\theta)$, (3) $[R(l)]^{-1} = R(l)$, (4) $[R(O)]^{-1} = R(O)$, (5) $[H(O,k)]^{-1} = H(O,1/k)$, (6) $[G(l,AB)]^{-1} = G(l,BA)$, (7) $[H(O,k,\theta)]^{-1} = H(O,1/k, -\theta)$.

3.2.17 THEOREM. (1) $T(AA) = I$, (2) $R(O,n360°) = I$, *where* n *is any integer*, (3) $H(O,l) = I$.

3.2.18 THEOREM. $R(l)$ *and* $R(O)$ *are involutoric transformations.*

3.2.19 THEOREM. *In the unextended plane* (1) *a translation of nonzero vector has no invariant points,* (2) *a rotation of an angle which is not a multiple of* 360° *has only its center as an invariant point,* (3) *a reflection in a line has only the points of its axis as invariant points,* (4) *a reflection in a point has only its center as an invariant point,* (5) *a homothety of ratio different from* 1 *has only its center as an invariant point.*

PROBLEMS

1. Let O, P, M, N, referred to a rectangular cartesian coordinate system, be the points $(0,0)$, $(1,1)$, $(1,0)$, $(2,0)$ respectively, and let l denote the x axis. Find the coordinates of the point P' obtained from the point P by the following transformations: (a) $T(OM)$, (b) $R(O,90°)$, (c) $R(l)$, (d) $R(M)$. (e) $R(O)$, (f) $H(O,2)$, (g) $H(N,-2)$, (h) $H(M,\frac{1}{2})$, (i) $G(l,MN)$, (j) $S(O,2,l)$, (k) $H(O,2,90°)$, (l) $H(N,2,45°)$.

2. (a) If $O_1 \neq O_2$, are the rotations $R(O_1,\theta_1)$ and $R(O_2,\theta_2)$ commutative?
   (b) Are $R(O)$ and $R(l)$ commutative?
   (c) Are $R(O_1O_2)$ and $R(O_1,\theta)$ commutative?
   (d) Are $T(AB)$ and $R(l)$ commutative?
   (e) Are $T(AB)$ and $R(O)$ commutative?

3. If AB is carried into $A'B'$ by a rotation, locate the center of the rotation.

4. Let P map into P' under a glide-reflection. (a) Show that PP' is bisected by the axis of the glide-reflection. (b) Show that the square of the glide-reflection is a translation of twice the vector of the glide-reflection.

5. Let $ABCD$ be a square with center O. Show that $R(B,90°) R(C,90°) = R(O)$.

6. Let S be the square whose vertices are $A:(1,1)$, $B:(-1,1)$, $C:(-1,-1)$, $D:(1,-1)$, and let O be the origin. (a) Show that S is carried into itself under each of the transformations: $R(x\text{ axis})$, $R(y\text{ axis})$, $R(AC)$, $R(BD)$, $R(O,90°)$, $R(O)$, $R(O,270°)$, I. (b) Show that the transformations of part (a) form a transformation group.

7. Show that each of the following sets of point transformations in a plane constitutes an abelian transformation group: (a) all translations, (b) all concentric rotations, (c) all concentric homotheties, (d) all concentric homologies.

8. Show that $R(O_2)R(O_1) = T(2O_1O_2)$.

9. (a) Show that $T(AB)R(O)$ is a reflection in point O' such that $\overline{OO'}$ is equal and parallel to $(\overline{AB})/2$.
   (b) Show that $R(O)T(AB)$ is a reflection in point O' such that $\overline{O'O}$ is equal and parallel to $(\overline{AB})/2$.
   (c) Show that $T(OO')R(O) = R(M)$, where M is the midpoint of OO'.

10. (a) Show that $R(O_3)R(O_2)R(O_1)$ is a reflection in point O such that $\overline{OO_3}$ is equal and parallel to $\overline{O_1O_2}$.
    (b) Show that $R(O_3)R(O_2)R(O_1) = R(O_1)R(O_2)R(O_3)$.

11. Show that $R(OO')R(O) = R(O)R(OO') = R(l)$, where l is the line through O perpendicular to OO'.

12. Show that if $O \neq O'$, then $R(O',-\theta)R(O,\theta)$ is a translation.

13. Where is a point P if its image under the homothety $H(O_1,k_1)$ coincides with its image under the homothety $H(O_2,k_2)$, $k_1 \neq k_2$?

14. Let l_1 and l_2 be two lines intersecting in a point O and let θ be the angle from l_1 to l_2. Show that $G(l_2,CD)G(l_1,AB) = T(CD)R(O,2\theta)T(AB)$. In particular, if $\theta = 90°$, then $G(l_2,CD)G(l_1,AB) = T(CD)R(O)T(AB)$.

3.3 APPLICATIONS OF THE HOMOTHETY TRANSFORMATION

Before continuing our study of geometrical transformations, we pause to consider a few applications of the homothety transformation.

We first describe a linkage apparatus, known as a *pantograph*, which was invented about 1603 by the German astronomer Christolph Scheiner (*ca.* 1575–1650) for mechanically copying a figure on an enlarged or reduced scale. The instrument is made in a variety of forms and can be purchased in a good stationery store. One form is pictured in Figure 3.3a, where the four equal rods are hinged by adjustable pivots at A, B, C, P, with $OA = AP$ and $PC = P'C = AB$. The instrument lies flat on the drawing paper and is

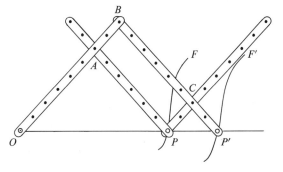

Figure 3.3a

fastened to the paper by a pointed pivot at O. Then if pencils are inserted at P and P', and P is made to trace a figure F, P' will trace the figure F' obtained from F by the homothety $H(O, \overline{OB}/\overline{OA})$. The reader can easily justify the working of the machine by showing that $APCB$ is a parallelogram, O, P, P' and collinear, and $\overline{OP'}/\overline{OP} = \overline{OB}/\overline{OA} = $ constant.

3.3.1 NOTATION. By the symbol $O(r)$ we mean the circle with center O and radius r.

3.3.2 DEFINITIONS. Let $A(a)$ and $B(b)$ be two nonconcentric circles and let I and E divide $\overline{AB}$ internally and externally in the ratio a/b. Then I and E are called the *internal* and the *external centers of similitude* of the two circles (see Figure 3.3b).

Figure 3.3b

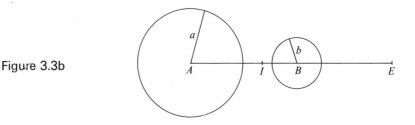

3.3.3 THEOREM. *Any two nonconcentric circles* A(a) *and* B(b) *with internal and external centers of similitude* I *and* E *are homothetic to each other under the homotheties* H(I, −b/a) *and* H(E,b/a).

Let P (see Figure 3.3c) be any point on $A(a)$ not collinear with A and B.

Figure 3.3c

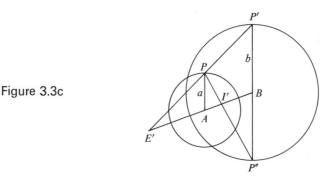

Let $P'BP''$ be the diameter of $B(b)$ parallel to AP, where $\overline{BP'}$ has the same direction as $\overline{AP}$. Let $P'P$ cut AB in E' and $P''P$ cut AB in I'. From similar triangles we find $\overline{E'P'}/\overline{E'P} = \overline{E'B}/\overline{E'A} = b/a$. Hence $E' = E$, the external center of similitude, and $B(b)$ is the image of $A(a)$ under the homothety

$H(E,b/a)$. Similarly, $\overline{I'P''}/\overline{I'P} = \overline{I'B}/\overline{I'A} = -b/a$, and $I' = I$, the internal center of similitude. It follows that $B(b)$ is the image of $A(a)$ under the homothety $H(I, -b/a)$.

3.3.4 THEOREM. *The orthocenter* H, *the circumcenter* O, *and the centroid* G *of a triangle* $A_1A_2A_3$ *are collinear and* $\overline{HG} = 2 \ \overline{GO}$.

Let M_1, M_2, M_3 (see Figure 3.3d) be the midpoints of the sides A_2A_3,

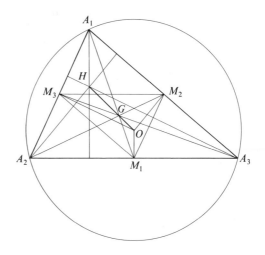

Figure 3.3d

A_3A_1, A_1A_2 of the triangle. Since $\overline{A_iG}/\overline{GM_i} = 2$ $(i = 1, 2, 3)$, triangle $M_1M_2M_3$ is carried into triangle $A_1A_2A_3$ by the homothety $H(G, -2)$. Therefore O, which is the orthocenter of triangle $M_1M_2M_3$, maps into the orthocenter H of triangle $A_1A_2A_3$. It follows that H, G, O are collinear and $\overline{HG} = 2 \ \overline{GO}$.

3.3.5 DEFINITION. The line of collinearity of the orthocenter, circumcenter, and centroid of a triangle is called the *Euler line* of the triangle.

3.3.6 THEOREM. *In triangle* $A_1A_2A_3$ *let* M_1, M_2, M_3 *be the midpoints of the sides* A_2A_3, A_3A_1, A_1A_2, H_1, H_2, H_3 *the feet of the altitudes on these sides*, N_1, N_2, N_3 *the midpoints of the segments* A_1H, A_2H, A_3H, *where* H *is the orthocenter of the triangle. Then the nine points* M_1, M_2, M_3, H_1, H_2, H_3, N_1, N_2, N_3 *lie on a circle whose center* N *is the midpoint of the segment joining the orthocenter* H *to the circumcenter* O *of the triangle, and whose radius is half the circumradius of the triangle.*

Referring to Figure 3.3e, we see that $\sphericalangle A_2A_3H_3 = 90° - \sphericalangle A_2 = \sphericalangle A_2A_1S_1 = \sphericalangle A_2A_3S_1$. Therefore right triangle HH_1A_3 is congruent to right triangle $S_1H_1A_3$, and H_1 is the midpoint of $\overline{HS_1}$. Similarly, H_2 is

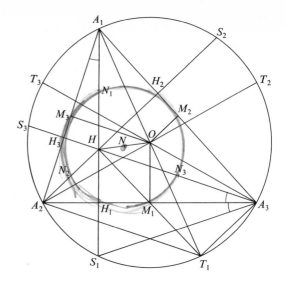

Figure 3.3e

the midpoint of HS_2 and H_3 is the midpoint of HS_3. Draw circumdiameter A_1T_1. Then T_1A_2 is parallel to A_3H_3 (since each is perpendicular to A_1A_2). Similarly, T_1A_3 is parallel to A_2H_2. Therefore $HA_3T_1A_2$ is a parallelogram and HT_1 and A_2A_3 bisect each other. That is, M_1 is the midpoint of HT_1. Similarly, M_2 is the midpoint of HT_2 and M_3 is the midpoint of HT_3. It now follows that the homothety $H(H,\frac{1}{2})$ carries A_1, A_2, A_3, S_1, S_2, S_3, T_1, T_2, T_3 into N_1, N_2, N_3, H_1, H_2, H_3, M_1, M_2, M_3, whence these latter nine points lie on a circle of radius half that of the circumcircle and with center N at the midpoint of HO.

3.3.7 DEFINITION. The circle of Theorem 3.3.6 is called the *nine-point circle* of triangle $A_1A_2A_3$.

It was O. Terquem who named this circle the *nine-point circle*, and this is the name commonly used in the English-speaking countries. Some French geometers refer to it as *Euler's circle*, and German geometers usually call it *Feuerbach's circle*.

3.3.8 DEFINITION. Let I and E be the internal and external centers of similitude of two given nonconcentric circles $A(a)$, $B(b)$ having unequal radii. Then the circle on IE as diameter is called the *circle of similitude* of the two given circles.

3.3.9 THEOREM. *Let* P *be any point on the circle of similitude of two nonconcentric circles* A(a), B(b) *having unequal radii. Then* B(b) *is the image of* A(a) *under the homology* H(P,b/a, $\angle \overline{APB}$).

Let I and E be the internal and external centers of similitude of the two given circles. If P coincides with I or E the theorem follows from Theorem 3.3.3. If P is distinct from I and E (see Figure 3.3f) then PI is perpendicular

Figure 3.3f

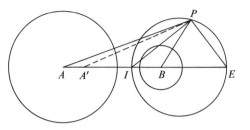

to PE and $(AB,IE) = -1$. Draw PA' so that PI bisects $\measuredangle A'PB$ internally. Then PE is the external bisector of the same angle, and it follows that $(A'B,IE) = -1$. Therefore $A' = A$ and $PB/PA = \overline{IB}/\overline{AI} = b/a$. The theorem now follows.

3.3.10 COROLLARY. *The locus of a point* P *moving in a plane such that the ratio of its distance from point* A *to its distance from point* B *of the plane is a positive constant* k ≠ 1 *is the circle on* IE *as diameter, where* I *and* E *divide the segment* $\overline{AB}$ *internally and externally in the ratio* k.

3.3.11 DEFINITION. The circle of Corollary 3.3.10 is called the *circle of Apollonius* of points A and B for the ratio k (see Section 1.6).

PROBLEMS

1. In Figure 3.3a show that O, P, P' are collinear and that $\overline{OP'}/\overline{OP} = \overline{OB}/\overline{OA}$.

2. In Figure 3.3g, AE, AB, BF represent three bars jointed at A and B. The bars AE and BF are attached at A and B to wheels of the same diameter, and around which goes a thin flexible steel band C. The result is that if bars AE and BF are so adjusted as to be parallel, they remain parallel however they are situated with respect to bar AB. The bars AE and BF are adjustable in length, and pencils are inserted at points E and F. D is a point adjustable along bar AB and about which the whole instrument can be rotated. Show that if D is fastened to AB so as to be collinear with E and F, then the pencil at E describes the map of a figure traced by the pencil at F under the homothety $H(D, -EA/BF)$.

3. Prove that if two circles have common external tangents, these tangents pass through the external center of similitude of the two circles, and if they have common internal tangents, these pass through the internal center of similitude of the two circles.

4. Let S be a center of similitude of two circles C_1 and C_2, and let one line through S cut C_1 in A and B and C_2 in A' and B', and a second line through S cut C_1

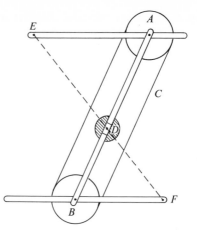

Figure 3.3g

in C and D and C_2 in C' and D', where the primed points are the maps of the corresponding unprimed points under the homothety having center S and carrying circle C_1 into circle C_2. Show that: (a) $B'D'$ is parallel to BD, (b) A', C', D, B are concyclic and A, C, D', B' are concyclic, (c) $(SA')(SB)$ $= (SA)(SB') = (SC')(SD) = (SC)(SD')$, (d) the tangents to C_1 and C_2 at B and A' intersect on the radical axis of C_1 and C_2.

5. Prove that the circle of similitude of two nonconcentric circles with unequal radii is the locus of points from which the two circles subtend equal angles.

6. Prove that two circles and their circle of similitude are coaxial.

7. (a) Show that the external centers of similitude of three circles with distinct centers taken in pairs are collinear.
   (b) Show that the external center of similitude of one pair of the circles and the internal centers of similitude of the other two pairs are collinear.

8. If a circle is tangent to each of two given nonconcentric circles, show that the line determined by the two points of tangency passes through a center of similitude of the two given circles.

9. If the distance between the centers of two circles $A(a)$ and $B(b)$ is c, locate the center of the circle of similitude of the two circles.

10. Show that any circle through the centers of two given nonconcentric circles of unequal radii is orthogonal to the circle of similitude of the two given circles.

11. In the notation of Theorems 3.3.4 and 3.3.6 show that:
    (a) $(HG,NO) = -1$.
    (b) The sum of the powers of the vertices A_1, A_2, A_3 with respect to the nine-point circle is $(A_2 A_3^2 + A_3 A_1^2 + A_1 A_2^2)/4$.

12. Prove that the circumcenter of the triangle formed by the tangents to the circumcircle of a given triangle at the vertices of the given triangle lies on the Euler line of the given triangle.

13. On the arc $M_1 H_1$ of the nine-point circle, take the point X_1 one-third the way from M_1 to H_1, Take similar points X_2 and X_3 on arcs $M_2 H_2$ and $M_3 H_3$. Show that triangle $X_1 X_2 X_3$ is equilateral.

14. If the Euler line is parallel to the side $A_2 A_3$, then $\tan A_2 \tan A_3 = 3$, and $\tan A_2$, $\tan A_1$, $\tan A_3$ are in arithmetic progression.

15. Show that the trilinear polar (see Problem 11 Section 2.4) of the orthocenter of a triangle is perpendicular to the Euler line of the triangle.

3.4 ISOMETRIES

In this section and the next we consider those point transformations of the unextended plane which preserve all lengths and those which preserve all shapes. These are known, respectively, as *isometries* and *similarities*. We commence with a formal definition of these concepts.

3.4.1 DEFINITIONS. A point transformation of the unextended plane onto itself which carries each pair of points A, B into a pair A', B' such that $A'B' = k(AB)$, where k is a fixed positive number, is called a *similarity* (or an *equiform transformation*), and the particular case where $k = 1$ is called an *isometry* (or a *congruent transformation*). A similarity is said to be *direct* or *opposite* according as $\triangle \overline{ABC}$ has or has not the same sense as $\triangle \overline{A'B'C'}$. (A direct similarity is sometimes called a *similitude*, and an opposite similarity an *antisimilitude*. A direct isometry is sometimes called a *displacement*, and an opposite isometry a *reversal*.)

It is very interesting that isometries and similarities can be factored into products of certain of the fundamental point transformations considered in Section 3.2. We proceed to obtain some of these factorizations for the isometries.

3.4.2 THEOREM. *There is a unique isometry that carries a given noncollinear triad of points* A, B, C *into a given congruent triad* A', B', C'.

Superimposing the plane (by sliding, or turning it over and then sliding) upon its original position so that triangle ABC coincides with triangle $A'B'C'$ induces an isometry of the plane onto itself in which the point triad A, B, C is carried into the point triad A', B', C'. There is only the one isometry, for if P is any point in the plane there is a unique point P' in the plane such that $P'A' = PA$, $P'B' = PB$, $P'C' = PC$.

3.4.3 THEOREM. *An isometry can be expressed as the product of at most three reflections in lines.*

Let an isometry carry the triad of points A, B, C into the congruent triad A', B', C'. We consider four cases. (1) If the two triads coincide, the isometry (by Theorem 3.4.2) is the identity I, which may be considered as the product of the reflection $R(l)$ with itself, where l is any line in the plane. (2) If A coincides with A' and B with B', but C and C' are distinct, the isometry (by Theorem 3.4.2) is the reflection $R(l)$, where l is the line AB. (3) If A

coincides with A', but B and B' and C and C' are distinct, the reflection $R(l)$, where l is the perpendicular bisector of BB', reduces this case to one of the two previous cases. (4) Finally, if A and A', B and B', C and C' are distinct, the reflection $R(l)$, where l is the perpendicular bisector of AA', reduces this case to one of the first three cases. In each case, the isometry is ultimately expressed as a product of no more than three reflections in lines.*

3.4.4 THEOREM. *An isometry with an invariant point can be represented as the product of at most two reflections in lines.*

Let A be an invariant point of the isometry and let B and C be two points not collinear with A. Then the point triad A, B, C is carried into the triad A', B', C' where A' coincides with A. The desired result now follows from the first three cases in the proof of Theorem 3.4.3.

3.4.5 THEOREM. *Let l_1 and l_2 be any two lines of the plane intersecting in a point O, and let θ be the directed angle from l_1 to l_2, then $R(l_2)R(l_1) = R(O,2\theta)$. Conversely, a rotation $R(O,2\theta)$ can be factored into the product $R(l_2)R(l_1)$ of reflections in two lines l_1 and l_2 through O, where either line may be arbitrarily chosen through O and then the other such that the directed angle from l_1 to l_2 is equal to θ.*

The proof is apparent from Figure 3.4a, since $OP' = OP$ and $\sphericalangle \overline{POP'} = \sphericalangle \overline{POQ} + \sphericalangle \overline{QOP'} = 2\ \sphericalangle \overline{L_1OQ} + 2\ \sphericalangle \overline{QOL_2} = 2\theta.$

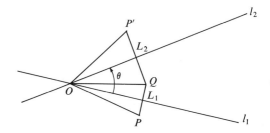

Figure 3.4a

3.4.6 THEOREM. *Let l_1 and l_2 be any two parallel (or coincident) lines of*

* One recalls the little test, that made the rounds of the mathematics meetings some years ago, for ferreting out incipient mathematicians. The test consists of two questions. (1) *You are in a room devoid of all furnishings except for a gas stove in one corner with one burner lit, and there is a kettle of water on the floor. What would you do to get the water in the kettle warm?* ANSWER: Place the kettle of water on the lighted burner. (2) *We have the same situation as in question (1) except that now there is also a table in the room and the kettle of water is on the table. What would you do to get the water in the kettle warm?* To give the same answer as before would be fatal. The correct answer is, "Remove the kettle from the table and place it on the floor, thus reducing the problem to one that has already been solved."

the plane, and let $\overline{A_1A_2}$ be the directed distance from line l_1 to line l_2, then $R(l_2)R(l_1) = T(2A_1A_2)$. Conversely, a translation $T(2A_1A_2)$ can be factored into the product $R(l_2)R(l_1)$ of reflections in two lines l_1 and l_2 perpendicular to A_1A_2, where either line may be arbitrarily chosen perpendicular to A_1A_2 and then the other such that the directed distance from l_1 to l_2 is equal to $\overline{A_1A_2}$.

The proof is apparent from Figure 3.4b, since PP' is parallel to A_1A_2 and $\overline{PP'} = \overline{PQ} + \overline{QP'} = 2\overline{L_1Q} + 2\overline{QL_2} = 2\overline{A_1A_2}$.

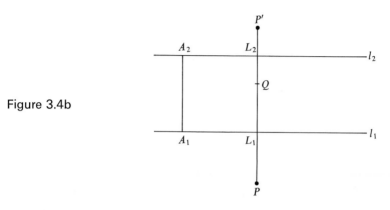

Figure 3.4b

3.4.7 THEOREM. *Any direct isometry is either a translation or a rotation.*

By Theorem 3.4.3, the isometry is a product of at most three reflections in lines. Since the isometry is direct, it must be a product of an *even* number of such reflections, and therefore of two such reflections. If the axes of the two reflections are parallel (or coincident), the isometry is a translation (by Theorem 3.4.6); otherwise the isometry is a rotation (by Theorem 3.4.5).

3.4.8 LEMMA. $R(O)R(l) = G(m, 2MO)$, *where* m *is the line through* O *perpendicular to* l *and cutting* l *in point* M.

For, see Figure 3.4c, $\overline{PN} = \overline{QQ''} + \overline{Q''Q'} = 2(\overline{MQ''} + \overline{Q''O}) = 2\overline{MO}$ and $\overline{NQ'} = \overline{Q'P'}$.

3.4.9 THEOREM. *An opposite isometry* T *is either a reflection in a line or a glide-reflection.*

By Theorem 3.4.4, if T has an invariant point, T must be a reflection in a line. Suppose T has no invariant point and let T carry point A into point A'. Let O be the midpoint of AA'. Then $R(O)T$ is an opposite isometry with invariant point A. Therefore $R(O)T = R(l)$, for some line l. That is, $T = [R(O)]^{-1}R(l) = R(O)R(l) = $ a glide-reflection, by Lemma 3.4.8.

3.4.10 THEOREM. *A product of three reflections in lines is either a reflection in a line or a glide-reflection.*

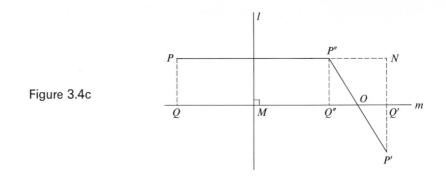

Figure 3.4c

For a product of three reflections in lines is an opposite isometry, and (by Theorem 3.4.9) an opposite isometry is either a reflection in a line or a glide-reflection.

PROBLEMS

1. Prove that an isometry maps straight lines into straight lines.

2. (a) Show that the product of two rotations is a rotation or a translation.
   (b) Show that a direct isometry which is not a translation has exactly one invariant point.

3. (a) Prove that any isometry with an invariant point is a rotation or a reflection in a line according as it is direct or opposite.
   (b) Prove that every opposite isometry with no invariant point is a glide-reflection.
   (c) Prove that if an isometry has more than one invariant point, it must be either the identity or a reflection in a line.

4. (a) Show that $R(l)T(AB)$ is a glide-reflection whose axis is a line m parallel to l at a distance equal to one-half the projection of $\overline{BA}$ on a line perpendicular to l, and whose vector is the projection of $\overline{AB}$ on l.
   (b) Show that $T(AB)R(l)$ is a glide-reflection whose axis is a line m parallel to l at a distance equal to one-half the projection of $\overline{AB}$ on a line perpendicular to l, and whose vector is the projection of $\overline{AB}$ on l.

5. Show that every opposite isometry is the product of a reflection in a line and a reflection in a point.

6. Show that $T(BA)R(O,\theta)T(AB) = R(O',\theta)$, where $\overline{O'O}$ is equal and parallel to $\overline{AB}$.

7. Show that $R(OO')R(O,\theta) = R(l)$, where l passes through O and the directed angle from l to OO' is $\theta/2$.

8. Let O, O' be two points on line l. Show that $G(l,2OO') = R(O')R(m)$, where m is the line through O perpendicular to l.

9. Let l_1, l_2, l_3 be the lines of the sides $A_2 A_3$, $A_3 A_1$, $A_1 A_2$ of a triangle $A_1 A_2 A_3$.
   (a) Show that $G(l_3, A_1 A_2)G(l_2, A_3 A_1)G(l_1, A_2 A_3)$ is a glide-reflection.

(b) Show that $[R(l_3)R(l_2)R(l_1)R(l_3)R(l_2)]^2$ is a translation along l_1.

(c) Show that $G(l_3,A_1A_2)G(l_2,A_3\,A_1)G(l_1,A_2A_3) = R(l_3)R(A_2\,,2\not{\angle}A_3)$.

(d) Show that $R(l_3)R(l_2)R(l_1) = R(l_3)R(A_3\,,2\not{\angle}A_3)$.

(e) If triangle $A_1A_2\,A_3$ is acute, show that $R(l_3)R(l_2)R(l_1)$ is a glide-reflection with axis H_1H_3 and having direction $\overline{H_1H_3}$ and length equal to the perimeter of the orthic triangle $H_1H_2\,H_3$ of triangle $A_1A_2\,A_3$.

10. If l_1, l_2, l_3 are concurrent in a point O, show that $R(l_3)R(l_2)R(l_1)$ is a reflection in a line through O.

11. If l_1, l_2, l_3 are parallel, show that $R(l_3)R(l_2)R(l_1)$ is a reflection in a line parallel to l_1, l_2, l_3.

3.5 SIMILARITIES

We now examine the similarities.

3.5.1 THEOREM. *There is a unique similarity that carries a given non-collinear triad of points* A, B, C *into a given similar triad* A', B', C'.

If the triads are congruent, the unique similarity is the unique isometry guaranteed by Theorem 3.4.2. If the triads are not congruent, choose a point O of the plane. Now there is a homothety T_1 with center O carrying the triad A, B, C into a triad A'', B'', C'' congruent to the triad A', B', C', and (by Theorem 3.4.2) an isometry T_2 carrying triad A'', B'', C'' into the triad A', B', C'. Therefore the similarity $T_2\,T_1$ carries triad A, B, C into triad A', B', C'. But this is the only similarity, for if P is any point of the plane there is a unique point P' such that $P'A' = kPA$, $P'B' = kPB$, $P'C' = kPC$, where $k = A'B'/AB$.

3.5.2 THEOREM. *Every nonisometric similarity has a unique invariant point.*

Let $k \neq 1$ be the ratio of the similarity S and let S carry A_0 into A_1. There is no loss in generality in assuming $k < 1$, for (since $k \neq 1$) either S or S^{-1} is actually a contraction, and S and S^{-1} have the same invariant points. If $A_1 = A_0$, then we have already found an invariant point. If $A_1 \neq A_0$, consider the sequence of points A_0, A_1, A_2, A_3, ..., where S carries A_i into A_{i+1}, $i = 0, 1, 2, \ldots$. If line segment A_0A_1 has length c, then A_1A_2 has length kc, A_2A_3 has length k^2c, etc. The circle of center A_0 and radius $c/(1-k)$ is carried into the circle of center A_1 and radius $kc/(1-k)$, then this circle into the circle of center A_2 and radius $k^2c/(1-k)$, etc. Since $c + kc/(1-k) = c/(1-k)$, these circles (see Figure 3.5a) form a nested sequence of circles whose radii tend to zero as i increases. By the theorem of nested sets of analysis, the sequence of circles converges to a point of accumulation O. Since S carries A_0, A_1, A_2, ... into A_1, A_2, A_3, ..., it follows that S leaves O invariant. Finally, S can have no more than one invariant point since the segment determined by two distinct invariant points would be mapped into itself, instead of being contracted by the ratio k.

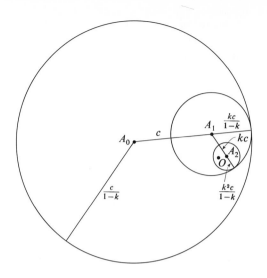

Figure 3.5a

3.5.3 THEOREM. *A direct similarity* S *is either a translation or a homology.*

If $k = 1$, S is a direct isometry and is then (by Theorem 3.4.7) either a translation or a rotation (which is a special homology). If $k \neq 1$, S has (by Theorem 3.5.2) an invariant point O. Then $S = TH(O,k)$, where T is a direct isometry with invariant point O. It follows that T is a rotation about O, and S is then a homology of center O.

3.5.4 THEOREM. *An opposite similarity* S *is either a glide-reflection or a stretch-reflection.*

If $k = 1$, S is an opposite isometry and is then (by Theorem 3.4.9) either a glide-reflection or a reflection in a line (which is a special glide-reflection). If $k \neq 1$, S has (by Theorem 3.5.2) an invariant point O. Then $S = TH(O,k)$, where T is an opposite isometry with invariant point O. It follows that T is a reflection in a line through O, and S is then a stretch-reflection of center O.

3.5.5 THEOREM. *A similarity that carries lines into parallel lines is either a translation or a homothety.*

The reader can easily show that this is a corollary of Theorems 3.5.3 and 3.5.4.

3.5.6 THEOREM. *If the line segments joining corresponding points of two*

given directly similar figures be divided proportionately, the locus of the dividing points is a figure directly similar to the given figures.

By Theorem 3.5.3, the two given figures are related by a translation or a homology. The case of a translation presents no difficulty; the locus of the dividing points is clearly a figure directly congruent to each of the two given figures. Suppose, then, that the two given figures are related by a homology $H(O,k,\theta)$. Let (see Figure 3.5b) A,A' be a fixed and P,P' a variable pair of

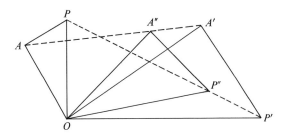

Figure 3.5b

corresponding points in the two given figures. Then $OP'/OP = OA'/OA = k$ and $\measuredangle \overline{POP'} = \measuredangle \overline{AOA'} = \theta$. Let P'' and A'' be taken on AA' and PP' such that $\overline{PP''}/\overline{P''P'} = \overline{AA''}/\overline{A''A'}$. Since triangles POP' and AOA' are directly similar, it follows that triangles POP'' and AOA'' are also directly similar, and $OP''/OP = OA''/OA = k'$, say, and $\measuredangle \overline{POP''} = \measuredangle \overline{AOA''} = \theta'$, say. It follows that the locus of P'' is the image of the locus of P under the homology $H(O,k',\theta')$. That is, the locus of P'' is a figure directly similar to the two given figures.

PROBLEMS

1. Prove that a similarity maps straight lines into straight lines and circles into circles.

2. (a) Prove that any direct similarity which is not a translation has an invariant point.
   (b) Prove that any opposite similarity which is not a glide-reflection has an invariant point.

3. (a) Show that if T is an opposite isometry, then T^2 is the identity or a translation.
   (b) Show that if T is an opposite similarity, then T^2 is a translation or a homothety.

4. If l_1 is perpendicular to l_2 and O is their point of intersection:
   (a) Show that $S(O,k,l_1) = S(O,-k,l_2)$.
   (b) Lines l_1 and l_2 are invariant under $S(O,k,l_1)$.

5. Prove Hjelmslev's Theorem: When all the points P on one line are related by an isometry to all the points P' on another line, the midpoints of the segments PP' are distinct and collinear, or else they all coincide.

6. If two maps of the same country on different scales are drawn on tracing paper and then superposed, show that there is just one place that is represented by the same spot on both maps.

7. What is the product of (a) two stretch-reflections? (b) a homology and a stretch-reflection?

8. Give a proof of Theorem 3.5.5 utilizing Desargues' Theorem.

9. Prove the following theorem, which is Problem 4025 of *The American Mathematical Monthly* (Feb. 1943):

Let A_1', A_2', ..., A_{2n}' be the vertices of equilateral triangles constructed externally (or internally) on the sides A_1A_2, A_2A_3, ..., $A_{2n}A_1$ of a plane polygon of $2n$ sides $(P) = A_1A_2 \cdots A_{2n}$, and M_1, M_2, ..., M_n be the midpoints of the principal diagonals A_1A_{n+1}, A_2A_{n+2}, ..., A_nA_{2n} of (P). The midpoints M_1', M_2', ..., M_n' of the principal diagonals $A_2'A_{n+1}'$, $A_2'A_{n+2}'$, ..., $A_n'A_{2n}'$ of the polygon $(P') = A_1'A_2' \cdots A_{2n}'$ are the vertices of equilateral triangles constructed upon the sides of the polygon $(p) = M_1M_2 \cdots M_n$.

Generalize by replacing the equilateral triangles by similar isosceles triangles.

10. (a) On the sides BC and CA of a triangle ABC, construct externally any two directly similar triangles, CBA_1 and ACB_1, Show that the midpoints of the three segments BC, A_1B_1, CA form a triangle directly similar to the two given triangles.

(b) On BC externally and on CA internally, construct any two directly similar triangles CBA_1 and CAB_1. Show that the midpoints of AB and A_1B_1 form with C a triangle directly similar to the two given triangles.

These two problems constitute Problem E 521 of *The American Mathematical Monthly* (Jan. 1943).

3.6 INVERSION

In this section we briefly consider the inversion transformation, which is perhaps the most useful transformation we have for simplifying plane figures. Use of this transformation will be made in many parts of our work.

The history of the inversion transformation is complex and not clear-cut. Inversely related points were known to François Vieta in the sixteenth century. Robert Simson, in his 1749 restoration of Apollonius' lost work *Plane Loci*, included (on the basis of commentary made by Pappus) one of the basic theorems of the theory of inversion, namely that the inverse of a straight line or a circle is a straight line or a circle. Simon A. J. L'Huilier (1750–1840) in his *Eléments d'analyse géométrique et d'analyse algébrique appliquées à la recherche des lieux géométriques* (Paris and Geneva, 1808) gave special cases of this theorem.

But inversion as a simplifying transformation for the study of figures is a product of more recent times, and was independently exploited by a number of writers. Bützberger has pointed out that Jacob Steiner disclosed, in an unpublished manuscript, a knowledge of the inversion transformation as early as 1824. It was refound in the following year by the Belgian astronomer and statistician Adolphe Quetelet. It was then found independently by L. I. Magnus, in a more general form, in 1831, by J. Bellavitis in 1836, then by J. W. Stubbs and J. R. Ingram, two Fellows of Trinity College, Dublin, in 1842 and 1843, and by Sir William Thomson (Lord Kelvin) in 1845. Thomson used inversion to give geometrical proofs of some difficult propositions in the mathematical theory of elasticity. In 1847 Liouville called inversion the *transformation by reciprocal radii*. Because of a property to be established shortly, inversion has also been called *reflection in a circle*.

3.6.1 DEFINITIONS AND NOTATION. If point P is not the center O of circle $O(r)$, the *inverse* of P in, or with respect to, circle $O(r)$ is the point P' lying on the line OP such that $(\overline{OP})(\overline{OP'}) = r^2$. Circle $O(r)$ is called the *circle of inversion*, point O the *center of inversion*, r the *radius of inversion*, and r^2 the *power of inversion*. We denote the inversion with center O and power $k > 0$ by the symbol $I(O,k)$.

From the above definition it follows that to each point P of the plane, other than O, there corresponds a unique inverse point P', and that if P' is the inverse of P, then P is the inverse of P'. Since there is no point corresponding, under the inversion, to the center O of inversion, we do not have a transformation of the set S of all points of the plane onto itself. In order to make inversion a transformation, as defined in Definition 3.1.2, we may do either of two things. We may let S' denote the set of all points of the plane except for the single point O, and then inversion will be a transformation of the "punctured plane" S' onto itself. Or we may add to the set S of all points in the plane a single ideal "point at infinity" to serve as the correspondent under the inversion of the center O of inversion, and then the inversion will be a transformation of this augmented set S'' onto itself. It turns out that the second approach is the more convenient one, and we accordingly adopt the following convention.

3.6.2 CONVENTION AND DEFINITIONS. When working with inversion, we add to the set S of all points of the plane a single ideal point at infinity, to be considered as lying on every line of the plane, and this ideal point, Z, shall be the image under the inversion of the center O of inversion, and the center O of inversion shall be the image under the inversion of this ideal point Z.* The plane, augmented in this way, will be referred to as the *inversive plane*.

Of course, Convention 3.6.2 is at variance with the earlier Convention 2.2.1. But conventions are made only for convenience, and no trouble will

---

* When inverting the center O of inversion into the ideal point Z at infinity, one is reminded of Stephen Leacock's line about the rider who "flung himself upon his horse and rode off madly in all directions."

arise if one states clearly which, if either, convention is being employed. For some investigations it is convenient to work in the ordinary plane, for others, in the extended plane, and for still others, in the inversive plane. It is to be understood that throughout the present section we shall be working in the inversive plane. The following theorem is apparent.

3.6.3 THEOREM. *Inversion is an involutoric transformation of the inversive plane onto itself which maps the interior of the circle of inversion onto the exterior of the circle of inversion and each point on the circle of inversion onto itself.*

One naturally wonders if there are any other self-inverse loci besides the circle of inversion. The next theorem deals with this matter. We recall that by "circle" is meant a straight line or a circle (see Definition 2.9.4).

3.6.4 THEOREM. *A "circle" orthogonal to the circle of inversion inverts into itself.*

This is obvious if the "circle" is a straight line, and the proof of Theorem 2.9.3 takes care of the case where the "circle" is a circle. Note that the "circle" inverts into itself as a whole and not point by point.

The following theorem suggests an easy way to construct the inverse of any given point distinct from the center of inversion.

3.6.5 THEOREM. *A point* D *outside the circle of inversion, and the point* C *where the chord of contact of the tangents from* D *to the circle of inversion cuts the diametral line* OD, *are inverse points.*

For (see Figure 3.6a), $(\overline{OD})(\overline{OC}) = (OT)^2 = r^2$.

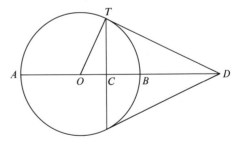

Figure 3.6a

The following four theorems relate the concept of inverse points with some earlier concepts.

3.6.6 THEOREM. *If* C, D *are inverse points with respect to circle* O(r), *then*

(AB,CD) $= -1$, *where* AB *is the diameter of* O(r) *through* C *and* D; *conversely, if* (AB,CD) $= -1$, *where* AB *is a diameter of circle* O(r), *then* C *and* D *are inverse points with respect to circle* O(r).

For (see Figure 3.6a), $(\overline{OC})(\overline{OD}) = r^2 = (OB)^2$ if and only if $(AB,CD) = -1$ (by Theorem 2.8.5).

3.6.7 THEOREM. *If* C, D *are inverse points with respect to circle* O(r), *then any circle through* C *and* D *cuts circle* O(r) *orthogonally; conversely, if a diameter of circle* O(r) *cuts a circle orthogonal to* O(r) *in* C *and* D, *then* C *and* D *are inverse points with respect to* O(r).

This, in view of Theorem 3.6.6, is merely an alternative statement of Theorem 2.9.3.

3.6.8 THEOREM. *If two intersecting circles are each orthogonal to a third circle, then the points of intersection of the two circles are inverse points with respect to the third circle.*

Let two circles intersect in points C and D and let O be the center of the third circle. Draw OC to cut the two given circles again in D′ and D″. Then D′ and D″ are each (by Theorem 3.6.7) the inverse of C with respect to the third circle. It follows that D′ = D″ = D, and C and D are inverse points with respect to the third circle.

It is Theorem 3.6.8 that has led some geometers to refer to inversion as *reflection in a circle.* For if two intersecting circles are each orthogonal to a straight line, then the points of intersection of the two circles are reflections of each other in the line. Therefore, using the terminology "reflection in a circle" for "inversion with respect to a circle," and recalling Convention 3.6.2, we may subsume both the above fact and Theorem 3.6.8 in the single statement: *If two intersecting "circles" are each orthogonal to a third "circle," then the points of intersection of the two "circles" are reflections of each other in the third "circle."* Of course the same end can be achieved by using the terminology "inversion in a line" for "reflection in a line," and some geometers do just this.

3.6.9 THEOREM. $I(O,k_2)I(O,k_1) = H(O,k_2/k_1)$.

Let P be any point and let $I(O,k_1)$ carry P into P′ and $I(O,k_2)$ carry P′ into P″. Then, if $P \neq O$, we have O, P, P′, P″ collinear and $(\overline{OP})(\overline{OP'}) = k_1$, $(\overline{OP'})(\overline{OP''}) = k_2$, whence O, P, P″ are collinear and $\overline{OP''}/\overline{OP} = k_2/k_1$. If $P = O$, then P′ = Z, P″ = O. The theorem now follows.

When a point P traces a given curve C, the inverse point P′ traces a curve C′ called the *inverse* of the given curve. The next four theorems investigate the nature of the inverses of straight lines and circles. The first of the theorems has already been established as part of Theorem 3.6.4.

3.6.10 THEOREM. *The inverse of a straight line l passing through the center O of inversion is the line l itself.*

3.6.11 THEOREM. *The inverse of a straight line l not passing through the center O of inversion is a circle C passing through O and having its diameter through O perpendicular to l.*

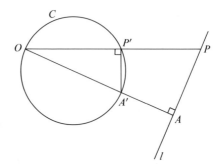

Figure 3.6b

Let point A (see Figure 3.6b) be the foot of the perpendicular dropped from O on l. Let P be any other ordinary point on l and let A', P' be the inverses of A, P. Then $(\overline{OA})(\overline{OA'}) = (\overline{OP})(\overline{OP'})$, whence $OP'/OA' = OA/OP$ and triangles $OP'A'$, OAP are similar. Therefore $\sphericalangle OP'A' = \sphericalangle OAP = 90°$. It follows that P' lies on the circle C having OA' as diameter. Conversely, if P' is any point on circle C other than O or A', let OP' cut line l in P. Then, by the above, P' must be the inverse of P. Note that point O on circle C corresponds to the point Z at infinity on l.

3.6.12 THEOREM. *The inverse of a circle C passing through the center O of inversion is a straight line l not passing through O and perpendicular to the diameter of C through O.*

Let point A (see Figure 3.6c) be the point on C diametrically opposite O, and let P be any point on circle C other than O and A. Let A', P' be the

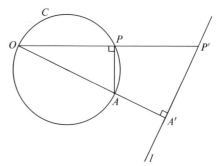

Figure 3.6c

inverses of A, P. Then $(\overline{OA})(\overline{OA'}) = (\overline{OP})(\overline{OP'})$, whence $OP'/OA' = OA/OP$ and triangles $OP'A'$, OAP are similar. Therefore $\sphericalangle OA'P' = \sphericalangle OPA = 90°$. It follows that P' lies on the line l through A' and perpendicular to OA. Conversely, if P' is any ordinary point on line l other than A', let OP' cut circle C in P. Then, by the above, P' must be the inverse of P. Note that the point Z at infinity on line l corresponds to the point O on circle C.

3.6.13 THEOREM. *The inverse of a circle C not passing through the center O of inversion is a circle C' not passing through O and homothetic to circle C with O as center of homothety.*

Let P (see Figure 3.6d) be any point on circle C. Let P' be the inverse of

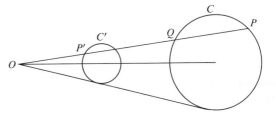

Figure 3.6d

P and let OP cut circle C again in Q, Q coinciding with P if OP is tangent to circle C. Let r^2 be the power of inversion and let k be the power of point O with respect to circle C. Then $(\overline{OP})(\overline{OP'}) = r^2$ and $(\overline{OP})(\overline{OQ}) = k$, whence $\overline{OP'}/\overline{OQ} = r^2/k$, a constant. It follows that P' describes the map of the locus of Q under the homothety $H(O,r^2/k)$. That is, P' describes a circle C' homothetic to circle C and having O as center of homothety. Since circle C does not pass through O, circle C' also does not pass through O.

3.6.14 DEFINITION. A point transformation of a plane onto itself that carries " circles " into " circles " is called a *circular*, or *Möbius, transformation*.

Combining Theorems 3.6.10 through 3.6.13 we have:

3.6.15 THEOREM. *Inversion is a circular transformation of the inversive plane.*

PROBLEMS

1. (a) Draw the figure obtained by inverting a square with respect to its center.
   (b) Draw the figure obtained by inverting a square with respect to one of its vertices.
2. (a) What is the inverse of a system of concurrent lines with respect to a point distinct from the point of concurrence?
   (b) What is the inverse of a system of parallel lines?

3. Prove that a coaxial system of circles inverts into a coaxial system of circles or into a set of concurrent or parallel lines.

4. (a) Let O be a point on a circle of center C, and let the inverse of this circle with respect to O as center of inversion intersect OC in B. If C' is the inverse of C, show that $OB = BC'$.
   (b) Show that the inverse C' of the center C of a given circle K is the inverse of the center O of inversion in the circle K' which is the inverse of the given circle K.
   (c) Calling reflection in a line inversion in the line, state the facts of parts (a) and (b) as a single theorem.
   (d) Prove that if two circles are orthogonal, the inverse of the center of either with respect to the other is the midpoint of their common chord.

5. Prove (for all cases): If two intersecting "circles" are each orthogonal to a third "circle," then the points of intersection of the two "circles" are reflections of each other in the third "circle."

6. Show that isometries and similarities are circular transformations of the ordinary plane.

3.7 PROPERTIES OF INVERSION

We first establish a very useful theorem (which will be generalized in a later chapter) concerning directed angles between two "circles." We need the following lemma, whose easy proof will be left to the reader.

3.7.1 LEMMA. *Let* C' *(see Figure 3.7a$_1$, and 3.7a$_2$) be the inverse of*

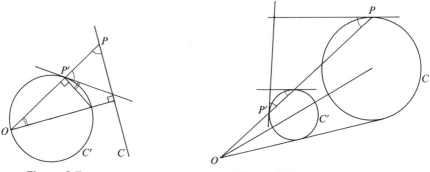

Figure 3.7a$_1$ Figure 3.7a$_2$

"circle" C, and let P, P' *be a pair of (perhaps coincident) corresponding points, under an inversion of center* O, *on C and C' respectively. Then the tangents (see Definition 2.7.4) to C and C' at P and P' are reflections of one another in the perpendicular to* OP *through the midpoint of* PP'.

3.7.2 THEOREM. *A directed angle of intersection of two "circles" is un-altered in magnitude but reversed in sense by an inversion.*

Let C and D be two "circles" intersecting in a point P, their inverses C' and D' intersecting in the inverse P' of P. Let c and d (see Figure 3.7b)

Figure 3.7b

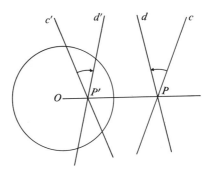

be the tangents to C and D at P, and let c' and d' be the tangents to C' and D' at P'. Since, by Lemma 3.7.1, c and c', as well as d and d', are reflections of one another in the perpendicular to OP at the midpoint of PP', it follows that the directed angle from c to d is equal but opposite to the directed angle from c' to d'.

In particular we have:

3.7.3 COROLLARY. (1) *If two "circles" are tangent, their inverses are tangent.* (2) *If two "circles" are orthogonal, their inverses are orthogonal.*

There are other things, besides the magnitudes of angles between "circles," which remain invariant under an inversion transformation. Theorems 3.7.4 and 3.7.6 give two useful invariants of this sort. Theorem 3.7.5 is an important metrical theorem which shows how inversion affects distances between points. Theorem 3.7.7 is a sample of a whole class of theorems which are valuable when using inversion in its role of a simplifying transformation. When employing a particular transformation in geometry it is of course important to know both the principal invariants of the transformation and some of the ways the transformation can simplify figures.

3.7.4 THEOREM. (1) *If a circle and two inverse points be inverted with respect to a center not on the circle, we obtain a circle and two inverse points.* (2) *If a circle and two inverse points be inverted with respect to a center on the circle, we obtain a straight line and two points which are reflections of one another in the straight line.*

Let points A and B (see Figure 3.7c) be inverse points with respect to a circle C and let K be any point. Draw circles C_1, C_2 through A and B but not through K. C_1, C_2 are orthogonal to C (by Theorem 3.6.7). Invert the figure with respect to center K.

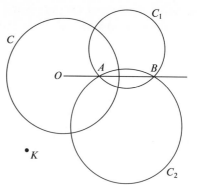

Figure 3.7c

(1) If K is not on C, we obtain circles C', C_1', C_2' and points A', B'. By Corollary 3.7.3, C_1' and C_2' are orthogonal to C', whence (by Theorem 3.6.8) A' and B' are inverse with respect to circle C'.

(2) If K is on C, we obtain a straight line C', circles C_1', C_2', and points A', B'. By Corollary 3.7.3, C_1' and C_2' are orthogonal to C'. It follows that A', B' are reflections of one another in line C'.

3.7.5 THEOREM. *If* P, P' *and* Q, Q' *are pairs of inverse points with respect to circle* O(r), *then* $P'Q' = (PQ)r^2/(OP)(OQ)$.

Suppose (see Figure 3.7d) O, P, Q are not collinear. Since $(\overline{OP})(\overline{OP'}) =$

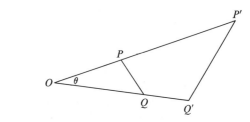

Figure 3.7d

$(\overline{OQ})(\overline{OQ'})$, triangle OPQ is similar to triangle $OQ'P'$, whence $P'Q'/PQ = OQ'/OP = (OQ')(OQ)/(OP)(OQ) = r^2/(OP)(OQ)$.

The case where O, P, Q are collinear follows from the above case by letting angle θ (see Figure 3.7d) approach zero. Or we may give a separate proof as follows (see Figure 3.7e):

$$(\overline{OP})(\overline{OP'}) = (\overline{OQ})(\overline{OQ'}),$$
$$(\overline{OQ} + \overline{QP})\overline{OP'} = \overline{OQ}(\overline{OP'} + \overline{P'Q'}),$$
$$(\overline{QP})(\overline{OP'}) = (\overline{OQ})(\overline{P'Q'}),$$
$$\overline{P'Q'} = (\overline{QP})(\overline{OP'})/\overline{OQ} = (\overline{QP})(\overline{OP'})(\overline{OP})/(\overline{OP})(\overline{OQ}) = (\overline{QP})r^2/(\overline{OP})(\overline{OQ}).$$

Figure 3.7e

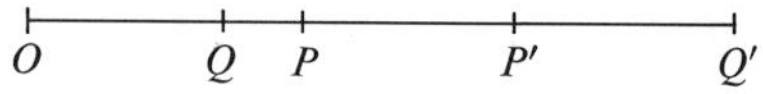

3.7.6 THEOREM. *The cross ratio of four distinct points on a "circle" is invariant under any inversion whose center is distinct from each of the four points.*

Let A, B, C, D be the four distinct points on a "circle" K, and let K invert into "circle" K'.

(1) If K and K' are both circles, then

$$
\begin{aligned}
(A'B',C'D') &= e(A'C'/C'B')/(A'D'/D'B') && \text{(by Theorem 2.6.7)}\\
&= e(AC/CB)/(AD/DB) && \text{(by Theorem 3.7.5)}\\
&= (AB,CD). && \text{(by Theorem 2.6.7)}
\end{aligned}
$$

(2) If K is a straight line and K' a circle, then

$$
\begin{aligned}
(A'B',C'D') &= e(A'C'/C'B')/(A'D'/D'B') && \text{(by Theorem 2.6.7)}\\
&= e(AC/CB)/(AD/DB) && \text{(by Theorem 3.7.5)}\\
&= (AB,CD). && \text{(see Problem 2, Section 2.5)}
\end{aligned}
$$

(3) If K is a circle and K' a straight line, then

$$
\begin{aligned}
(A'B',C'D') &= e(A'C'/C'B')/(A'D'/D'B') && \text{(see Problem 2, Section 2.5)}\\
&= e(AC/CB)/(AD/DB) && \text{(by Theorem 3.7.5)}\\
&= (AB,CD). && \text{(by Theorem 2.6.7)}
\end{aligned}
$$

(4) If K and K' are both straight lines, then

$$
\begin{aligned}
(A'B',C'D') &= e(A'C'/C'B')/(A'D'/D'B') && \text{(see Problem 2, Section 2.5)}\\
&= e(AC/CB)/(AD/DB) && \text{(by Theorem 3.7.5)}\\
&= (AB,CD). && \text{(see Problem 2, Section 2.5)}
\end{aligned}
$$

3.7.7 THEOREM. *Two nonintersecting circles can always be inverted into a pair of concentric circles.*

Let C_1 and C_2 be a pair of nonintersecting circles, and let l be their radical axis. Using two points on l as centers, draw two circles D_1 and D_2 each orthogonal to both C_1 and C_2. Then (by Theorem 2.10.7 (1)) D_1 and D_2 intersect in two points, P_1 and P_2. Choose either of these two points, say P_1, as a center of inversion and invert the entire figure. C_1 and C_2 (by Theorem 3.6.13) become circles C_1' and C_2'. D_1 and D_2 (by Theorem 3.6.12) become straight lines D_1' and D_2', each of which (by Corollary 3.7.3 (2)) cuts circles C_1' and C_2' orthogonally. This means that C_1' and C_2' are concentric.

PROBLEMS

1. Two circles intersect orthogonally at P; O is any point on any circle touching the former circles at Q and R. Prove that the circles OPQ, OPR intersect at an angle of $45°$.

2. If PQ, RS are common tangents to two circles PAR, QAS, prove that the circles PAQ, RAS are tangent to each other.

3. (a) Invert with respect to the center of the semicircle the theorem: An angle inscribed in a semicircle is a right angle.

   (b) Invert with respect to A the theorem: If A, B, C, D are concyclic points, then angles ABD, ACD are equal or supplementary.

4. Prove that the circles having for diameters the three chords AB, AC, AD of a given circle intersect by pairs in three collinear points.

5. Given a triangle ABC and a point M. Draw the circles MBC, MCA, MAB, and then draw the tangents to these circles at M to cut BC, CA, AB in R, S, T. Prove that R, S, T are collinear.

6. If A, B, C, D are four coplanar points with no three collinear, prove that circles ABC and ADC intersect at the same angle as the circles BDA and BCD.

7. Circles K_1, K_2 touch each other at T, and a variable circle through T cuts K_1, K_2 orthogonally in X_1, X_2 respectively. Prove that X_1X_2 passes through a fixed point.

8. A variable circle K touches a fixed circle K_1 and is orthogonal to another fixed circle K_2. Show that K touches another fixed circle coaxial with K_1 and K_2.

9. A, B, C, D are four concyclic points. If a circle through A and B touches one through C and D, prove that the locus of the point of contact is a circle.

10. What is the locus of the inverse of a given point in a system of tangent coaxial circles?

11. AC is a diameter of a given circle, and chords AB, CD intersect (produced if necessary) in a point O. Prove that circle OBD is orthogonal to the given circle.

12. Solve Problem 4, Section 3.6, parts (a) and (b), by means of Theorem 3.7.4.

3.8 APPLICATIONS OF INVERSION

We give a few illustrations of inversion as a simplifying transformation. We first emphasize the *transform-solve-invert* procedure, described in the introduction to this chapter, by an informal discussion of the problem (see Figure 3.8a$_1$):

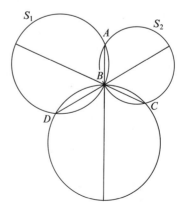

Figure 3.8a$_1$

Let two circles S_1 *and* S_2 *intersect in* A *and* B, *and let the diameters of* S_1 *and* S_2 *through* B *cut* S_2 *and* S_1 *in* C *and* D. *Show that line* AB *passes through the center of circle* BCD.

The figure of the problem involves three lines and three circles, all passing through a common point *B*. This suggests that we *transform* the figure into a simpler one by an inversion having center *B*, for under such an inversion the three lines will (by Theorem 3.6.10) map into themselves, and the three circles will (by Theorem 3.6.12) map into three lines. For convenience, we sketch the appearance of the simplified figure, not upon the first figure, as it would naturally appear, but to the right of the first figure (see Figure $3.8a_2$). Note that circles *BCD*, $S_1 = ABD$, $S_2 = ABC$ have become straight

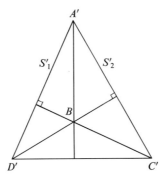

Figure $3.8a_2$

lines $D'C'$, $A'D'$, $A'C'$, respectively, and the straight lines *AB*, *CB*, *DB* have become the straight lines $A'B$, $C'B$, $D'B$, respectively. Since line *BC*, being a diametral line of circle S_1, cuts S_1 orthogonally, we have (by Corollary 3.7.3 (2)) that BC' is perpendicular to $S_1' = A'D'$. Similarly, we have that BD' is perpendicular to $S_2' = A'C'$.

Now it is our desire to show that *AB* is a diametral line of circle *BCD*, or, in other words, to show that *AB* is orthogonal to circle *BCD*. We therefore attempt to *solve*, in the simplified figure, the allied problem: *Show that* A'B *is perpendicular to* D'C'. But this is easily accomplished, for, since BC' and BD' are perpendicular to $A'D'$ and $A'C'$ respectively, *B* is the orthocenter of triangle $A'D'C'$, and $A'B$ must be perpendicular to $D'C'$.

Since $A'B$ is perpendicular to $D'C'$, if we *invert* the transformation that carried the first figure into the simplified one, we discover that *AB* is orthogonal to circle *BCD*, and our original problem is now solved.

The three-part procedure, *transform-solve-invert*, has carried us through. Our problem has turned out to be nothing but the inverse, with respect to the orthocenter as center of inversion, of the fact that the three altitudes of a triangle are concurrent, and the relation of the problem might well have been first discovered in just this way.

The five applications of inversion which now follow will be sketched only briefly, and the reader is invited to supply any missing details.

Ptolemy's Theorem

The following proposition was brilliantly employed by Claudius Ptolemy (85?–165?) for the development of a table of chords in the first book of his *Almagest*, the great definitive Greek work on astronomy (see Problem 9, section 1.6). In all probability the proposition was known before Ptolemy's time, but his proof of it is the first that has come down to us. It is interesting that a very simple demonstration of the proposition—indeed, of an extension of the proposition—can be given by means of the inversion transformation.

3.8.1 PTOLEMY'S THEOREM. *In a cyclic convex quadrilateral the product of the diagonals is equal to the sum of the products of the two pairs of opposite sides.*

Referring to Figure 3.8b, subject the quadrilateral and its circumcircle

Figure 3.8b

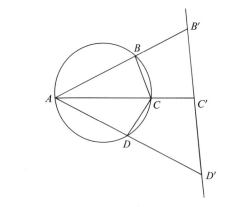

to the inversion $I(A,1)$. The vertices B, C, D map into points B', C', D' lying on a straight line. It follows that $B'C' + C'D' = B'D'$, whence (by Theorem 3.7.5)

$$BC/(AB \cdot AC) + CD/(AC \cdot AD) = BD/(AB \cdot AD),$$

or

$$BC \cdot AD + CD \cdot AB = BD \cdot AC.$$

If quadrilateral $ABCD$ is not cyclic, then B', C', D' will not be collinear and $B'C' + C'D' > B'D'$. Using this fact the reader can easily supply a proof, fashioned after the above, for the following:

3.8.2 EXTENSION OF PTOLEMY'S THEOREM. *In a convex quadrilateral ABCD,*

$$BC \cdot AD + CD \cdot AB \geq BD \cdot AC,$$

with equality if and only if the quadrilateral is cyclic.

Pappus' Ancient Theorem

In Book IV of Pappus' *Collection* appears the following beautiful proposition, referred to by Pappus as being already ancient in his time. The proof of the proposition by inversion is singularly attractive.

3.8.3 PAPPUS' ANCIENT THEOREM. *Let* X, Y, Z *be three collinear points with* Y *between* X *and* Z, *and let* C, C_1, K_0 *denote semicircles, all lying on the same side of* XZ, *on* XZ, XY, YZ *as diameters. Let* K_1, K_2, K_3, . . . *denote circles touching* C *and* C_1, *with* K_1 *also touching* K_0, K_2 *also touching* K_1, K_3 *also touching* K_2, *and so on. Denote the radius of* K_n *by* r_n, *and the distance of the center of* K_n *from* XZ *by* h_n. *Then* $h_n = 2nr_n$.

Subject the figure to the inversion $I(X, t_n^2)$, where t_n is the tangent length from X to circle K_n. Then (see Figure 3.8c) K_n inverts into itself, C and C_1

Figure 3.8c

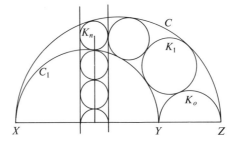

invert into a pair of parallel lines tangent to K_n and perpendicular to XZ. K_0, K_1, K_2, . . ., K_{n-1} invert into a semicircle and circles, all of the same radius, tangent to the two parallel lines and such that K_1' touches K_0', K_2' touches K_1', . . ., K_{n-1}' touches K_{n-2}' and also K_n. It is now clear that $h_n = 2nr_n$.

Feuerbach's Theorem

Geometers universally regard the so-called Feuerbach's Theorem as undoubtedly one of the most beautiful theorems in the modern geometry of the triangle. The theorem was first stated and proved by Karl Wilhelm Feuerbach (1800–1834) in a work of his published in 1822; his proof was of a computational nature and employed trigonometry. A surprising number of proofs of the theorem have been given since, but probably none is as neat as the following proof employing the inversion transformation. We first state two definitions.

3.8.4 DEFINITION. A circle tangent to one side of a triangle and to the other two sides produced is called an *excircle* of the triangle. (There are four circles touching all three side lines of a triangle—the incircle and three excircles.)

3.8.5 DEFINITION. Two lines are said to be *antiparallel* relative to two transversals if the quadrilateral formed by the four lines is cyclic.

3.8.6 FEUERBACH'S THEOREM. *The nine-point circle of a triangle is tangent to the incircle and to each of the excircles of the triangle.*

Figure 3.8d shows the incircle (I) and one excircle (I') of a triangle ABC.

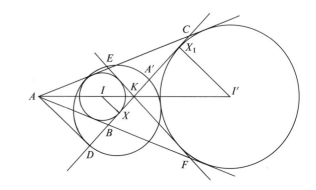

Figure 3.8d

The four common tangents to these two circles determine the homothetic centers A and K lying on the line of centers II', and we have $(AK, II') = -1$. If D, X, X_1 are the feet of the perpendiculars from A, I, I' on line BC, we then have $(DK, XX_1) = -1$. Now the line segments BC and XX_1 have a common midpoint A', and

$$(1) \qquad (A'K)(A'D) = (A'X)^2 = (A'X_1)^2.$$

Subject the figure to the inversion $I(A', \overline{A'X}^2)$. The circles (I) and (I') invert into themselves. Since the nine-point circle passes through A' and D, it follows that this circle inverts into a straight line through K, the inverse of D by (1). Also, the angle which this line makes with BC is equal but opposite to the angle which the tangent to the nine-point circle at D makes with BC, or therefore equal in both magnitude and sign to the angle which the tangent to the nine-point circle at A' makes with BC. But it is easily shown that this latter tangent is parallel to the opposite side of the orthic triangle, and therefore antiparallel to BC relative to AB and AC. But EF is antiparallel to BC relative to AB and AC. It follows that line EF is the inverse of the nine-point circle. Since this line is tangent to both (I) and (I'), we have that the nine-point circle is tangent to both (I) and (I'). That the nine-point circle is tangent to each of the other excircles can be shown in a like manner.

Steiner's Porism

Consider a circle C_1 lying entirely within another circle C_2, and a sequence of circles, K_1, K_2, ..., each having external contact with C_1 and internal contact with C_2, and such that K_2 touches K_1, K_3 touches K_2, and so on. A number of interesting questions suggest themselves in connection with such a figure. For example, can it ever be that the sequence K_1, K_2, ... is finite, in the sense that finally a circle K_n of the sequence is reached which touches both K_{n-1} and K_1? Do the points of contact of the circles K_1, K_2, ... lie on a circle? Do their centers lie on a circle? Etc. The figure was a dear one to Jacob Steiner, and he proved a number of remarkable properties of it. It seems that the best way to study the figure is by the inversion transformation. We content ourselves here with a proof of just one of Steiner's theorems. We first formulate a definition.

3.8.7 DEFINITION. A *Steiner chain of circles* is a sequence of circles, finite in number, each tangent to two fixed nonintersecting circles and to two other circles of the sequence.

3.8.8 STEINER'S PORISM. *If two given nonintersecting circles admit a Steiner chain, they admit an infinite number, all of which contain the same number of circles, and any circle tangent to the two given circles and surrounding either none or both of them is a member of such a chain.*

The proof is simple. By Theorem 3.7.7, the two given circles may be inverted into a pair of concentric circles, the circles of the Steiner chain then becoming a Steiner chain of equal circles for the two concentric circles. Since the circles of this associated Steiner chain may each be advanced cyclically in the ring in which they lie to form a similar chain, and this can be done in infinitely many ways, the theorem follows.

Theorem 3.8.8 is representative of a whole class of propositions in which there is a condition for a certain relation to subsist, but if the condition holds then the relation subsists infinitely often. Such propositions are called *porisms*. Three books of porisms by Euclid have been lost.

Peaucellier's Cell

An outstanding geometrical problem of the last half of the nineteenth century was to discover a linkage mechanism for drawing a straight line. A solution was finally found in 1864 by a French army officer, A. Peaucellier (1832–1913), and an announcement of the invention was made by A. Mannheim (1831–1906), a brother officer of engineers and inventor of the so-called Mannheim slide rule, at a meeting of the Paris Philomathic Society in 1867. But the announcement was little heeded until Lipkin, a young student of the celebrated Russian mathematician Chebyshev (1821–1894), independently reinvented the mechanism in 1871. Chebyshev had been trying to demonstrate the impossibility of such a mechanism. Lipkin received a substantial reward from the Russian Government, whereupon Peaucellier's merit was finally recognized and he was awarded the great mechanical prize of the

Institut de France. Peaucellier's instrument contains 7 bars. In 1874 Harry Hart (1848–1920) discovered a 5-bar linkage for drawing straight lines, and no one has since been able to reduce this number of links or to prove that a further reduction is impossible. Both Peaucellier's and Hart's linkages are based upon the fact that the inverse of a circle through the center of inversion is a straight line.

The subject of linkages became quite fashionable among geometers, and many linkages were found for constructing special curves, such as conics, cardioids, lemniscates, and cissoids. In 1933, R. Kanayama published (in the *Tôhoku Mathematics Journal*, v. 37, 1933, pp. 294–319), a bibliography of 306 titles of papers and works on linkage mechanisms written between 1631 and 1931. It has been shown (in *Scripta Mathematica*, v. 2, 1934, pp. 293–294) that this list is far from complete, and of course many additional papers have appeared since 1931.

It has been proved that there exists a linkage for drawing any given algebraic curve, but that there cannot exist a linkage for drawing any transcendental curve. Linkages have been devised for mechanically solving algebraic equations.

3.8.9 PEAUCELLIER'S CELL. *In Figure 3.8e, let the points* A *and* B *of the*

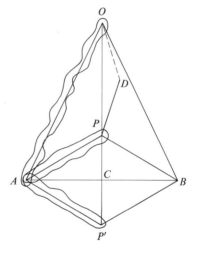

Figure 3.8e

jointed rhombus PAP'B *be joined to the fixed point* O *by means of equal bars* OA *and* OB, OA > PA. *Then, if all points of the figure are free to move except point* O, *the points* P *and* P' *will describe inverse curves under the inversion* I(O,OA² − PA²). *In particular, if a seventh bar* DP, DP > OP/2, *is attached to* P *and the point* D *fixed so that* DO = DP, *then* P' *will describe a straight line.*

For we have

$$(OP)(OP') = (OC - PC)(OC + PC)$$
$$= OC^2 - PC^2$$
$$= (OC^2 + CA^2) - (PC^2 + CA^2)$$
$$= OA^2 - PA^2.$$

PROBLEMS

1. Prove that if a circle C_1 inverts into a circle C_2, then the circle of similitude of C_1 and C_2 inverts into the radical axis of C_1 and C_2.

2. Circles OBC, OCA, OAB are cut in P, Q, R respectively by another circle through O. Prove that $(BP)(CQ)(AR) = (CP)(AQ)(BR)$.

3. Let $T_1 T_2 T_3 T_4$ be a convex quadrilateral inscribed in a circle C. Let C_1, C_2, C_3, C_4 be four circles touching circle C externally at T_1, T_2, T_3, T_4 respectively. Show that

$$t_{12}t_{34} + t_{23}t_{41} = t_{13}t_{24},$$

where t_{ij} is the length of a common external tangent to circles C_i and C_j. (This is a special case of a more general theorem due to Casey. It can be considered as a generalization of Ptolemy's Theorem.)

4. (a) If $A(a)$ and $B(b)$ are two orthogonal circles, show that $I(B,b^2)I(A,a^2) = I(A,a^2)I(B,b^2)$.
   (b) If K_1 K_2 are two orthogonal circles, A, A' inverse points in K_1, B and B' the inverses of A and A' in K_2, show that B, B' are inverse points in K_1.

5. If A, B, C, D are four concyclic points in the order A, C, B, D, and if p, q, r are the lengths of the perpendiculars from D to the lines AB, BC, CA respectively, show that

$$AB/p = BC/q + CA/r.$$

6. Let C' be the inverse of circle C under the inversion $I(O,r^2)$, and let p and p' be the powers of O with respect to C and C' respectively. Show that $pp' = r^4$.

7. Prove that the product of three inversions in three circles of a coaxial system is an inversion in a circle of that system.

8. We call the product $R(O)I(O,r^2)$ an *antinversion* in circle $O(r)$.
   (a) Show that an antinversion is a circular transformation of the inversive plane.
   (b) Show that a circle through a pair of antinverse points for a circle K cuts K diametrically.

9. Show that if two point triads inscribed in the same circle are copolar at a point C, then the inverse with respect to C of either triad is homothetic to the other triad.

10. (a) Show that two circles can be inverted into themselves from any point on their radical axis and outside both circles.
    (b) When can three circles be inverted into themselves?

11. Show that a nonintersecting coaxial system of circles can be inverted into a system of concentric circles.

12. Show that any three circles can be inverted into three circles whose centers are collinear.

13. Show that any three points can, in general, be inverted into the vertices of a triangle similar to a given triangle.

14. Show that any three noncollinear points can be inverted into the vertices of an equilateral triangle of given size.

15. Circle C_1 inverts into circle C_2 with respect to circle C. Show that C_1 and C_2 invert into equal circles with respect to any point on circle C.

16. If a quadrilateral with sides a, b, c, x is inscribed in a semicircle of diameter x, show that

$$x^3 - (a^2 + b^2 + c^2)x - 2abc = 0.$$

(This is Problem E 574 of *The American Mathematical Monthly*, Feb. 1944.)

17. Prove Theorem 3.8.2.

18. Prove Ptolemy's Second Theorem: If $ABCD$ is a convex quadrilateral inscribed in a circle, then

$$AC/BD = (AB \cdot AD + CB \cdot CD)/(BA \cdot BC + DA \cdot DC).$$

19. (a) If A', B' are the inverses of A, B, then show that AA', BB' are antiparallel relative to AB and $A'B'$.
    (b) If A, B, C, D are four points such that AB and CD are antiparallel relative to AD and BC, show that the four points can be inverted into the vertices of a rectangle.

20. Fill in the details of the proof of Theorem 3.8.6.

21. Show that the points of contact of the circles of a Steiner chain of circles all lie on a circle.

22. A linkage, called *Hart's contraparallelogram* (invented by H. Hart in 1874) is pictured in Figure 3.8f. The four rods AB, CD, BC, DA, $AB = CD$, BC

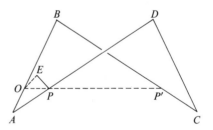

Figure 3.8f

$= DA$, are hinged at A, B, C, D. If O, P, P' divide AB, AD, CB proportionately, and the linkage is pivoted at O, show that O, P, P' are always collinear and that P and P' describe inverse curves with respect to O as center of the inversion. Hence if a fifth rod $EP > OP/2$ be pivoted at E with $OE = EP$, P' will describe a straight line.

23. Let four lines through a point V cut a circle in A, A'; B, B'; C, C'; D, D' respectively. Show that $(AB,CD) = (A'B',C'D')$. (This is Problem 13, Section 2.5.)

3.9 RECIPROCATION

We now consider a remarkable transformation of the set S of all points of the extended plane onto the set T of all straight lines of the extended plane.

3.9.1 DEFINITIONS AND NOTATION. Let $O(r)$ be a fixed circle (see Figure 3.9a) and let P be any ordinary point other than O. Let P' be the inverse

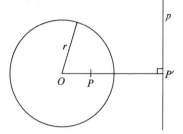

Figure 3.9a

of P in circle $O(r)$. Then the line p through P' and perpendicular to OPP' is called the *polar* of P for the circle $O(r)$. The *polar of* O is taken as the line at infinity, and the *polar of an ideal point* P is taken as the line through O perpendicular to the direction OP.

If line p is the polar point P, then point P is called the *pole* of line p.

The pole-polar transformation set up by circle $K = O(r)$ will be denoted by $P(K)$ or $P(O(r))$ and will be called *reciprocation* in circle K.

Some nascent properties of reciprocation may be found in the works of Apollonius and Pappus. The theory was considerably developed by Desargues in his treatise on conic sections of 1639, and by his student Philippe de La Hire (1640–1718), and then greatly elaborated in the first half of the nineteenth century in connection with the study of the conic sections in projective geometry. The term *pole* was introduced in 1810 by the French mathematician F. J. Servois, and the corresponding term *polar* by Gergonne two to three years later. Gergonne and Poncelet developed the idea of poles and polars into a regular method out of which grew the elegant *principle of duality* of projective geometry. We shall look into the projective aspects of poles and polars in a later chapter.

The easy proof of the following theorem is left to the reader.

3.9.2 THEOREM. (1) *The polar of a point for a circle intersects the circle, is tangent to the circle at the point, or does not intersect the circle, according as the point is outside, on, or inside the circle.* (2) *If point* P *is outside a circle, then its polar for the circle passes through the points of contact of the tangents to the circle from* P.

The next theorem is basic in applications of reciprocation.

3.9.3 THEOREM. (1) *If, for a given circle, the polar of* P *passes through* Q, *then the polar of* Q *passes through* P. (2) *If, for a given circle, the pole of line* p *lies on line* q, *then the pole of* q *lies on* p. (3) *If, for a given circle,* P *and* Q *are the poles of* p *and* q, *then the pole of line* PQ *is the point of intersection of* p *and* q.

(1) Suppose *P* and *Q* are ordinary points. Let *P'* (see Figure 3.9b) be the

Figure 3.9b

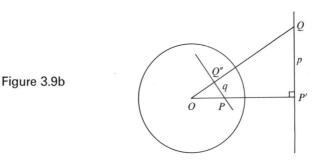

inverse of *P* and *Q'* the inverse of *Q* in the given circle, and suppose *P'* and *Q* are distinct. Then $\overline{OP} \cdot \overline{OP'} = \overline{OQ} \cdot \overline{OQ'}$, whence *P*, *P'*, *Q*, *Q'* are concyclic and $\measuredangle PP'Q = \measuredangle PQ'O$. But, since *Q* lies on the polar of *P*, $\measuredangle PP'Q = 90°$. Therefore $\measuredangle PQ'O = 90°$, and *P* lies on the polar of *Q*. If *P'* = *Q*, the theorem is obvious. The cases where *P*, or *Q*, or both *P* and *Q* are ideal points are easily handled.

(2) Let *P* and *Q* be the poles of *p* and *q*. It is given that *q* (the polar of *Q*) passes through *P*. It follows, by (1), that *p* (the polar of *P*) passes through *Q*.

(3) Let *p* and *q* intersect in *R*. Then the polar of *P* passes through *R*, whence the polar of *R* passes through *P*. Similarly, the polar of *R* passes through *Q*. Therefore line *PQ* is the polar of *R*.

3.9.4 COROLLARY. *The polars, for a given circle, of a range of points constitute a pencil of lines; the poles, for a given circle, of a pencil of lines constitute a range of points.*

3.9.5 DEFINITIONS. Two points such that each lies on the polar of the other, for a given circle, are called *conjugate points* for the circle; two lines such that each passes through the pole of the other, for a given circle, are called *conjugate lines* for the circle.

The reader should find no difficulty in establishing the following facts.

3.9.6 THEOREM. *For a given circle:* (1) *Each point of a line has a conjugate point on that line.* (2) *Each line through a point has a conjugate line through* *that point.* (3) *Of two distinct conjugate points on a line that cuts the circle, one is inside and the other outside the circle.* (4) *Of two distinct conjugate lines that intersect outside the circle, one cuts the circle and the other does not.* (5) *Any point on the circle is conjugate to all the points on the tangent to the circle at the point.* (6) *Any tangent to the circle is conjugate to all lines through its point of contact with the circle.*

The next few theorems will be found important in the projective theory of poles and polars.

3.9.7 TheoreM. *If, for a given circle, two conjugate points lie on a line which intersects the circle, they are harmonically separated by the points of intersection.*

Let A and B (see Figure 3.9c) be two such points, and let A' be the inverse

Figure 3.9c

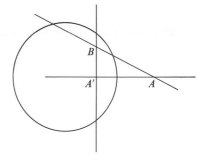

of A in the circle. If $B = A'$, the desired result follows immediately. If $B \neq A'$, then $A'B$ is the polar of A and $\angle AA'B = 90°$. The circle on AB as diameter, since it passes through A', is (by Theorem 3.6.7) orthogonal to the given circle. It follows (by Theorem 2.9.3) that A and B are harmonically separated by the points in which their line intersects the circle.

3.9.8 Corollary. *If a variable line through a given point intersects a circle, the harmonic conjugates of the point with respect to the intersections of the line and circle all lie on the polar of the given point.*

3.9.9 TheoreM. *If, for a given circle, two conjugate lines intersect outside the circle, they are harmonically separated by the tangents to the circle from their point of intersection.*

Let a and b (see Figure 3.9d) be two such lines and let S be their point of intersection. Since a and b are conjugate lines for the circle, the pole A of a lies on b, and the pole B of b lies on a. Then (by Theorem 3.9.3(3)) line AB is the polar of S and must pass through the points P and Q where the

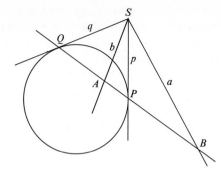

Figure 3.9d

tangents from S touch the circle. Now A and B are conjugate points, whence (by Theorem 3.9.7) $(AB,PQ) = -1$. It follows that $(ba,pq) = -1$.

3.9.10 THEOREM. *If A, B, C, D are four distinct collinear points, and a, b, c, d are their polars for a given circle, then* $(AB,CD) = (ab,cd)$.

Referring to Figure 3.9e, the polars a, b, c, d all pass through S, the pole

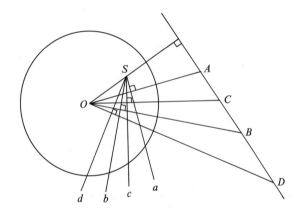

Figure 3.9e

of the line $ABCD$. Since each polar is perpendicular to the line joining its pole to the center O of the circle, we see that

$$(ab,cd) = O(AB,CD) = (AB,CD).$$

PROBLEMS

1. Establish Theorem 3.9.2.

2. Establish Theorem 3.9.6.

3. If P and Q are conjugate points for a circle, show that $(PQ)^2$ is equal to the sum of the powers of P and Q with respect to the circle.

4. If PR is a diameter of a circle K_1 orthogonal to a circle K_2 of center O, and if OP meets K_1 in Q, prove that line QR is the polar of P for K_2.

5. (a) If P and Q are conjugate points for circle K, prove that the circle on PQ as diameter is orthogonal to K.
   (b) If two circles are orthogonal, prove that the extremities of any diameter of one are conjugate points for the other.

6. (a) Let K_1, K_2, K_3 be three circles having a radical circle R, and let P be any point on R. Show that the polars of P for K_1, K_2, K_3 are concurrent.
   (b) A common tangent to two circles K_1 and K_2 touches them at P and Q respectively. Show that P and Q are conjugate points for any circle coaxial with K_1 and K_2.
   (c) The tangent to the circumcircle of triangle ABC at vertex A intersects line BC at T and is produced to U so that $AT = TU$. Prove that A and U are conjugate points for any circle passing through B and C.
   (d) Let ABC be a right triangle with right angle at B, and let B' be the midpoint of AC. Prove that A and C are conjugate points for any circle which touches BB' at B.
   (e) If a pair of opposite vertices of a square are conjugate points for a circle, prove that the other pair of opposite vertices are also conjugate points for the circle.

7. Prove Salmon's Theorem: If P, Q are two points, and PX, QY are the perpendiculars from P, Q to the polars of Q, P respectively, for a circle of center O, then $OP/OQ = PX/QY$.

3.10 APPLICATIONS OF RECIPROCATION

In this section we give a few applications of the reciprocation transformation. The first is a proof of Brianchon's Theorem for a circle (see Problem 4, Section 2.6). This theorem was discovered and published in a paper by C. J. Brianchon (1785–1864) in 1806, when still a student at the École Polytechnique in Paris, over 150 years after Pascal had stated his famous mystic hexagram theorem. Brianchon's paper was one of the first publications to employ the theory of poles and polars to obtain new geometrical results, and his theorem played a leading role in the recognition of the far-reaching principle of duality. The following proof of Brianchon's theorem is essentially that given by Brianchon himself.

3.10.1 BRIANCHON'S THEOREM FOR A CIRCLE. *If a hexagon (not necessarily convex) is circumscribed about a circle, the three lines joining pairs of opposite vertices are concurrent.*

Let $ABCDEF$ (see Figure 3.10a) be a hexagon circumscribed about a circle. The polars for the circle, of the vertices A, B, C, D, E, F, form the sides a, b, c, d, e, f of an inscribed hexagon. Since a is the polar of A and d the polar of D, the point ad is (by Theorem 3.9.3(3)) the pole of line AD. Similarly, the point be is the pole of line BE, and the point cf is the pole of

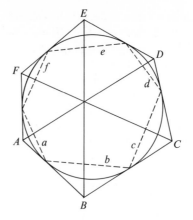

Figure 3.10a

line *CF.* Now, by Pascal's mystic hexagram theorem, the points *ad, be, cf* are collinear. It follows (by Theorem 3.9.4) that the polars *AD, BE, CF* are concurrent.

3.10.2 THEOREM. *Let* ABCD *be a complete quadrangle inscribed in a circle. Then each diagonal point of the quadrangle is the pole, for the circle, of the line determined by the other two diagonal points.*

For we have (see Figure 3.10b) $(AD,MQ) = (BC,NQ) = -1$, whence (by

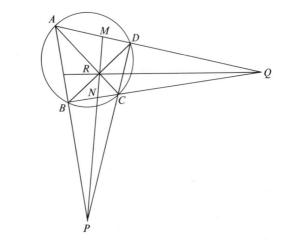

Figure 3.10b

Corollary 3.9.8) line *PR* is the polar of point *Q.* Similarly line *QR* is the polar of point *P.* It then follows (by Theorem 3.9.3 (3)) that line *PQ* is the polar of point *R.*

3.10.3 THE BUTTERFLY THEOREM. *Let* O *be the midpoint of a given chord*

of a circle, let two other chords TU *and* VW *be drawn through* O, *and let* TW *and* VU *cut the given chord in* E *and* F *respectively. Then* O *is the midpoint of* FE.

If the given chord is a diameter of the circle, the theorem is obvious. Otherwise (see Figure 3.10c) produce *TW* and *VU* to intersect in *R*, and

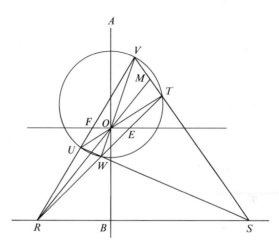

Figure 3.10c

VT and *UW* to intersect in *S*. Then (by Theorem 3.10.2) *RS* is the polar of *O*, whence *RS* is perpendicular to the diametral line *AOB*. But *FE* is perpendicular to *AOB*. Therefore *FE* is parallel to *RS*. Now (by Theorem 2.8.11) $R(VT,MS) = -1$, whence $(FE,O\infty) = -1$, and *O* bisects *FE*.

Theorem 3.10.3 has received its name from the fancied resemblance of the figure of the theorem to a butterfly with outspread wings. It is a real stickler if one is limited to the use of only high school geometry.

3.10.4   THEOREM.   *Let* PQR *be a triangle and let* P′, Q′, R′ *be the poles of* QR, RP, PQ *for a circle* K. *Then* P, Q, R *are the poles of* Q′R′, R′P′, P′Q′ *for circle* K.

The easy proof is left to the reader.

3.10.5   DEFINITIONS.   Two triangles are said to be *conjugate*, or *polar*, for a circle if each vertex of one triangle is the pole of a side of the other triangle. If the triangle is conjugate to itself—that is, each vertex is the pole of the opposite side—the triangle is said to be *self-conjugate*, or *self-polar*, for the circle.

We merely sketch the proofs of the next two theorems, leaving it to the reader to draw the accompanying figures and to complete details.

3.10.6 THEOREM. *Two triangles which are conjugate for a circle are co-polar to one another.*

Let ABC and $A'B'C'$ be a pair of conjugate triangles. Let BC and $B'C'$ meet in M and let AA' cut BC in N and $B'C'$ in N'. Then (by Theorem 3.9.10) $(BC,NM) = A'(C'B',MN) = (B'C',N'M)$. It now follows that BB', CC', NN' are concurrent (by Corollary 2.6.2).

3.10.7 HESSE'S THEOREM FOR THE CIRCLE (1840). *If two pairs of opposite vertices of a complete quadrilateral are conjugate points for a circle, then the third pair of opposite vertices are also conjugate points for the circle.*

Let A, A'; B, B'; C, C' be the three pairs of opposite vertices of the quadrilateral, and suppose A, A'; B, B' are conjugate pairs of points for a circle. Let $A''B''C''$ be the triangle conjugate to triangle ABC. Now $B''C''$ passes through A', and $A''C''$ passes through B'. But triangles ABC, $A''B''C''$ are copolar (by Theorem 3.10.6), and therefore corresponding sides meet in three collinear points, two of which are seen to be A' and B'. It follows that C' must be the third point. That is, $B''A''$ passes through C', and C and C' are conjugate points.

PROBLEMS

1. A variable chord PQ of a given circle K passes through a fixed point T. Prove that the tangents at P and Q intersect on a fixed line t.

2. Assume the theorem: "The feet of the perpendiculars from any point on the circumcircle of a triangle to the sides of the triangle are collinear." (The line of collinearity is called the *Simson line* of the point for the triangle.)
   Let I be the incircle of a triangle ABC, and let perpendiculars through I to IA, IB, IC meet a given tangent to the incircle in P, Q, R. Prove that AP, BQ, CR are concurrent.

3. Consider the two propositions: (1) The lines joining the vertices of a triangle to the points of contact of the opposite sides with the incircle of the triangle are concurrent. (2) The tangents to the circumcircle of a triangle at the vertices of the triangle meet the opposite sides of the triangle in three collinear points.
   Show that Proposition (2) can be obtained from Proposition (1) by subjecting Proposition (1) to a reciprocation in the incircle of the triangle.

4. Draw a figure for Theorem 3.10.6 and verify on it the proof of the theorem given in the text.

5. If P, Q are two conjugate points for a circle, and R is the pole of PQ, show that PQR is a self-conjugate triangle.

6. If a triangle is self-conjugate for a circle, prove that its orthocenter is the center of the circle.

7. (a) Given an obtuse triangle, prove that there exists one and only one circle for which the triangle is self-conjugate. (The circle for which an obtuse triangle is self-conjugate is called the *polar circle* of the triangle.)

(b) Prove that the polar circle of an obtuse triangle ABC has its center at the orthocenter of the triangle, and its radius equal to $[(\overline{HA})(\overline{HD})]^{1/2}$, where D is the foot of the altitude from A.

(c) Prove that the inverse of the circumcircle of an obtuse triangle with respect to its polar circle is the nine-point circle of the triangle.

8. AB, CD are conjugate chords of a circle. (a) Show that $(AB,CD) = -1$.
   (b) Show that $(AC)(BD) = (BC)(AD) = (AB)(CD)/2$.

9. (a) Prove, in Figure 3.10b, that triangle PQR is self-conjugate for the circle $ABCD$.
   (b) Show that the circles on PQ, QR, RP as diameters are orthogonal to circle $ABCD$.

10. Draw a figure for Theorem 3.10.7 and verify on it the proof of the theorem given in the text.

11. Let X, Y, Z be the points of contact of the incircle of triangle ABC with the sides BC, CA, AB respectively, and let P be the point on the incircle diametrically opposite point X.
    (a) If A' is the midpoint of BC, show that AA', PX, YZ are concurrent.
    (b) If PA, PY, PZ cut line BC in M, Q, R respectively, show that $RM = MQ$.

3.11   SPACE TRANSFORMATIONS

The elementary point transformations of unextended (three-dimensional) space are: (1) translation, (2) rotation about an axis, (3) reflection in a point, (4) reflection in a line, (5) reflection in a plane, and (6) homothety. In (1), the set S of all points of unextended space is mapped onto itself by carrying each point P of S into a point P' of S such that $\overline{PP'}$ is equal and parallel to a given directed segment $\overline{AB}$ of space. There are no invariant points under a translation of nonzero vector $\overline{AB}$. In (2), each point P of S is carried into a point P' of S by rotating P about a fixed line in space through a given angle. The fixed line is called the axis of the rotation, and the points of the axis are the invariant points of the rotation. In (3), each point P of S is carried into the point P' of S such that PP' is bisected by a fixed point O of space. The fixed point O is the only invariant point of the transformation. In (4), each point P of S is carried into the point P' of S such that PP' is perpendicularly bisected by a fixed line l of space. This transformation is obviously equivalent to a rotation of $180°$ about the line l. In (5), each point P of S is carried into the point P' of S such that PP' is perpendicularly bisected by a fixed plane p of space, and the points of p are the invariant points of the transformation. In (6), each point P of S is carried into the point P' of S collinear with P and a fixed point O of space, and such that $\overline{OP'}/\overline{OP} = k$, where k is a nonzero real number. If $k \neq 1$, the point O is the only invariant point of the transformation.

Certain compounds of the above elementary point transformations of space are basic in a study of similarities and isometries in space. These compounds are: (7) screw-displacement, (8) glide-reflection, (9) rotatory-reflection, and (10) space homology. A *screw-displacement* is the product of a rotation and a translation along the axis of rotation; a *glide-reflection* is the product of a reflection in a plane and a translation of vector $\overline{AB}$, where $\overline{AB}$ lies in the plane; a *rotatory-reflection* is the product of a reflection in a plane and a rotation about a fixed axis perpendicular to the plane; a *space homology* is the product of a homothety and a rotation about an axis passing through the center of the homothety.

Similarities and isometries in space are defined exactly as they were defined in a plane. Thus, a point transformation of unextended space onto itself which carries each pair of points A, B into a pair A', B' such that $A'B' = k(AB)$, where k is a fixed positive number, is called a *similarity*, and the particular case where $k = 1$ is called an *isometry*. A similarity is said to be *direct* or *opposite* according as tetrahedron $ABCD$ has or has not the same sense as tetrahedron $A'B'C'D'$.

We lack the space and time to develop the theory of similarities and isometries of space, and accordingly list the following interesting theorems without proof.

3.11.1 THEOREM. *Every isometry in space is the product of at most four reflections in planes.*

3.11.2 THEOREM. *Every isometry in space containing an invariant point is the product of at most three reflections in planes.*

3.11.3 THEOREM. *Every direct isometry in space is the product of two reflections in lines.*

3.11.4 THEOREM. *Any direct isometry is either a rotation, a translation, or a screw-displacement.*

3.11.5 THEOREM. *Any opposite isometry is either a rotatory-reflection or a glide-reflection.*

3.11.6 THEOREM. *Any nonisometric similarity is a space homology.*

The inversion transformation of Section 3.6 is also easily generalized to space.

3.11.7 DEFINITIONS AND NOTATION. We denote the sphere of center O and radius r by the symbol $O(r)$. If point P is not the center O of sphere $O(r)$, the *inverse* of P in, or with respect to, sphere $O(r)$ is the point P' lying on the line OP such that $(\overline{OP})(\overline{OP'}) = r^2$. Sphere $O(r)$ is called the *sphere of inversion*, point O the *center of inversion*, r the *radius of inversion*, and r^2 the *power of inversion*.

3.11.8 CONVENTION AND DEFINITIONS. When working with inversion in space, we add to the set S of all points of space a single ideal *point at infinity*, to be considered as lying on every plane of space, and this ideal point, Z, shall be the image under the inversion of the center O of inversion, and the center O of inversion shall be the image under the inversion of the ideal point Z. Space, augmented in this way, will be referred to as *inversive space*.

It is now apparent that space inversion is a transformation of inversive space onto itself. Much of the theory of planar inversion can easily be extended to space inversion. For example, if "sphere" (with the quotation marks) denotes either a plane or a sphere, one can prove the following theorem.

3.11.9 THEOREM. *In a space inversion, "spheres" invert into "spheres," and "circles" invert into "circles."*

In particular, one can show that: (1) a plane through the center of inversion inverts into itself, (2) a plane not through the center of inversion inverts into a sphere through the center of inversion, (3) a sphere through the center of inversion inverts into a plane not through the center of inversion, (4) a sphere not through the center of inversion inverts into a sphere not through the center of inversion.

The pole-polar relation of Section 3.9 can be generalized to extended space.

3.11.10 DEFINITIONS. Let $O(r)$ be a fixed sphere and let P be any ordinary point other than O. Let P' be the inverse of P in the sphere $O(r)$. Then the plane p through P' and perpendicular to OPP' is called the *polar* of P for the sphere $O(r)$. The *polar of* O is taken as the plane at infinity, and the *polar of an ideal point* P is taken as the plane through O perpendicular to the direction OP. If plane p is the polar of point P, then point P is called the *pole* of plane p.

In the pole-polar relation of extended space we have a transformation of the set of all points of extended space onto the set of all planes of extended space. It can be shown that this transformation carries a range of points into a pencil of planes, and we accordingly have a straight line (considered as the base of a range of points) associated with a straight line (considered as the axis of a pencil of planes). This pole-polar transformation of extended space gives rise to a remarkable principle of duality of extended space.

Besides point transformations mapping a whole three-dimensional space onto itself, there are point transformations which map a part of space onto another part of space. Consider, for example, two planes p_1 and p_2 in unextended space, and a fixed direction not parallel to either plane. We may induce a transformation of the set of all points of p_1 onto the set of all points of p_2 by the simple procedure of associating with each point of P_1 of p_1 the point P_2 of p_2 such that P_1P_2 is parallel to the given direction. If we consider p_1 and p_2 as immersed in extended space, then the transformation carries the line at infinity of p_1 into the line at infinity of p_2. As another example, again consider two ordinary planes p_1 and p_2 in extended space, and a point O not on either plane. We can induce a transformation of the set of all points of p_1 onto the set of all points of p_2 by associating with each point P_1 of p_1 the point P_2 of p_2 such that O, P_1, P_2 are collinear. These particular transformations that we have been describing are very important in geometry and they will be discussed in Chapter 6.

We conclude the section with an outline of a point transformation of a part of inversive space onto another part of itself. The transformation, known as stereographic projection, was known to the ancient Greeks and affords a simple and useful method of transferring figures from a plane to the surface of a sphere, or vice versa.

3.11.11 DEFINITION AND NOTATION. Let K be a sphere of diameter d, and p a plane tangent to K at its south pole S; let N be the north pole of K. If P is any point of the sphere other than N, we associate with it the point P' of p such that N, P, P' are collinear. With N we associate the point at infinity on the inversive plane p. This transformation of the set of all points of the sphere K onto the set of all points of the inversive plane p is called *stereographic projection.*

The proofs of the first two of the following theorems are left to the reader.

3.11.12 THEOREM. *The meridians on* K *correspond to the straight lines on* p *through* S, *and the circles of latitude on* K *correspond to the circles on* p *having center* S. *In particular, the equator of* K *corresponds to the circle on* p *of center* S *and radius* d.

3.11.13 THEOREM. *The circles on* K *through* N *correspond to the straight lines on* p.

3.11.14 THEOREM. *If points* P *and* Q *on the sphere* K *are reflections of one another in the equatorial plane of* K, *then their images* P', Q' *on plane* p *are inverses of one another in the circle* S(d) *of* p.

For (see Figure 3.11a) $SP' = d \tan PNS$, $SQ' = d \tan QNS$. But angles PNS and QNS are complementary, whence $(SP')(SQ') = d^2$. Clearly S, P', Q' are collinear.

Theorem 3.11.14 exhibits an interesting interpretation of a planar inversion.

Figure 3.11a

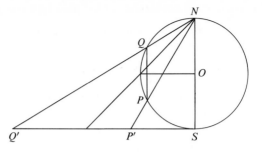

To perform a planar inversion, we may first project the figure stereographically onto a sphere, then interchange the hemispheres by reflection in the equatorial plane, and finally project stereographically back onto the plane.

3.11.15 THEOREM. *Stereographic projection is a space inversion in the sphere* N(d).

For (see Figure 3.11a) $NP = d \cos PNS$, $NP' = d \sec PNS$, whence $(NP)(NP') = d^2$, and of course N, P, P' are collinear.

3.11.16 THEOREM. *Circles on* K *correspond to "circles" on* p.

This follows from Theorems 3.11.15 and 3.11.9.

3.11.17 THEOREM. *The angle between two lines of* p *is equal to the angle between their stereographic projections on* K.

For two lines AB, AC of p map into circles $NA'B'$, $NA'C'$ of K. The tangents to these circles at N are parallel to AB and AC respectively. But the circles intersect at equal angles at N and A'. It follows that an angle between the circles at A' is equal to the corresponding angle between the straight lines at A.

Considered as a mapping of the sphere onto the plane, stereographic projection furnishes a satisfactory representation on a plane of a limited region of the sphere. In fact, stereographic projection is one of the most commonly used methods of constructing geographical maps. Particularly useful in such maps is the preservation of angles guaranteed by Theorem 3.11.17.

Stereographic projection also furnishes an elegant and timesaving way of obtaining the formulas of spherical trigonometry from those of plane trigonometry. This approach to spherical trigonometry was developed by the crystallographer and mineralogist Giuseppe Cesàro (1849–1939). For an exposition in English see J. D. H. Donnay, *Spherical Trigonometry after the Cesàro Method* (New York: Interscience Publishers, Inc., 1945).

PROBLEMS

1. (a) Which of the elementary point transformations of unextended space are involutoric? (b) Which are direct isometries? (c) Which are opposite isometries?

2. (a) Are the rotation and the translation of a screw-displacement commutative?
   (b) Are the reflection and the translation of a glide-reflection commutative?
   (c) Are the reflection and the rotation of a rotatory-reflection commutative?

3. (a) Give a proof of Theorem 3.11.1 patterned after that of Theorem 3.4.3.
   (b) Give a proof of Theorem 3.11.2 patterned after that of Theorem 3.4.4.

4. State a space analogue of Theorem 3.4.2.

5. (a) Prove that a translation in space can be factored into a product of reflections in two parallel planes.
   (b) Prove that a rotation about an axis can be factored into a product of reflections in two planes passing through the axis of rotation.

6. Show that the product of reflections in two perpendicular planes is a reflection in the line of intersection of the two planes.

7. Show that if a direct isometry leaves a point O fixed, it leaves some line through O fixed, that is, the isometry is a rotation.

8. Prove Euler's Theorem on Rotations (1776): The product of two rotations about axes through a point O is a rotation about an axis through O.

9. (a) Show that the product of two rotations of 180° about two given intersecting lines that form an angle θ is a rotation of 2θ about a line perpendicular to the two given lines at their point of intersection.
   (b) Show that the product of rotations of 180° about three mutually perpendicular concurrent lines is the identity.

10. In a space homology H let k denote the ratio of homothety and θ the angle of rotation. Show that:
    (a) If $\theta = 0$, $k = 1$, then H is the identity.
    (b) If $\theta = 180°$, $k = 1$, then H is a reflection in a line.
    (c) If $\theta = \theta$, $k = 1$, then H is a rotation about an axis.
    (d) If $\theta = 0$, $k = -1$, then H is a reflection in a point.
    (e) If $\theta = 180°$, $k = -1$, then H is a reflection in a plane.
    (f) If $\theta = \theta$, $k = -1$, then H is a rotatory-reflection.
    (g) If $\theta = 0$, $k = k$, then H is a homothety.

11. Show that a "sphere" orthogonal to the sphere of inversion inverts into itself.

12. Show that a sphere through a pair of inverse points is orthogonal to the sphere of inversion.

13. Prove Theorem 3.11.9.

14. Given a tetrahedron $ABCD$ and a point M, prove that the tangent planes, at M, to the four spheres $MBCD$, $MCDA$, $MDAB$, $MABC$, meet the respective faces BCD, CDA, DAB, ABC in four coplanar lines. (This is Problem E 493 of *The American Mathematical Monthly*, June–July 1942).

15. Prove Frederick Soddy's Hexlet Theorem (1936): Let S_1, S_2, S_3 be three spheres all touching one another. Let $K_1, K_2, \ldots$ be a sequence of spheres touching one another successively and all touching S_1, S_2, S_3. Show that K_6 touches K_1.

16. (a) Show that if, for a given sphere, the polar plane of point P passes through point Q, then the polar plane of point Q passes through point P.

(b) Show that if, for a given sphere, the pole of plane p lies on plane q, the pole of q lies on p.

(c) Show that if, for a given sphere, the noncollinear points P, Q, R are the poles of p, q, r, then the pole of plane PQR is the point of intersection of p, q, r.

17. Show that the polars, for a given sphere, of a range of points form a pencil of planes.

18. Define *conjugate points* and *conjugate planes* for a sphere.

19. Define *conjugate tetrahedra* and *self-conjugate tetrahedron* of a sphere.

20. (a) Establish Theorem 3.11.12.

(b) Establish Theorem 3.11.13.

BIBLIOGRAPHY

See the Bibliography of Chapter 2; also:

BARRY, E. H., *Introduction to Geometrical Transformations*. Boston, Mass.: Prindle, Weber & Schmidt, Inc., 1966.

CHOQUET, GUSTAVE, *Geometry in a Modern Setting*. Boston, Mass.: Houghton Mifflin Company, 1969.

COXETER, H. S. M., *Introduction to Geometry*. New York: John Wiley and Sons, Inc., 1961.

COXFORD, A. F., and Z. P. USISKIN, *Geometry, a Transformation Approach*. River Forest, Illinois: Laidlaw Brothers, 1971.

DONNAY, J. D. H., *Spherical Trigonometry after the Cesàro Method*. New York: Interscience Publishers, Inc., 1945.

ECCLES, F. M., *An Introduction to Transformational Geometry*. Reading, Mass.: Addison-Wesley, 1971.

FORDER, H. G., *Geometry*. New York: Hutchinson's University Library, 1950,

GRAUSTEIN, W. C., *Introduction to Higher Geometry*. New York: The Macmillan Company, 1930.

JEGER, MAX, *Transformation Geometry*, tr. by A. W. Deicke and A. G. Howson. London: George Allen and Unwin, Ltd., 1964.

KLEIN, FELIX, *Elementary Mathematics from an Advanced Standpoint, Geometry*, tr. by E. R. Hedrick and C. A. Noble. New York: Dover Publications, Inc., 1939. First German edition published in 1908.

LEVI, HOWARD, *Topics in Geometry*. Boston, Mass.: Prindle, Weber & Schmidt, Inc., 1968.

LEVY, L. S., *Geometry: Modern Mathematics via the Euclidean Plane*. Boston, Mass.: Prindle, Weber & Schmidt, Inc., 1970.

MODENOV, P. S., and A. S. PARKHOMENKO, *Geometric Transformations* (2 vols.), tr. by M. B. P. Slater. New York: Academic Press, 1965.

YAGLOM, I. M., *Geometric Transformations*, tr. by Allen Shields. New York: Random House, Inc., New Mathematical Library, No. 8, 1962.

4

Euclidean Constructions

There is much to be said in favor of a game which you play alone. It can be played or abandoned whenever you wish. There is no bother about securing a willing and suitable opponent, nor do you annoy anyone if you suddenly decide to desist play. Since you are, in a sense, your own opponent, the company is most congenial and perfectly matched in skill and intelligence, and there is no embarrassing sarcastic utterance should you make a stupid play. The game is particularly good if it is truly challenging and if it possesses manifold variety. It is still better if also the rules of the game are very few and simple. And little more can be asked if, in addition, the game requires no highly specialized equipment, and so can be played almost anywhere and at almost any time.

The Greek geometers of antiquity devised a game—we might call it *geometrical solitaire*—which, judged on all the above points, must surely stand at the very top of any list of games to be played alone. Over the ages it has attracted hosts of players, and though now well over 2000 years old, it seems not to have lost any of its singular charm and appeal. This chapter is concerned with a few facets of this fascinating game, and with some of its interesting modern variants.

4.1 THE EUCLIDEAN TOOLS

For convenience we repeat here the first three postulates of Euclid's *Elements*:

1. A straight line can be drawn from any point to any point.
2. A finite straight line can be produced continuously in a straight line.
3. A circle may be described with any center and distance.

These postulates state the primitive constructions from which all other constructions in the *Elements* are to be compounded. They constitute, so to speak, the rules of the game of Euclidean construction. Since they restrict constructions to only those that can be made in a permissible way with straightedge and compass,* these two instruments, so limited, are known as the *Euclidean tools*.

The first two postulates tell us what we can do with a Euclidean straightedge; we are permitted to draw as much as may be desired of the straight line determined by any two given points. The third postulate tells us what we can do with the Euclidean compass; we are permitted to draw the circle of given center and having any straight line segment radiating from that center as a radius—that is, we are permitted to draw the circle of given center and passing through a given point. Note that neither instrument is to be used for transferring distances. This means that the straightedge cannot be marked, and the compass must be regarded as having the characteristic that if either leg is lifted from the paper, the instrument immediately collapses. For this reason, a Euclidean compass is often referred to as a *collapsing compass*; it differs from a *modern compass*, which retains its opening and hence can be used as a divider for transferring distances.

It would seem that a modern compass might be more powerful than a collapsing compass. Curiously enough, such turns out not to be the case; any construction performable with a modern compass can also be carried out (in perhaps a longer way) by means of a collapsing compass. We prove this fact as our first theorem, right after introducing the following convenient notation.

4.1.1 NOTATION. The circle with center O and passing through a given point C will be denoted by $O(C)$, and the circle with center O and radius equal to a given segment AB will be denoted by $O(AB)$.

4.1.2 THEOREM. *The collapsing and modern compasses are equivalent.*

To prove the theorem it suffices to show that we may, with a collapsing compass, construct any circle $O(AB)$. This may be accomplished as follows (see Figure 4.1a). Draw circles $A(O)$ and $O(A)$ to intersect in D and E; draw circles $D(B)$ and $E(B)$ to intersect again in F; draw circle $O(F)$. It is an easy matter to prove that $OF = AB$, whence circle $O(F)$ is the same as circle $O(AB)$.

---

* Though contrary to common English usage, we shall use this word in the singular.

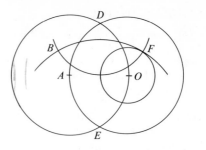

Figure 4.1a

In view of Theorem 4.1.2, we may dispense with the Euclidean, or collapsing compass, and in its place employ the more convenient modern compass. We are assured that the set of constructions performable with straightedge and Euclidian compass is the same as the set performable with straightedge and modern compass. As a matter of fact, in all our construction work, we shall not be interested in actually and exactly carrying out the constructions, but merely in assuring ourselves that such constructions are possible. To use a phrase of Jacob Steiner, we shall do our constructions "simply by means of the tongue," rather than with actual instruments on paper. We seek then, at least for the time being, the easiest construction to describe rather than the simplest or best construction actually to carry out with the instruments.

If one were asked to find the midpoint of a given line segment using only the straightedge, one would be justified in exclaiming that surely the Euclidean straightedge alone will not suffice, and that some additional tool or permission must be furnished. The same is true of the combined Euclidean tools; there are constructions which cannot be performed with these tools alone, at least under the restrictions imposed upon them. Three famous problems of this sort, which originated in ancient Greece, are:

1. *The duplication of the cube*, or the problem of constructing the edge of a cube having twice the volume of a given cube.
2. *The trisection of an angle*, or the problem of dividing a given arbitrary angle into three equal parts.
3. *The quadrature of the circle*, or the problem of constructing a square having an area equal to that of a given circle.

The fact that there are constructions beyond the Euclidean tools adds a certain zest to the construction game. It becomes desirable to obtain a criterion for determining whether a given construction problem is or is not within the power of our tools. Synthetic geometry has not been able to cope with this problem, and we accordingly reserve a discussion of this interesting facet of Euclidean constructions for a later chapter, where analytical methods will be employed.

But, in spite of the limited power of our instruments, one is surprised at the really intricate constructions that can be accomplished with them. Thus, though with our instruments we cannot, for example, solve the seemingly simple problem of drawing the two lines trisecting an angle of 60°, we can draw all the circles which touch three given circles (the *problem of Apollonius*); we can draw three circles in the angles of a triangle such that each circle touches the other two and also the two sides of the angle (the *problem of Malfatti*); we can inscribe in a given circle a triangle whose sides, produced if necessary, pass through three given points (the *Castillon-Cramer problem*).

As a concluding remark of this section we point out that Euclid used constructions in the sense of existence theorems—to prove that certain entities actually exist. Thus one may define a *bisector* of a given angle as a line in the plane of the angle, passing through the vertex of the angle, and such that it divides the given angle into two equal angles. But a definition does not establish the existence of the thing being defined; this requires proof. To show that a given angle does possess a bisector, we show that this entity can actually be constructed. Existence theorems are very important in mathematics, and actual construction of an entity is the most satisfying way of proving its existence. One might define a *square circle* as a figure which is both a square and a circle, but one would never be able to prove that such an entity exists; the class of square circles is a class without any members. In mathematics it is nice to know that the set of entities satisfying a certain definition is not just the empty set.

PROBLEMS

1. In the proof of Theorem 4.1.2, show that $OF = AB$.

2. A student reading Euclid's *Elements* for the first time might experience surprise at the three opening propositions of Book I (for statements of these propositions see Appendix 1). These three propositions are constructions, and are trivial with straightedge and *modern* compass, but require some ingenuity with straightedge and *Euclidean* compass.
   (a) Solve I 1 with Euclidean tools.
   (b) Solve I 2 with Euclidean tools.
   (c) Solve I 3 with Euclidean tools.
   (d) Show that I 2 proves that the straightedge and Euclidean compass are equivalent to the straightedge and modern compass.

3. Consider the following two arguments:

   I. PROPOSITION. *Of all triangles inscribed in a circle, the equilateral is the greatest.*

   1. If ABC is a nonequilateral triangle inscribed in a circle, so that $AB \neq AC$, say, construct triangle XBC, where X is the intersection of the perpendicular bisector of BC with arc BAC.
   2. Then triangle XBC > triangle ABC.
   3. Hence, if we have a nonequilateral triangle inscribed in a circle, we can always construct a greater inscribed triangle.
   4. Therefore, of all triangles inscribed in a circle, the equilateral is the greatest.

   II. PROPOSITION. *Of all natural numbers, 1 is the greatest.*

   1. If m is a natural number other than 1, construct the natural number m^2.

2. Then $m^2 > m$.
3. Hence, if we have a natural number other than 1, we can always construct a greater natural number.
4. Therefore, of all natural numbers, 1 is the greatest.

Now the conclusion in argument I is true, and that in argument II is false. But the two arguments are formally identical. What, then, is wrong?

4.2 THE METHOD OF LOCI

In this section we very briefly consider what is perhaps the most basic method in the solution of geometric construction problems. It can often be used alone, and often in combination with some other method. It may be considered as a fundamental maneuver in the construction game.

The solution of a construction problem very often depends upon first finding some key point. Thus the problem of drawing a circle through three given points is essentially solved once the center of the circle is located. Again, the problem of drawing a tangent to a circle from an external point is essentially solved once the point of contact of the tangent with the circle has been found. Now the key point satisfies certain conditions, and each condition considered alone generally restricts the position of the key point to a certain locus. The key point is thus found at the intersections of certain loci. This method of solving a construction problem is aptly referred to as the *method of loci*.

To illustrate, denote the three given points in our first problem above by *A*, *B*, *C*. Now the sought center *O* of the circle through *A*, *B*, *C* must be equidistant from *A* and *B* and also from *B* and *C*. The first condition places *O* on the perpendicular bisector of *AB*, and the second condition places *O* on the perpendicular bisector of *BC*. The point *O* is thus found at the intersection, if it exists, of these two perpendicular bisectors. If the three given points are not collinear, there is exactly one solution; otherwise there is none.

Suppose, in our second problem above, we denote the center of the given circle by *O*, the external point by *E*, and the sought point of contact of the tangent from *E* to the circle by *T*. Now *T*, first of all, lies on the given circle. Also, since $\angle OTE = 90°$, *T* lies on the circle having *OE* as diameter. The sought point *T* is thus found at an intersection of these two circles. There are always two solutions to the problem.

In order to apply the method of loci to the solution of geometric constructions, it is evidently of great value to know a considerable number of loci that are constructible straight lines and circles. Here are a few such loci.

1. The locus of points at a given distance from a given point is the circle having the given point as center and the given distance as radius.
2. The locus of points at a given distance from a given line consists of the two lines parallel to the given line and at the given distance from it.

3. The locus of points equidistant from two given points is the perpendicular bisector of the segment joining the two given points.
4. The locus of points equidistant from two given intersecting lines consists of the bisectors of the angles formed by the two given lines.
5. The locus of points from which lines drawn to the endpoints of a given line segment enclose a given angle consists of a pair of congruent circular arcs having the given segment as a chord, the two arcs lying on opposite sides of the given segment. In particular, if the given angle is a right angle, the two arcs are the semicircles having the given segment as diameter.
6. The locus of points whose distances from two given points A and B have a given ratio $k \neq 1$ is the circle on IE as diameter, where I and E divide AB internally and externally in the given ratio. (This is the circle of Apollonius for the given segment and the given ratio.)
7. The locus of points whose distances from two given intersecting lines have a given ratio k is a pair of straight lines through the point of intersection of the given lines. Locus (4) is the special case where $k = 1$.
8. The locus of points for which the difference of the squares of the distances from two given points is a constant is a straight line perpendicular to the line determined by the two given points.
9. The locus of points for which the sum of the squares of the distances from two given points is a constant is a circle having its center at the midpoint of the segment joining the two given points.

Not only should the reader verify the correctness of each of the above loci, but he should assure himself that each one can actually be constructed with compass and straightedge. For example, to construct locus (5) he might lay off the supplement of the given angle at one end A of the given segment AB and then find the center of one of the desired arcs as the intersection of the perpendicular bisector of AB and the perpendicular at A to the other side of the layed-off angle. Locus (6) is easily constructed once the points I and E are found, and the finding of these was the subject matter of Problem 7, Section 2.1. Each of the lines of locus (7) can be found once one point, other than the intersection of the two given lines, has been found, and such a point can be found by (2), any two distances having the given ratio being chosen. To construct locus (8), let the given points be A and B and denote the given difference of squares of distances by d^2. Construct any right triangle having a leg d and the other leg greater than half of AB. Using A and B as centers and radii equal to the hypotenuse and other leg, respectively, of the constructed right triangle, find a point P on the sought locus. We are here assuming that P has its greater distance from A. To find the points M' and N' where locus (9) cuts the segment joining the two given points A and B, draw angle $BAD = 45°$ and cut line AD in M and N with circle $B(s)$, where s^2 is the given sum of squares of distances. Then M' and N' are the feet of the perpendiculars from M and N on AB.

Let us now consider a few construction problems solvable by the method of loci.

4.2.1 PROBLEM. *Draw a circle passing through two given points and subtending a given angle at a third point.*

Referring to Figure 4.2a, let A and B be the two given points and let C

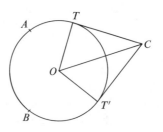

Figure 4.2a

be the third point. Denote the center of the sought circle by O and let T and T' denote the points of contact of the tangents to the circle from point C. Since $\angle TCT'$ is given, the form of right triangle OTC is known. That is, we know the ratio OT/OC. But $OA/OC = OB/OC = OT/OC$. It follows that O lies on the circle of Apollonius for A and C and ratio OT/OC, and on the circle of Apollonius for B and C and ratio OT/OC. Point O is thus found at the intersections, if any exist, of these two circles of Apollonius. The details are left to the reader.

4.2.2 PROBLEM. *Find a point the distances of which from three given lines have given ratios.*

Use locus (7).

4.2.3 PROBLEM. *In a triangle find a point the distances of which from the three vertices have given ratios.*

Use locus (6).

4.2.4 PROBLEM. *Construct a triangle given one side, the altitude on that side, and the sum of the squares of the other two sides.*

Find the vertex opposite the given side by using loci (2) and (9).

4.2.5 PROBLEM. *Construct a triangle given one side, the opposite angle, and the difference of the squares of the other two sides.*

Find the vertex opposite the given side by using loci (5) and (8).

PROBLEMS

1. Establish the constructions given in the text for loci (1) through (9).

2. Complete the details of Problem 4.2.1.

3. Construct a triangle given one side and the altitude and median to that side.

4. Draw a circle touching two given parallel lines and passing through a given point.

5. Construct a triangle given one side, the opposite angle, and the median to the given side.

6. Find a point at which three given circles subtend equal angles.

7. Two balls are placed on a diameter of a circular billiard table. How must one ball be played in order to hit the other after its recoil from the circumference?

8. Through two given points of a circle draw two parallel chords whose sum shall have a given length.

9. Draw a circle of given radius touching a given circle and having its center on a given line.

10. Inscribe a right triangle in a given circle so that each leg will pass through a given point.

11. Draw a circle of given radius, passing through a given point, and cutting off a chord of given length on a given line.

12. Draw a circle tangent to a given line at a given point and also tangent to a given circle.

13. Draw a tangent to a given circle so that a given line cuts off on the tangent a given distance from the point of contact.

14. Construct a cyclic quadrilateral given one angle, an adjacent side, and the two diagonals.

15. Through a given point draw a line intersecting a given circle so that the distances of the points of intersection from a given line have a given sum.

16. Find a point from which three parts AB, BC, CD of a given line are seen under equal angles.

17. Find the locus of points for which the distances from two given lines have a given sum.

18. (a) On the circumference of a given circle find a point for which the sum of the distances from two given lines is given.
(b) Find the point on the given circle for which the sum of the distances from the two given lines is a minimum.

4.3   THE METHOD OF TRANSFORMATION

There are many geometric construction problems that can be solved by applying one of the transformations discussed in the previous chapter. We illustrate this *method of transformation* by the following sequence of problems, in which certain details are left to the reader.

4.3.1   PROBLEM.   *Draw a line in a given direction on which two given circles cut off equal chords.*

In Figure 4.3a, let C_1 and C_2 be the given circles and let t be a line in

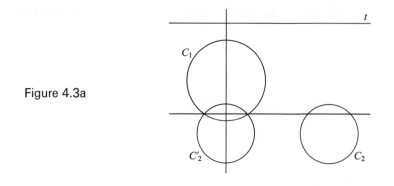

Figure 4.3a

the given direction. Translate C_2 parallel to t to position C_2' in which the line of centers of C_1 and C_2' is perpendicular to t. Then the line through the points of intersection, if such exist, of C_1 and C_2' is the sought line.

4.3.2 PROBLEM. *Through one of the points of intersection of two given intersecting circles, draw a line on which the two circles cut off equal chords.*

In Figure 4.3b, let C_1 and C_2 be the two given intersecting circles, and let

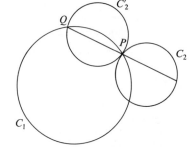

Figure 4.3b

P be one of the points of intersection. Reflect C_2 in point P into position C_2' and let Q be the other intersection of C_1 and C_2'. Then QP is the sought line.

4.3.3 PROBLEM. *Inscribe a square in a given triangle, so that one side of the square lies on a given side of the triangle.*

In Figure 4.3c, let ABC be the given triangle and BC the side on which the required square is to lie. Choose any point D' on side AB and construct the square $D'E'F'G'$ as indicated in the figure. If E' falls on AC, the problem is solved. Otherwise we have solved the problem for a triangle $A'BC'$ which

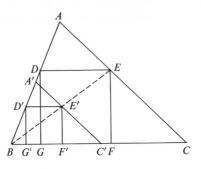

Figure 4.3c

is homothetic to triangle ABC with B as center of homothety. It follows that line BE' cuts AC in vertex E of the sought square inscribed in triangle ABC.

It is interesting, in connection with the last problem, to contemplate that from a sheer guess of the position of the sought square we were able to find the actual position of the square. The method employed, often called the *method of similitude*, is a geometric counterpart of the *rule of false position* used by the ancient Egyptians to solve linear equations in one unknown. Suppose, for example, we are to solve the simple equation $x + x/5 = 24$. Assume any convenient value of x, say $x = 5$. Then $x + x/5 = 6$, instead of 24. Since 6 must be multiplied by 4 to give the required 24, the correct value of x must be 4(5), or 20. From a sheer guess, and without employment of algebraic procedures, we have obtained the correct answer.

4.3.4   A GENERAL PROBLEM.   *Given a point* O *and two curves* C_1 *and* C_2. *Locate a triangle* OP_1P_2, *where* P_1 *is on* C_1 *and* P_2 *is on* C_2, *similar to a given triangle* $O'P_1'P_2'$.

In Figure 4.3d, let C_2' be the map of C_2 under the homology

$$H(O,\ \measuredangle P_2'O'P_1',\ O'P_1'/O'P_2').$$

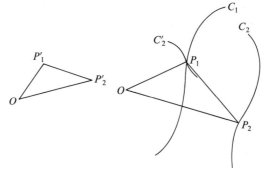

Figure 4.3d

Then C_1 and C_2' intersect in the possible positions of P_1. If C_1 and C_2 are "circles," the problem can be solved with Euclidean tools.

4.3.5 PROBLEM. *Draw an equilateral triangle having its three vertices on three given parallel lines.*

In Figure 4.3e, choose any point O on one of the three given parallel lines,

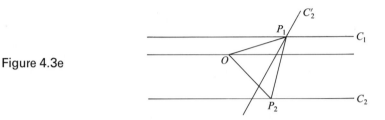

Figure 4.3e

and denote the other two parallel lines by C_1 and C_2. We may now apply the above General Problem 4.3.4 by subjecting line C_2 to the homology $H(O, 60°, 1)$.

4.3.6 PROBLEM. *Draw a "circle" touching three given concurrent non-coaxial circles.*

In Figure 4.3f, let the three given circles C_1, C_2, C_3 intersect in point O.

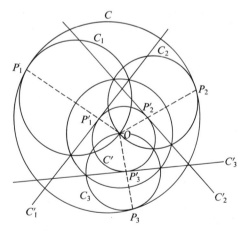

Figure 4.3f

Subject the figure to any convenient inversion of center O. Then C_1, C_2, C_3 become three straight lines C_1', C_2', C_3', which are easily constructed with Euclidean tools (they are actually the common chords of C_1, C_2, C_3 with the circle of inversion). Draw a circle C' touching all three of the lines C_1',

C_2', C_3'. The inverse C of this circle, which can be constructed with Euclidean tools (if C' touches C_1', C_2', C_3' at P_1', P_2', P_3', respectively, then C touches C_1, C_2, C_3 at the points P_1, P_2, P_3 where OP_1', OP_2', OP_3' cut C_1, C_2, C_3 again), is a "circle" touching C_1, C_2, C_3. Note that there are four solutions to the problem.

The *method of homology*, illustrated in Problem 4.3.5, and the *method of inversion*, illustrated in Problem 4.3.6, are powerful methods, and many construction problems that would otherwise be very difficult yield to these methods. They are, of course, instances of the general method of transformation. Note that Problem 4.3.6, is a special case of the Problem of Apollonius.

PROBLEMS

1. Place a line segment equal and parallel to a given line segment and having its extremities on two given circles.

2. From a vessel two known points are seen under a given angle. The vessel sails a given distance in a known direction, and now the same two points are seen under another known angle. Find the position of the vessel.

3. In a given quadrilateral inscribe a parallelogram the center of which is at a given point.

4. Solve the general problem: To a given line draw a perpendicular on which two given curves will cut off equal lengths measured from the foot of the perpendicular.

5. Place a square with two opposite vertices on a given line and the other two vertices on two given circles.

6. Draw a triangle given the positions of three points which divide the three sides in given ratios.

7. Inscribe a quadrilateral of given shape in a semicircle, a specified side of the quadrilateral lying along the diameter of the semicircle.

8. Given the focus and directrix of a parabola, find the points of intersection of the parabola with a given line.

9. Draw a circle passing through a given point and touching two given lines.

10. Find points D and E on sides AB and AC of a triangle ABC so that $BD = DE = EC$.

11. Draw a triangle given A, $a + b$, $a + c$.

12. Solve the general problem: Through a given point O draw a line intersecting two given curves C_1 and C_2 in points P_1 and P_2 so that OP_1 and OP_2 shall be in a given ratio to one another.

13. Through a given point O within a circle draw a chord which is divided by O in a given ratio.

14. Through a given point O on a given circle draw a chord which is bisected by another given chord.

15. In a given triangle ABC inscribe another, $A'B'C'$, which shall have its sides parallel to three given lines.

16. Two radii are drawn in a circle. Draw a chord which will be trisected by the radii.

17. In a given parallelogram inscribe an isosceles triangle of given vertex angle and with its vertex at a given vertex of the parallelogram.

18. Draw an equilateral triangle having its three vertices on three given concentric circles.

19. Solve the general problem: Through a given point O draw a line cutting two given curves C_1 and C_2 in points P_1 and P_2 such that $(OP_1)(OP_2)$ is a given constant.

20. Through a given point O draw a line cutting two given lines in points A and B so that $(OA)(OB)$ is given.

21. Through a point of intersection of two circles draw a line on which the two circles intercept chords having a given product.

22. Draw a circle passing through a given point P and touching two given circles.

23. Draw a circle through two given points and tangent to a given circle.

24. Draw a circle tangent externally to three given mutually external circles.

4.4 THE DOUBLE POINTS OF TWO COAXIAL HOMOGRAPHIC RANGES

In this section we describe another, and very clever, maneuver in the construction game. Some writers call this maneuver the *method of trial and error*. We need the following basic construction.

4.4.1 PROBLEM. *Find the common, or double, points of two distinct coaxial homographic ranges.*

A number of ingenious solutions have been devised for this problem. We give one here which is based upon simple cross-ratio properties of a circle. In Figure 4.4a, let A and A', B and B', C and C' be three pairs of corresponding points of two distinct coaxial homographic ranges. Draw any circle in the plane and on it take any point O. Draw the six lines OA, OB, OC, OA', OB', OC' and let them cut the circle again in A_1, B_1, C_1, A_1', B_1', C_1' respectively. Let A_1B_1' and $A_1'B_1$ intersect in P, and A_1C_1' and $A_1'C_1$ in Q. Draw line PQ. If this line cuts the circle in D_1 and E_1, then the lines OD_1 and OE_1 will intersect the common base of the homographic ranges in their double points D and E.

To prove the above construction, suppose PQ does cut the circle in points D_1 and E_1. Let S be the intersection of PQ and A_1A_1'. Then

$$(AB,CD) = O(A_1B_1,C_1D_1) = A_1'(A_1B_1,C_1D_1) = (SP,QD_1)$$
$$= A_1(A_1'B_1',C_1'D_1) = O(A_1'B_1',C_1'D_1) = (A'B',C'D),$$

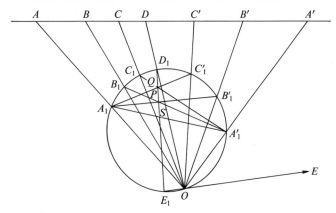

Figure 4.4a

and D is a common, or double, point of the two coaxial homographic ranges. In a similar way it can be shown that E is also a double point of the two ranges.

Conversely, suppose D is a double point of the two coaxial homographic ranges. Let D_1 be the point where OD cuts the circle again, and let $A_1'D_1$ and A_1D_1 cut the line PQ in M and N respectively. Then

$$(AB,CD) = O(A_1B_1,C_1D_1) = A_1'(A_1B_1,C_1D_1) = (SP,QM).$$

Similarly, $(A'B',C'D) = (SP,QN)$. But, since D is a double point, $(AB,CD) = (A'B',C'D)$. Therefore $(SP,QM) = (SP,QN)$, or $M = N = D_1$, and PQ passes through the point D_1. It follows that our construction gives all the double points of the two distinct coaxial homographic ranges.

4.4.2 REMARKS. (1) Since PQ may intersect, touch, or fail to intersect the circle, two distinct coaxial homographic ranges have two, one, or no double points.

(2) The above construction also solves two other important problems: that of finding the common, or double, rays of two distinct copunctual homographic pencils, and that of finding the common, or double, points of two distinct concyclic homographic ranges.

(3) Note that PQ is the Pascal line of the hexagon $A_1B_1'C_1A_1'B_1C_1'$, whence B_1C_1' and $B_1'C_1$ also intersect on it.

We now illustrate our new maneuver in a solution of the following problem.

4.4.3 PROBLEM. *Construct a triangle inscribed in one given triangle and circumscribed about another given triangle.*

Referring to Figure 4.4b, let ABC and $A'B'C'$ be the two given triangles. We wish to construct a triangle PQR whose vertices P, Q, R lie on the

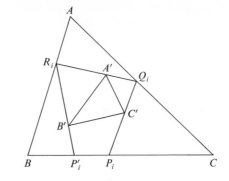

Figure 4.4b

sides *BC*, *CA*, *AB* of triangle *ABC* and whose sides *QR*, *RP*, *PQ* pass through the vertices *A'*, *B'*, *C'* of triangle *A'B'C'*. Take an arbitrary point P_i on *BC* and draw $P_i C'$ to cut *AC* in Q_i; next draw $Q_i A'$ to cut *AB* in R_i; then draw $R_i B'$ to cut *BC* in P_i'. It is easily seen that range P_i is homographic to range P_i', and the desired point *P* is a double point of these two coaxial homographic ranges. We may find *P* by applying Problem 4.4.1.

We now see why the new method is sometimes referred to as the *method of trial and error*; from three *guesses* of the position of a point we find its actual position.

PROBLEMS

1. Through a given point *P* draw a line intersecting two given lines *m* and *m'* in corresponding points of two homographic ranges lying on *m* and *m'*.

2. Given two homographic pencils, find the pairs of corresponding rays which intersect on a given line *m*.

3. Given two homographic pencils, find a pair of corresponding rays which intersect at a given angle.

4. Given two homographic pencils, find a pair of corresponding rays which are parallel.

5. Construct a line segment whose extremities lie one each on two given intersecting lines, and which subtends a given angle at each of two given points.

6. Find two points on a given line which are isogonal conjugates with respect to a given triangle.

7. In a given triangle *ABC* inscribe another, *A'B'C'*, which shall have its sides parallel to three given lines. (This is Problem 15, Section 4.3.)

8. Given four coplanar lines, *p*, *q*, *r*, *s*. Find points *A* and *B* on *p* and *q* such that the projections of *AB* on *r* and *s* shall have given lengths.

9. Through a given point draw two lines which cut off segments of given lengths on two given lines.

10. Given two fixed points O and O' on two fixed lines m and m'. Through a fixed point P draw a line cutting m and m' in points A and A' such that $OA/O'A'$ is a given constant. (This problem was the subject matter of Apollonius' treatise *On Proportional Section*.)

11. Given two fixed points O and O' on two fixed lines m and m'. Through a fixed point P draw a line cutting m and m' in points A and A' such that $(OA)(O'A')$ is a given constant. (This problem was the subject matter of Apollonius' treatise *On Spatial Section*.)

12. Through a given point draw a line to include with two given intersecting lines a triangle of given area.

13. Inscribe a triangle in a given triangle such that its sides will subtend given angles at given points.

14. Solve the Castillon-Cramer Problem: Inscribe in a given circle a triangle whose sides, produced if necessary, pass through three given points.

15. Circumscribe a triangle about a given circle such that each vertex shall lie on a given line.

4.5 THE MOHR-MASCHERONI CONSTRUCTION THEOREM

The eighteenth-century Italian geometer and poet, Lorenzo Mascheroni (1750–1800), made the surprising discovery that all Euclidean constructions, insofar as the given and required elements are points, can be made with the compass alone, and that the straightedge is thus a redundant tool. Of course, straight lines cannot be drawn with the compass, but any straight line arrived at in a Euclidean construction can be determined by the compass by finding two points on the line. This discovery appeared in 1797 in Mascheroni's *Geometria del compasso*. Generally speaking, Mascheroni established his results by using the idea of reflection in a line. In 1890, the Viennese geometer, August Adler (1863–1923), published a new proof of Mascheroni's results, using the inversion transformation.

Then an unexpected thing happened. Shortly before 1928, a student of the Danish mathematician Johannes Hjelmslev (1873–1950), while browsing in a bookstore in Copenhagen, came across a copy of an old book, *Euclides Danicus*, published in 1672 by an obscure writer named Georg Mohr. Upon examining the book, Hjelmslev was surprised to find that it contained Mascheroni's discovery, with a different solution, arrived at a hundred and twenty-five years before Mascheroni's publication had appeared.

The present section will be devoted to a proof of the Mohr-Mascheroni discovery. We shall employ the Mascheroni approach, and shall relegate Adler's approach to the problems at the end of the section. We first introduce a compact and elegant way of describing any given construction. The method will become clear from an example, and we choose the construction appearing in Theorem 4.1.2. That construction can be condensed into the following table:

A(O), O(A)	D(B), E(B)	O(F)
D, E	F	

The first line of the table tells us what "circles" we are to draw, and the second line labels the points of intersection so obtained. The table is divided vertically into steps. Reading the above table we have: Step 1. Draw circles $A(O)$ and $O(A)$ to intersect in points D and E. Step 2. Draw circles $D(B)$ and $E(B)$ to intersect in point F. Step 3. Draw circle $O(F)$. It will be noted that this is precisely the construction appearing in Theorem 4.1.2.

We are now ready to proceed.

4.5.1 PROBLEM. *Given points* A, B, C, D, *construct, with a modern compass alone, the points of intersection of circle* C(D) *and line* AB.

Case 1. C not on AB (see Figure 4.5a).

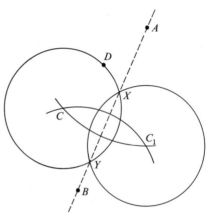

Figure 4.5a

A(C), B(C)	C(D), C₁(CD)
C₁	X, Y

Case 2. C on AB (see Figure 4.5b).

A(D), C(D)	C(DD₁), D(C)	C(DD₁), D₁(C)	F(D₁), F₁(D)	F(CM), C(D)
D₁	F, 4th vertex of □ CD₁DF	F₁, 4th vertex of □ CDD₁F₁	M	X, Y

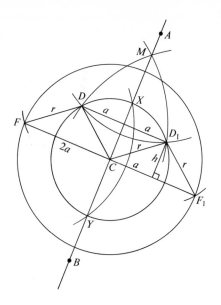

Figure 4.5b

The proof of case 1 is easy. In case 2, observe (see Figure 4.5b) that

$$(CM)^2 = (FM)^2 - 4a^2 = (FD_1)^2 - 4a^2 = (9a^2 + h^2) - 4a^2$$
$$= 9a^2 + r^2 - a^2 - 4a^2 = 4a^2 + r^2 = (FX)^2.$$

4.5.2 PROBLEM. *Given points A, B, C, D, construct, with a modern compass alone, the points of intersection of the lines* AB *and* CD *(see Figure 4.5c).*

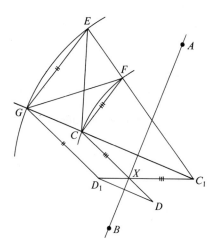

Figure 4.5c

The proof of the construction, given on the next page, is easy. We have C_1D_1GE similar to C_1XCF. But $GD_1 = GE$. Therefore $CX = CF$.

$A(C), B(C)$	$A(D), B(D)$	$C(DD_1),$ $D_1(CD)$	$C_1(G), G(D_1)$	$C_1(C), G(CE)$	$C(F),$ $C_1(CF)$
C_1	D_1	G, collinear with C, C_1	E, either intersection	F, collinear with C_1, E	X

4.5.3 THE MOHR-MASCHERONI CONSTRUCTION THEOREM. *Any Euclidean construction, insofar as the given and required elements are points, may be accomplished with the Euclidean compass alone.*

For in a Euclidean construction, every new point is determined as an intersection of two circles, of a straight line and a circle, or of two straight lines, and the construction, no matter how complicated, is a succession of a finite number of these processes. Because of the equivalence of modern and Euclidean compasses (see Theorem 4.1.2), it is then sufficient to show that with a modern compass alone we are able to solve the following three problems:

    I. Given A, B, C, D, find the points of intersection of $A(B)$ and $C(D)$.
    II. Given A, B, C, D, find the points of intersection of AB and $C(D)$.
    III. Given A, B, C, D, find the point of intersection of AB and CD.

But I is obvious, and we have solved II and III in Problem 4.5.1 and Problem 4.5.2, respectively.

It is to be noted that our proof of the Mohr-Mascheroni construction theorem is more than a mere existence proof, for not only have we shown the existence of a construction using only the Euclidean compass which can replace any given Euclidean construction, but we have shown how such a construction can actually be obtained from the given Euclidean construction. It must be confessed, though, that the resulting construction using only the Euclidean compass would, in all likelihood, be far more complicated than is necessary. The task of finding a "simplest" construction employing only the Euclidean compass, or even only the modern compass, is usually very difficult indeed, and requires considerable ingenuity on the part of the solver.

PROBLEMS

1. Solve or establish, as the case may be, the following constructions using only the *Euclidean* compass:
    (a) Given points A and B, find point C on AB produced such that $AB = BC$.

$B(A), A(B)$	$B(A), R(B)$	$B(A), S(B)$
R	S	C

(b) Given points A and B, and a positive integer n, find point C on AB produced such that $AC = n(AB)$.

(c) Given points O, A, B, find the reflection of O in line AB.

(d) Given points O, D, M, find the inverse M' of M in circle $O(D)$.

Case 1. $OM > (OD)/2$.

O(D), M(O)	A(O), B(O)
A, B	M'

Case 2. $OM \leqq (OD)/2$.

(e) Given noncollinear points A, B, O, find the center Q of the inverse of line AB in circle $O(D)$.

Find P, the reflection of O in AB, and then Q, the inverse of P in $O(D)$.

(f) Given points O, D and a circle k not through O, find the center M' of the inverse k' of k in circle $O(D)$.

Find M, the inverse of O in k, and then M', the inverse of M in $O(D)$.

(g) Given points A, B, C, D, construct, with a Euclidean compass alone, the points X, Y of intersection of circle $C(D)$ and line AB.

(h) Given points A, B, C, D, construct, with a Euclidean compass alone, the point X of intersection of lines AB and CD.

The above steps essentially constitute Adler's proof of the Mohr-Mascheroni construction theorem. Note that Mascheroni's approach exploits reflections in lines whereas Adler's approach exploits the inversion transformation.

2. Establish the following solution, using a Euclidean compass alone, of the problem of finding the center D of the circle through three given noncollinear points A, B, C.

Draw circle $A(B)$. Find (by Problem 1, (d)) the inverse C' of C in $A(B)$. Find (by Problem 1, (c)) the reflection D' of A in BC'. Find (by Problem 1, (d)) the inverse D of D' in $A(B)$.

3. Given points A and B, find, with a Euclidean compass alone, the midpoint M of segment AB.

4. On page 268 of Cajori's *A History of Mathematics* we read: "Napoleon proposed to the French mathematicians the problem, to divide the circumference of a circle into four equal parts by the compasses only. Mascheroni does this by applying the radius three times to the circumference; he obtains the arcs AB, BC, CD; then AD is a diameter; the rest is obvious." Complete the "obvious" part of the construction.

5. Look up the following constructions with compass alone which have appeared in *The American Mathematical Monthly*: Problem 3000, Apr. 1924; Problem 3327, June-July 1929; Problem 3706, June-July 1936; Problem E 100, Jan. 1935; Problem E 567, Jan. 1944.

4.6 THE PONCELET-STEINER CONSTRUCTION THEOREM

Though all Euclidean constructions, insofar as the given and required elements are points, are possible with a Euclidean compass alone, it is an easy matter to assure ourselves that not all Euclidean constructions are similarly possible with a Euclidean straightedge alone. To see this, consider the problem of finding the point M midway between two given points A and B, and suppose the problem can be solved with straightedge alone. That is, suppose there exists a finite sequence of lines, drawn according to the restrictions of Euclid's first two postulates, that finally leads from the two given points A and B to the desired midpoint M. Choose a point O outside the plane of construction, and from O project the entire construction upon a second plane not through O. Points A, B, M of the first plane project into points A', B', M' of the second plane, and the sequence of lines leading to the point M in the first plane projects into a sequence of lines leading to the point M' in the second plane. The description of the straightedge construction in the second plane, utilizing the projected sequence of lines, of the point M' from the points A' and B' is exactly like the description of the straightedge construction in the first plane, utilizing the original sequence of lines, of the point M from the points A and B. Since M is the midpoint of AB, it follows, then, that M' must be the midpoint of $A'B'$. But this is absurd, for the midpoint of a line segment need not project into the midpoint of the projected segment. It follows that the simple Euclidean problem of finding the point M midway between two given points A and B is not possible with the straightedge alone.

Our inability to solve all Euclidean constructions with a straightedge alone shows that the straightedge must be assisted with the compass, or with some other tool. It is natural to wonder if the compass can be replaced by some kind of compass less powerful than the Euclidean and modern compasses. As early as the tenth century, the Arab mathematician, Abû'l-Wefâ (940–998), considered constructions carried out with a straightedge and a so-called *rusty compass*, or a compass of fixed opening. Constructions of this sort appeared in Europe in the late fifteenth and early sixteenth centuries and engaged the attention, among others, of the great artists Albrecht Dürer (1471–1528) and Leonardo da Vinci (1452–1519). In this early work, the motivation was a practical one, and the radius of the rusty compass was chosen as some length convenient for the problem at hand. In the middle of the sixteenth century, a new viewpoint on the matter was adopted by Italian mathematicians. The motivation became a purely academic one, and the radius of the rusty compass was considered as arbitrarily assigned at the start. A number of writers showed how all the constructions in Euclid's *Elements* can be carried out with a straightedge and a given rusty compass. Some real ingenuity is required to accomplish this, as becomes evident to anyone who tries to construct with a straightedge and a rusty compass the triangle whose three sides are given.

But the fact that all the constructions in Euclid's *Elements* can be carried out with a straightedge and a given rusty compass does not prove that a straightedge and a rusty compass are together equivalent to a straightedge and a Euclidean compass. This equivalence was first indicated in 1822 by Victor Poncelet, who stated, with a suggested method of proof, that all Euclidean constructions can be carried out with the straightedge alone in the presence of a single circle and its center. This implies that all Euclidean constructions can be carried out with a straightedge and a rusty compass, and that, moreover, the rusty compass need be used *only once*, and thenceforth discarded. In 1833, Jacob Steiner gave a complete and systematic treatment of Poncelet's theorem. It is our aim in this section to develop a proof of the above Poncelet-Steiner theorem.

4.6.1 PROBLEM.   *Given points* A, B, U, P, *where* U *is the midpoint of* AB *and* P *is not on line* AB, *construct, with a straightedge alone, the line through* P *parallel to line* AB.

Draw (see Figure 4.6a) lines *AP*, *BP*, and an arbitrary line *BC* through

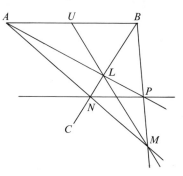

Figure 4.6a

B cutting *AP* in *L*. Draw *UL* to cut *BP* in *M*. Draw *AM* to cut *BC* in *N*. Then *PN* is the sought parallel. The proof follows from the fact that in the quadrangle *ABPN*, *PN* must cut *AB* in the harmonic conjugate of *U* for *A* and *B*; since *U* is the midpoint of *AB*, the harmonic conjugate is the point at infinity on *AB*.

4.6.2 PROBLEM.   *Given a circle* k *with its center* O, *a line* AB, *and a point* P *not on* AB, *construct, with straightedge alone, the line through* P *parallel to line* AB.

Referring to Figure 4.6b, draw any two diameters *RS*, *UV* of k not parallel to *AB*. Draw (by Problem 4.6.1) *PC* parallel to *RS* to cut *AB* in *C*, *PD* parallel to *UV* to cut *AB* in *D*, *DE* and *CE* parallel to *RS* and *UV* respectively to cut in *E*. Let *PE* cut *DC* in *M*. We now have a bisected segment on line *AB* and we may proceed as in Problem 4.6.1.

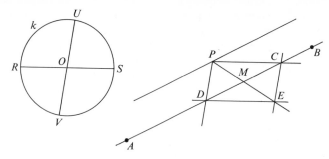

Figure 4.6b

4.6.3 PROBLEM. *Given a circle* k *with its center* O, *a line* AB, *and a point* P *not on* AB, *construct, with straightedge alone, the reflection of* P *in* AB.

Referring to Figure 4.6c, draw any diameter *CD* of *k* not parallel to *AB*.

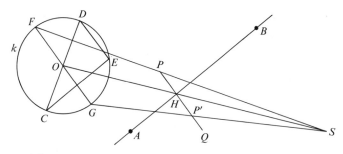

Figure 4.6c

Draw (by Problem 4.6.2) chord *CE* parallel to *AB*, and then diameter *FOG* and line *PQ* parallel to *DE*. Let *PQ* cut *AB* in *H* and let *FP* and *OH* meet in *S*. Then *GS* cuts *PQ* in the sought point *P'*. If *FP* and *OH* are parallel, draw *GS* through *G* parallel to *FP* to cut *PQ* in the sought point *P'*. In either case, the proof follows from the fact that *PHP'* is homothetic to *FOG*.

If *PQ* collines with *FG* (see Figure 4.6d) the above construction fails. In this case connect *F*, *O*, *G* with any point *S'* not on *FG* and cut these joins by a line parallel to *FOG*, obtaining the points *F'*, *O'*, *G'*. We may now proceed as before.

4.6.4 PROBLEM. *Given a circle* k *with its center* O, *and points* A, B, C, D, *construct, with straightedge alone, the points of intersection of line* AB *and circle* C(D).

Referring to Figure 4.6e, draw radius *OR* parallel to *CD* and let *OC*,

176   *Euclidean Constructions*

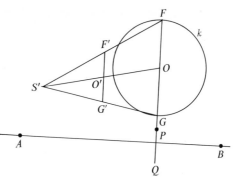

Figure 4.6d

RD meet in *S*. Draw any line *DL* and then *RM* parallel to *DL* to cut *SL* in *N*. Draw *NP* parallel to *AB* to cut *k* in *U* and *V*. Now draw *US*, *VS* to cut *AB* in the sought points *X* and *Y*.

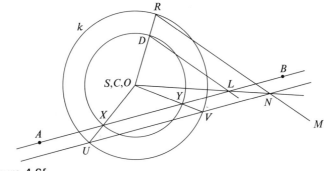

Figure 4.6e

If *O*, *D*, *C* colline, find (by Problem 4.6.3) the reflection *D'* of *D* in any line through *C*, and proceed as before.

If *CD* = *OR*, then *S* is at infinity, but a construction can be carried out similar to the above by means of parallels.

If *C* = *O*, take *S* = *O* (see Figure 4.6f).

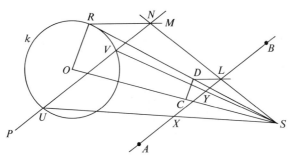

Figure 4.6f

In all cases, a proof is easy by simple homothety considerations.

4.6.5 PROBLEM. *Given a circle* k *with its center* O, *and points* A, B, C, D, *construct, with straightedge alone, the radical axis of circles* A(B) *and* C(D).

Referring to Figure 4.6g, draw $\overline{AQ}$ parallel to $\overline{CD}$. Find (by Problem

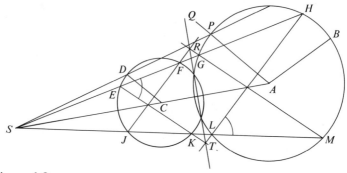

Figure 4.6g

4.6.4) the point P where AQ cuts $A(B)$. AC and PD determine the external center of similitude S of $A(B)$ and $C(D)$. Draw any two lines through S to cut the circles in E, F, G, H and J, K, L, M, which can be found by Problem 4.6.4. Let JF and MG intersect in R and EK and HL in T. Then RT is the sought radical axis.

A proof may be given as follows. Since arcs FK and HM have the same angular measure, $\angle HEK = \angle HLM$. Therefore $HEKL$ is concyclic, whence $(TK)(TE) = (TL)(TH)$, and T is on the radical axis of $A(B)$ and $C(D)$. R is similarly on the radical axis of $A(B)$ and $C(D)$.

4.6.6 PROBLEM. *Given a circle* k *with its center* O, *and points* A, B, C, D, *construct, with straightedge alone, the points of intersection of circles* A(B) *and* C(D).

Find, by Problem 4.6.5, the radical axis of $A(B)$ and $C(D)$. Then, by Problem 4.6.4, find the points of intersection of this radical axis with $A(B)$.

4.6.7 THE PONCELET-STEINER CONSTRUCTION THEOREM. *Any Euclidean construction, insofar as the given and required elements are points, may be accomplished with straightedge alone in the presence of a given circle and its center.*

We leave the proof, which can be patterned after that of Theorem 4.5.3, to the reader.

PROBLEMS

1. It has recently been shown* that the Georg Mohr mentioned in Section 4.5 was the author of an anonymously published booklet entitled *Compendium Euclidis Curiosi*, which appeared in 1673 and which in effect shows that all the constructions of Euclid's *Elements* are possible with a straightedge and a given rusty compass. Solve, with a straightedge and a given rusty compass, the following first fourteen constructions found in Mohr's work (see the second reference in the footnote).

   (1) To divide a given line into two equal parts.
   (2) To erect a perpendicular to a line from a given point in the given line.
   (3) To construct an equilateral triangle on a given side.
   (4) To erect a perpendicular to a line from a given point off the given line.
   (5) Through a given point to draw a line parallel to a given line.
   (6) To add two given line segments.
   (7) To subtract a shorter segment from a given segment.
   (8) Upon the end of a given line to place a given segment perpendicularly.
   (9) To divide a line into any number of equal parts.
   (10) Given two lines, to find the third proportional.
   (11) Given three lines, to find the fourth proportional.
   (12) To find the mean proportional to two given segments.
   (13) To change a given rectangle into a square.
   (14) To draw a triangle, given the three sides.

2. With a straightedge alone find the polar p of a given point P for a given circle k.

3. (a) With a straightedge alone construct the tangents to a given circle k from a given external point P.
   (b) With a straightedge alone construct the tangent to a given circle k at a given point P on k.

4. Given three collinear points A, B, C with $AB = BC$, construct, with a straightedge alone, (a) the point X on line AB such that $AX = n(AB)$, where n is a given positive integer, (b) the point Y on AB such that $AY = (AB)/n$.

5. Supply the proof of Theorem 4.6.7.

6. Look up the following constructions with straightedge alone which have appeared in *The American Mathematical Monthly*: Problem 3089, Jan. 1929; Problem 3137, June-July 1926; Problem E 539, June-July 1943 and Aug.-Sept. 1948; Problem E 793, Nov. 1948.

4.7 SOME OTHER RESULTS

There are other construction theorems that are of great interest to devotees of the construction game, but space forbids us from doing much more than briefly to comment on a few of them.

It will be recalled that the Poncelet-Steiner theorem states that any Euclidean construction can be carried out with a straightedge alone in the

---

* See A. E. Hallerberg, "The geometry of the fixed-compass," *The Mathematics Teacher*, Apr. 1959, pp. 230–244, and A. E. Hallerberg, "Georg Mohr and Euclidis Curiosi," *The Mathematics Teacher*, Feb. 1960, pp. 127–132.

presence of a circle *and its center*. It can be shown (by a method similar to that employed at the start of Section 4.6 to show that the point midway between two given points cannot be found with straightedge alone) that the center of a given circle cannot be found by straightedge alone. It follows that, in the Poncelet-Steiner theorem, the center of the given circle is a necessary piece of data. But the interesting question arises: How many circles in the plane do we need in order to find their centers with a straightedge alone? It has been shown that two circles suffice provided they intersect, are tangent, or are concentric; otherwise three noncoaxial circles are necessary. It also can be shown that the centers of any two given circles can be found with straightedge alone if we are given either a center of similitude of the two circles, or a point on their radical axis, and if only one circle is given, its center can be found with straightedge alone if we are also given a parallelogram somewhere in the plane of construction. Perhaps the most remarkable finding in connection with the Poncelet-Steiner theorem is that not all of the given circle is needed, but that *all Euclidean constructions are solvable with straightedge alone in the presence of a circular arc, no matter how small, and its center.*

Adler and others have shown that *all Euclidean constructions are solvable with a double-edged ruler, be the two edges parallel or intersecting at an angle.* Examples of the latter type of two-edged ruler are a carpenter's square and a draughtsman's triangle. It is interesting that while an increase in the number of compasses will not enable us to solve anything more than the Euclidean constructions, two carpenter squares make it possible for us to duplicate a given cube and to trisect an arbitrary angle. These latter problems are also solvable with compass and a *marked* straightedge—that is, a straightedge bearing two marks along its edge. Various tools and linkages have been invented which will solve certain problems beyond those solvable with Euclidean tools alone. Another interesting discovery is that *all Euclidean constructions, insofar as the given and required elements are points, can be solved without any tools whatever, by simply folding and creasing the paper representing the plane of construction.*

Suppose we consider a point of intersection of two loci as *ill-defined* if the two loci intersect at the point in an angle less than some given small angle θ, and that otherwise the point of intersection will be considered as *well-defined*. The question arises: Can any given Euclidean construction be accomplished with the Euclidean tools by using only well-defined intersections? Interestingly enough, the answer is an affirmative one, and the result can be strengthened. In addition to the small angle θ, let us be given a small linear distance d, and suppose we define a Euclidean construction to be *well-defined* if it utilizes only well-defined intersections, straight lines determined by pairs of points which are a greater distance than d apart, and circles with radii greater than d. Then it can be shown that *any Euclidean construction can be accomplished by a well-defined one.* Another, and easy to prove, theorem is the following concerning constructions in a limited space: *Let a Euclidean construction be composed of given, auxiliary, and required loci* G, A, R *respectively, and let* S *be a convex region placed on the construction such that* S *contains at least a part* G', A', R' *of each locus of* G, A, R. *Then there exists a Euclidean construction determining* R', *insofar as each locus of* R' *is considered determined by points, from the loci of* G', *the construction being performed entirely inside* S.

Euclid's first postulate presupposes that our straightedge is as long as we wish, so that we can draw the line determined by *any* two given points *A* and *B*. Suppose we have a straightedge of small finite length ε, such that it will not span the distance between *A* and *B*. Can we, with our little ε-straightedge alone, still draw the line joining *A* and *B*? This would surely seem to be impossible but, curiously enough, after a certain amount of trial and error, the join can be drawn. Let us show how this may be accomplished. After a finite amount of trial and error it is always possible (see Figure 4.7a) to obtain, with our ε-straightedge, two lines 1 and 2 through

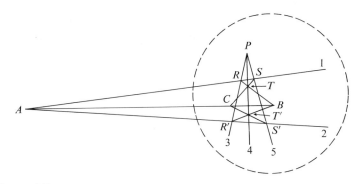

Figure 4.7a

A and lying so close to *B* that the ε-straightedge will comfortably span lines 1, 2, and *AB*. Choose a point *P* sufficiently close to *B* and draw three lines 3, 4, 5 through *P*. Let line 3 cut lines 1 and 2 in points *R* and *R'*, and let line 5 cut lines 1 and 2 in points *S* and *S'*. Draw *BR* and *BR'* to cut line 4 in *T* and *T'*. Finally, draw *ST* and *S'T'* to intersect in *C*. Then, since triangles *RST*, *R'S'T'* are copolar at *P*, they are (by Desargues' Theorem) coaxial, and points *A*, *B*, *C* must be collinear. But our ε-straightedge will span *B* and *C*, and thus the line *BA* can be drawn.

The problem of finding the "best" Euclidean solution to a required construction has also been considered, and a science of *geometrography* was developed in 1907 by Émile Lemoine for quantitatively comparing one construction with another. To this end, Lemoine considered the following five operations:

S_1: to make the straightedge pass through one given point,
S_2: to rule a straight line,

C_1: to make one compass leg coincide with a given point,

C_2: to make one compass leg coincide with an arbitrary point of a given locus,

C_3: to describe a circle.

If the above operations are performed m_1, m_2, n_1, n_2, n_3 times in a construction, then $m_1 S_1 + m_2 S_2 + n_1 C_1 + n_2 C_2 + n_3 C_3$ is regarded as the *symbol* of the construction. The total number of operations, $m_1 + m_2 + n_1 + n_2 + n_3$, is called the *simplicity* of the construction, and the total number of coincidences, $m_1 + n_1 + n_2$, is called the *exactitude* of the construction. The total number of loci drawn is $m_2 + n_3$, the difference between the simplicity and the exactitude of the construction.

As simple examples, the symbol for drawing the straight line through A and B is $2S_1 + S_2$, and that for drawing the circle with center C and radius AB is $3C_1 + C_3$.

Of course the construction game has been extended to three-dimensional space, but we shall not consider this extension here.

PROBLEMS

1. Prove that the center of a given circle cannot be found with straightedge alone.

2. (a) Prove that, given a parallelogram somewhere in the plane of construction, we can, with straightedge alone, draw a line parallel to any given line AB through any given point P.
   (b) Prove that, given a parallelogram somewhere in the plane of construction, we can, with straightedge alone, find the center of any given circle k.

3. (a) Prove that, with straightedge alone, we can find the common center of two given concentric circles.
   (b) Prove that, with straightedge alone, we can find the centers of two given tangent circles.
   (c) Prove that, with straightedge alone, we can find the centers of two given intersecting circles.

4. Prove that the centers of any two given circles can be found with straightedge alone if we are given a center of similitude of the two circles.

5. Solve the following problems using a ruler having two parallel edges:
   (a) To obtain a bisected segment on a given line AB.
   (b) Through a given point P to draw a line parallel to a given line AB.
   (c) Through a given point P to draw a line perpendicular to a given line AB.
   (d) Given points A, B, C, D, to find the points of intersection of line AB and circle $C(D)$.
   (e) Given points A, B, C, D, to find the points of intersection of circles $A(B)$ and $C(D)$.

6. Show that Problem 5 establishes the theorem: *All Euclidean constructions, insofar as the given and required elements are points, can be accomplished with a ruler having two parallel edges.*

7. Look up Euclidean constructions by paper folding in R. C. Yates, *Geometrical Tools*, Educational Publishers, Inc., St. Louis (revised 1949).

8. Look up well-defined Euclidean constructions in Howard Eves and Vern Hoggatt, "Euclidean constructions with well-defined intersections," *The Mathematics Teacher*, vol. 44, no. 4 (April 1951), pp. 262–3.

9. Establish the theorem of Section 4.7 concerning Euclidean constructions in a limited space.

10. Let us be given two curves *m* and *n*, and a point *O*. Suppose we permit ourselves to mark, on a given straightedge, a segment *MN*, and then to adjust the straightedge so that it passes through *O* and cuts the curves *m* and *n* with *M* on *m* and *N* on *n*. The line drawn along the straightedge is then said to have been drawn by "the insertion principle." Problems beyond the Euclidean tools can often be solved with these tools if we also permit ourselves to use the insertion principle. Establish the correctness of the following constructions, each of which uses the insertion principle.
(a) Let *AB* be a given segment. Draw $\sphericalangle ABM = 90°$ and $\sphericalangle ABN = 120°$. Now draw *ACD* cutting *BM* in *C* and *BN* in *D* and such that $CD = AB$. Then $(AC)^3 = 2(AB)^3$. Essentially this construction for duplicating a cube was given in publications by Vieta (1646) and Newton (1728).
(b) Let *AOB* be any central angle in a given circle. Through *B* draw a line *BCD* cutting the circle again in *C*, *AO* produced in *D*, and such that $CD = OA$, the radius of the circle. Then $\sphericalangle ADB = (\sphericalangle AOB)/3$. This solution of the trisection problem is implied by a theorem given by Archimedes (*ca.* 240 B.C.).

11. Over the years many mechanical contrivances, linkage machines, and compound compasses, have been devised to solve the trisection problem. An interesting and elementary implement of this kind is the so-called *tomahawk*. The inventor of the tomahawk is not known, but the instrument was described in a book in 1835. To construct a tomahawk, start with a line segment *RU* trisected at *S* and *T* (see Figure 4.7b). Draw a semicircle on *SU* as diameter and draw *SV* perpendicular to *RU*. Complete the instrument as indicated in the figure. To trisect an angle *ABC* with the tomahawk place the implement on

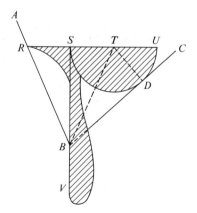

Figure 4.7b

the angle so that R falls on BA, SV passes through B, and the semicircle touches BC, at D say. Show that BS and BT then trisect the angle.

The tomahawk may be constructed with straightedge and compass on tracing paper and then adjusted on the given angle. By this subterfuge we may trisect an angle with straightedge and compass.

12. A construction using Euclidean tools but requiring an infinite number of operations is called an *asymptotic Euclidean construction*. Establish the following two constructions of this type for solving the trisection and the quadrature problems. (For an asymptotic Euclidean solution of the duplication problem, see T. L. Heath, *History of Greek Mathematics*, vol. 1, pp. 268–270.)

(a) Let OT_1 be the bisector of $\angle AOB$, OT_2 that of $\angle AOT_1$, OT_3 that of $\angle T_2 OT_1$, OT_4 that of $\angle T_3 OT_2$, OT_5 that of $\angle T_4 OT_3$, and so forth. Then $\lim\limits_{i \to \infty} OT_i = OT$, one of the trisectors of $\angle AOB$. (This construction was given by Fialkowski, 1860.)

(b) On the segment AB_1 produced mark off $B_1 B_2 = AB_1$, $B_2 B_3 = 2(B_1 B_2)$, $B_3 B_4 = 2(B_2 B_3)$, and so forth. With B_1, B_2, B_3, ... as centers draw the circles $B_1(A)$, $B_2(A)$, $B_3(A)$, ... Let M_1 be the midpoint of the semicircle on AB_2. Draw $B_2 M_1$ to cut circle $B_2(A)$ in M_2, $B_3 M_2$ to cut circle $B_3(A)$ in M_3, ... Let N_i be the projection of M_i on the common tangent of the circles at A. Then $\lim\limits_{i \to \infty} AN_i =$ quadrant of circle $B_1(A)$.

13. Find the symbol, simplicity, and exactitude for the following familiar constructions of a line through a given point A and parallel to a given line MN.

(a) Through A draw any line to cut MN in B. With any radius r draw the circle $B(r)$ to cut MB in C and AB in D. Draw circle $A(r)$ to cut AB in E. Draw circle $E(CD)$ to cut circle $A(r)$ in X. Draw AX, obtaining the required parallel.

(b) With any suitable point D as center draw circle $D(A)$ to cut MN in B and C. Draw circle $C(AB)$ to cut circle $D(A)$ in X. Draw AX.

(c) With any suitable radius r draw the circle $A(r)$ to cut MN in B. Draw circle $B(r)$ to cut MN in C. Draw circle $C(r)$ to cut circle $A(r)$ in X. Draw AX.

14. The Arabians were interested in constructions on a spherical surface. Consider the following problems, to be solved with Euclidean tools and appropriate planar constructions.

(a) Given a material sphere, find its diameter.

(b) On a given sphere locate the vertices of an inscribed cube.

(c) On a given material sphere locate the vertices of an inscribed regular tetrahedron.

15. J. S. Mackay, in his *Geometrography*, regarded the following construction as probably the simplest way of finding the center of a given circle; the construction is credited to a person named Swale, in about 1830.

With any point O on the circumference of the given circle as center and any convenient radius describe an arc PQR, cutting the circle in P and Q. With Q as center, and with the same radius, describe an arc OR cutting arc PQR in R. Draw PR, cutting the circle again in L. Then LR is the radius of the given circle, and the center K may now be found by means of two intersecting arcs.

(a) Establish the correctness of Swale's construction.

(b) Find the simplicity and exactitude of the construction.

4.8 THE REGULAR SEVENTEEN-SIDED POLYGON

In Book IV of Euclid's *Elements* are found constructions, with straightedge and compass, of regular polygons of three, four, five, six, and fifteen sides. By successive angle, or arc, bisections, we may then with Euclidean tools construct regular polygons having $3(2^n)$, $4(2^n)$, $5(2^n)$, and $15(2^n)$ sides, where $n = 0, 1, \ldots$. Not until almost the nineteenth century was it suspected that any other regular polygons could be constructed with these limited tools.

In 1796, the eminent German mathematician Karl Friedrich Gauss, then only 19 years old, developed the theory that shows that a regular polygon having a *prime* number of sides can be constructed with Euclidean tools if and only if that number is of the form

$$f(n) = 2^{2^n} + 1.$$

For $n = 0, 1, 2, 3, 4$ we find $f(n) = 3, 5, 17, 257, 65537$, all prime numbers. Thus, unknown to the Greeks, regular polygons of 17, 257, and 65537 sides can be constructed with straightedge and compass. For no other value of n, than those listed above, is it known that $f(n)$ is a prime number. Indeed, after considerable effort, $f(n)$ has actually been shown to be composite for a number of values of $n > 4$, and the general feeling among number theorists today is that $f(n)$ is probably composite for all $n > 4$. The numbers $f(n)$ are usually referred to as *Fermat numbers*, because the great French number theorist, Pierre de Fermat (1601–1665), about 1640 conjectured (incorrectly, as it has turned out) that $f(n)$ is prime for all nonnegative integral n.

Many Euclidean constructions of the regular polygon of 17 sides (the regular heptadecagon) have been given. In 1832, F. J. Richelot published an investigation of the regular polygon of 257 sides, and a Professor Hermes of Lingen gave up ten years of his life to the problem of constructing a regular polygon of 65537 sides. It has been reported that it was Gauss's discovery, at the age of 19, that a regular polygon of 17 sides can be constructed with straightedge and compass that decided him to devote his life to mathematics. His pride in this discovery is evidenced by his request that a regular polygon of 17 sides be engraved on his tombstone. Although the request was never fulfilled, such a polygon is found on a monument to Gauss erected at his birthplace in Braunschweig (Brunswick).

We will not here take up Gauss's general theory of polygonal construction, but will merely give an elementary synthetic treatment of the regular 17-sided polygon. Some of the easy details will be left to the reader, who may find it interesting to apply an analogous treatment to the construction of a regular pentagon. We commence with some preliminary matter.

4.8.1 LEMMA. *If* C *and* D *are two points on a semicircumference* ACDB *of radius* R, *with* C *lying between* A *and* D, *and if* C′ *is the reflection of* C *in the diameter* AB, *then:* (1) (AC)(BD) = R(C′D − CD), (2) (AD)(BC) = R(C′D + CD), (3) (AC)(BC) = R(CC′).

(1) On DC' mark off $DL = DC$ and connect A with L, A with D, A with C', and D with the center O of the semicircle (see Figure 4.8a$_1$). In triangles

Figure 4.8a$_1$

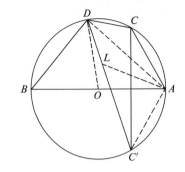

DCA and DLA, $DC = DL$, $DA = DA$, and $\angle CDA = \angle LDA$. Therefore triangles DCA and DLA are congruent, whence $AL = AC = AC'$, and triangles BOD and $C'AL$ are isosceles. But $\angle ABD = \angle AC'D$, whence triangles BOD and $C'AL$ are similar and $BD/C'L = OB/AC'$. That is, $BD/(C'D - CD) = R/AC$, or $(AC)(BD) = R(C'D - CD)$.

(2) On $C'D$ produced, mark off $DL' = DC$ and connect B with L', B with D, B with C', and D with O (see Figure 4.8a$_2$). In triangles DCB and $DL'B$,

Figure 4.8a$_2$

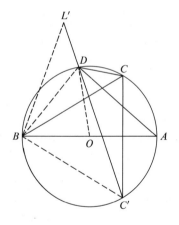

$DC = DL'$, $DB = DB$, and $\angle CDB = (1/2)(\text{arc } CAC'B) = (1/2)(\text{arc } BDCC')$ $= (1/2)(\text{arc } BD + \text{arc } DCC') = \angle L'DB$. Therefore triangles DCB and $DL'B$ are congruent, whence $BL' = BC = BC'$, and triangles AOD and $C'BL'$ are isosceles. But $\angle BAD = \angle BC'D$, whence triangles AOD and $C'BL'$ are similar and $AD/C'L' = OA/BC'$. That is, $AD/(C'D + CD) = R/BC$, or $(AD)(BC) = R(C'D + CD)$.

(3) Since $\angle BCA = 90°$, we have $(AC)(BC) = 2(\text{area triangle } BCA) = (BA)(CC')/2 = R(CC')$.

4.8.2 THEOREM. *Let the circumference of a circle of radius R be divided into seventeen equal parts and let* AB *be the diameter through one of the points of section,* A, *and the midpoint* B *of the opposite arc. Let the points of section on each side of the diameter* AB *be consecutively named* $C_1, C_2, \ldots,$ C_8 *and* $C_1', C_2', \ldots, C_8'$ *beginning next to* A *(see Figure 4.8b). Then*

$$(BC_1)(BC_2)(BC_3)(BC_4)(BC_5)(BC_6)(BC_7)(BC_8) = R^8.$$

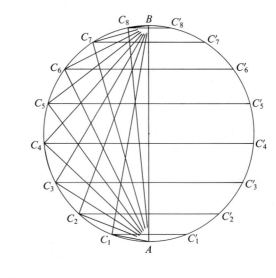

Figure 4.8b

For, by Lemma 4.8.1 (3), we have $(AC_i)(BC_i) = R(C_iC_i')$, $i = 1, \ldots, 8$. But, since the circumference is divided into equal parts, $AC_1 = C_8 C_8'$, $AC_2 = C_1 C_1'$, $AC_3 = C_7 C_7'$, $AC_4 = C_2 C_2'$, $AC_5 = C_6 C_6'$, $AC_6 = C_3 C_3'$, $AC_7 = C_5 C_5'$, $AC_8 = C_4 C_4'$. The theorem now readily follows.

4.8.3 COROLLARY. $(BC_1)(BC_2)(BC_4)(BC_8) = R^4.$

4.8.4 COROLLARY. $(BC_3)(BC_5)(BC_6)(BC_7) = R^4.$

4.8.5 THEOREM. *In the notation of Theorem* 4.8.2, *we have*

$$(BC_1)(BC_4) = R(BC_3 + BC_5), \qquad (BC_2)(BC_8) = R(BC_6 - BC_7),$$
$$(BC_3)(BC_5) = R(BC_2 + BC_8), \qquad (BC_6)(BC_7) = R(BC_1 - BC_4).$$

Let us denote the midpoints of arcs $AC_1, C_1C_2, C_2C_3, \ldots, C_7C_8$ by $C_{0.5}, C_{1.5}, C_{2.5}, \ldots, C_{7.5}$, and the midpoints of arcs $AC_1', C_1'C_2', C_2'C_3',$ $\ldots, C_7'C_8'$ by $C_{0.5}', C_{1.5}', C_{2.5}', \ldots, C_{7.5}'$.

Now, by Lemma 4.8.1 (2),

$$(AC_{4.5})(BC_1) = R(C_{4.5}' C_1 + C_{4.5} C_1).$$

But $AC_{4.5} = BC_4$, $C_{4.5}' C_1 = BC_3$, $C_{4.5} C_1 = BC_5$, whence, by substitution,

$$(BC_1)(BC_4) = R(BC_3 + BC_5).$$

By Lemma 4.8.1 (1),

$$(AC_{0.5})(BC_2) = R(C_2' C_{0.5} - C_2 C_{0.5}).$$

But $AC_{0.5} = BC_8$, $C_2' C_{0.5} = BC_6$, $C_2 C_{0.5} = BC_7$, whence, by substitution,

$$(BC_2)(BC_8) = R(BC_6 - BC_7).$$

In like manner, using Lemma 4.8.1 (1) and (2), we can show that

$$(BC_3)(BC_5) = R(BC_2 + BC_8), \quad (BC_6)(BC_7) = R(BC_1 - BC_4).$$

4.8.6 NOTATION. Set $M = BC_3 + BC_5$, $N = BC_6 - BC_7$, $P = BC_2 + BC_8$, $Q = BC_1 - BC_4$.

4.8.7 THEOREM. $MN = PQ - R^2$.

By Theorem 4.8.5 we have

$$M = BC_3 + BC_5 = (BC_1)(BC_4)/R,$$
$$N = BC_6 - BC_7 = (BC_2)(BC_8)/R,$$

whence, by Corollary 4.8.3,

$$MN = (BC_1)(BC_2)(BC_4)(BC_8)/R^2 = R^4/R^2 = R^2.$$

Similarly, using Corollary 4.8.4, we can show that $PQ = R^2$.

4.8.8 THEOREM. $BC_1 - BC_2 + BC_3 - BC_4 + BC_5 - BC_6 + BC_7 - BC_8 = R$.

We have, by Theorem 4.8.7,

$$R^2 = MN = (BC_3 + BC_5)(BC_6 - BC_7)$$
$$= (BC_3)(BC_6) + (BC_5)(BC_6) - (BC_3)(BC_7) - (BC_5)(BC_7).$$

But, by Lemma 4.8.1 (1), we have

$$(BC_3)(BC_6) = R(BC_3 - BC_8), \quad (BC_5)(BC_6) = R(BC_1 - BC_6),$$
$$(BC_3)(BC_7) = R(BC_4 - BC_7), \quad (BC_5)(BC_7) = R(BC_2 - BC_5).$$

Substituting, we obtain

$$BC_1 - BC_2 + BC_3 - BC_4 + BC_5 - BC_6 + BC_7 - BC_8 = R.$$

4.8.9 THEOREM. (1) $(M - N) - (P - Q) = R$, (2) $(M - N)(P - Q) = 4R^2$.

(1) We have (using Theorem 4.8.8)

$$(M - N) - (P - Q)$$
$$= (BC_3 + BC_5) - (BC_6 - BC_7) - (BC_2 + BC_8) + (BC_1 - BC_4)$$
$$= BC_1 - BC_2 + BC_3 - BC_4 + BC_5 - BC_6 + BC_7 - BC_8 = R.$$

(2) By Theorem 4.8.5,

$$(M - N)(P - Q)$$
$$= [(BC_3 + BC_5) - (BC_6 - BC_7)][(BC_2 + BC_8) - (BC_1 - BC_4)]$$
$$= - (BC_1)(BC_3) - (BC_1)(BC_5) + (BC_1)(BC_6) - (BC_1)(BC_7)$$
$$+ (BC_2)(BC_3) + (BC_2)(BC_5) - (BC_2)(BC_6) + (BC_2)(BC_7)$$
$$+ (BC_4)(BC_3) + (BC_4)(BC_5) - (BC_4)(BC_6) + (BC_4)(BC_7)$$
$$+ (BC_8)(BC_3) + (BC_8)(BC_5) - (BC_8)(BC_6) + (BC_8)(BC_7).$$

But, by Lemma 4.8.1 (1) and (2), we can show that

$$(BC_1)(BC_3) = R(BC_2 + BC_4), \qquad (BC_1)(BC_5) = R(BC_4 + BC_6),$$
$$(BC_1)(BC_6) = R(BC_5 + BC_7), \qquad (BC_1)(BC_7) = R(BC_6 + BC_8),$$
$$(BC_2)(BC_3) = R(BC_1 + BC_5), \qquad (BC_2)(BC_5) = R(BC_3 + BC_7),$$
$$(BC_2)(BC_6) = R(BC_4 + BC_8), \qquad (BC_2)(BC_7) = R(BC_5 - BC_8),$$
$$(BC_4)(BC_3) = R(BC_1 + BC_7), \qquad (BC_4)(BC_5) = R(BC_1 - BC_8),$$
$$(BC_4)(BC_6) = R(BC_2 - BC_7), \qquad (BC_4)(BC_7) = R(BC_3 - BC_6),$$
$$(BC_8)(BC_3) = R(BC_5 - BC_6), \qquad (BC_8)(BC_5) = R(BC_3 - BC_4),$$
$$(BC_8)(BC_6) = R(BC_2 - BC_3), \qquad (BC_8)(BC_7) = R(BC_1 - BC_2).$$

Substituting we obtain

$$(M - N)(P - Q)$$
$$= 4R(BC_1 - BC_2 + BC_3 - BC_4 + BC_5 - BC_6 + BC_7 - BC_8) = 4R^2.$$

4.8.1 PROBLEM. *To construct a regular seventeen-sided polygon.*

From the above we have

$$(M - N)(P - Q) = 4R^2, \qquad (M - N) - (P - Q) = R,$$
$$MN = R^2, \qquad PQ = R^2,$$
$$(BC_2)(BC_8) = RN, \qquad BC_2 + BC_8 = P.$$

Knowing the product and difference of $M - N$ and $P - Q$, these two quantities can be constructed. Then, knowing the product and difference of M and N, N can be constructed. Similarly, knowing the product and difference of P and Q, P can be constructed. Finally, knowing the product and sum of BC_2 and BC_8, these quantities can be constructed, and thence the regular 17-sided polygon inscribed in a circle of radius R.

The following sequence of figures shows the actual step-by-step construction (see Problem 9, Section 1.4).

(1) Let $ab = R$.

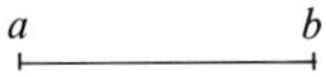

$$\overset{a}{\vert}\rule{2cm}{0.4pt}\overset{b}{\vert}$$

Figure 4.8c

(2) Take $cd = R$, $ce = 2R$.
Then $eg = M - N$ and
$ef = P - Q$.

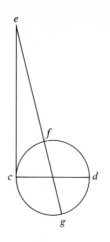

Figure 4.8d

(3) Take $hi = M - N$,
$hj = R$. Then
$jl = M$, $jk = N$.

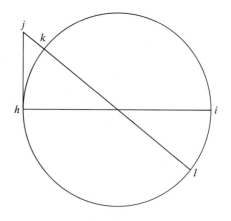

Figure 4.8e

(4) Take $mn = P - Q$,
$mo = R$. Then
$oq = P$, $op = Q$.

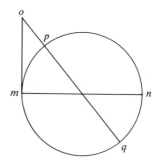

Figure 4.8f

(5) Find *tu*, the mean proportional between $rt = R$ and $ts = N$.

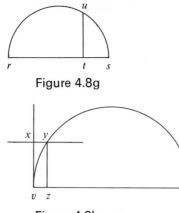

Figure 4.8g

(6) Take $vw = P$, $vx = tu$. Draw *xy* parallel to *vw* cutting semicircle on *vw* at *y*. Drop *yz* perpendicular to *vw*. Then $vz = BC_8$, $zw = BC_2$.

Figure 4.8h

PROBLEMS

1. In 1732, Euler showed that $f(5)$ has the factor 641. Verify this.

2. Suppose $n = rs$, where *n*, *r*, *s* are positive integers. Show that if a regular *n*-gon is constructible with Euclidean tools, then so also are a regular *r*-gon and a regular *s*-gon.

3. Suppose *r* and *s* are relatively prime positive integers and that a regular *r*-gon and a regular *s*-gon are constructible with Euclidean tools. Show that a regular *rs*-gon is also so constructible.

4. It can be shown that the only regular polygons that can be constructed with Euclidean tools are those the number *n* of whose sides can be expressed in the form

$$2^{\alpha} f(\alpha_1) f(\alpha_2) \cdots f(\alpha_k),$$

where $\alpha, \alpha_1, \alpha_2, \ldots, \alpha_k$ are distinct integers and each $f(\alpha_i)$ is a prime. On the basis of this theorem, list those regular *n*-gons, $n < 100$, that can be constructed with Euclidean tools.

5. Complete the proof of Theorem 4.8.5.

6. Complete the proof of Theorem 4.8.6.

7. Verify the details in the proof of Theorem 4.8.8.

8. Verify the details in the proof of Theorem 4.8.9.

9. Draw a circle and actually, with Euclidean tools, divide its circumference into 17 equal parts.

10. Show that

$$BC_8 = R\left[-1 + \sqrt{17} + \sqrt{34 - 2\sqrt{17}}\right]\Big/8$$

$$+ R\left[\sqrt{68 + 12\sqrt{17} - 16\sqrt{34 + 2\sqrt{17}} - 2(1 - \sqrt{17})\sqrt{34 - 2\sqrt{17}}}\right]\Big/8.$$

11. Show that Theorem 4.8.2 can be generalized to hold for any regular polygon of an odd number of sides.

12. (a) Carry out a treatment of the construction of a regular pentagon analogous to the treatment in the text of the construction of a regular 17-sided polygon.

(b) Show that for the regular pentagon, $BC_2 = R(\sqrt{5} - 1)/2$.

13. Try to verify the following construction of a regular 17-gon (H. W. Richmond, "To construct a regular polygon of seventeen sides," *Mathematische Annalen*, vol. 67 (1909), p. 459).

Let OA and OB be two perpendicular radii of a given circle with center O. Find C on OB such that $OC = OB/4$. Now find D on OA such that angle $OCD = $ (angle OCA)/4. Next find E on AO produced such that angle $DCE = 45°$. Draw the circle on AE as diameter, cutting OB in F, and then draw the circle $D(F)$, cutting OA and AO produced in G_4 and G_6. Erect perpendiculars to OA at G_4 and G_6, cutting the given circle in P_4 and P_6. These last points are the fourth and sixth vertices of the regular 17-gon whose first vertex is A.

BIBLIOGRAPHY

ALTSHILLER-COURT, NATHAN, *College Geometry, an Introduction to the Modern Geometry of the Triangle and the Circle.* New York: Barnes and Noble, Inc., 1952.

COOLIDGE, J. L., *A Treatise on the Circle and the Sphere.* New York: Oxford University Press, 1916.

———, *A History of Geometrical Methods.* New York: Oxford University Press, 1940.

DAUS, P. M., *College Geometry.* Englewood Cliffs, N.J.: Prentice-Hall, Inc., 1941.

DAVIS, D. R., *Modern College Geometry.* Reading, Mass.: Addison-Wesley Publishing Company, Inc., 1949.

DICKSON, L. E., *New First Course in the Theory of Equations.* New York: John Wiley and Sons, Inc., 1939.

EVES, HOWARD, *An Introduction to the History of Mathematics.* New York: Holt, Rinehart and Winston, Inc., 3rd ed., 1969.

———, and C. V. NEWSOM, *An Introduction to the Foundations and Fundamental Concepts of Mathematics.* New York: Holt, Rinehart and Winston, Inc., rev. ed., 1965.

HUDSON, H. P., *Ruler and Compasses.* New York: Longmans, Green and Company, 1916. Reprinted in *Squaring the Circle, and Other Monographs*, Chelsea Publishing Company, 1953.

JARDIN, DOV., *Constructions with the Bi-Ruler and with the Double-Ruler.* Jerusalem, Israel: privately printed, 1964.

KAZARINOFF, N. D., *Ruler and the Round.* Boston: Prindle, Weber & Schmidt, Inc., 1970.

KEMPE, A. B., *How to Draw a Straight Line; a Lecture on Linkages.* New York: The Macmillan Company, 1887. Reprinted in *Squaring the Circle, and Other Monographs*, Chelsea Publishing Company, 1953.

KLEIN, FELIX, *Famous Problems of Elementary Geometry*, tr. by W. W. Beman and D. E. Smith. New York: G. E. Stechert and Company, 1930. Reprinted in *Famous Problems, and Other Monographs*, Chelsea Publishing Company, 1955. Original German volume published in 1895.

KOSTOVSKII, A. N., *Geometrical Constructions Using Compasses Only*, tr. by Halina Moss. New York: Blaisdell Publishing Company, 1961. Original Russian volume published in 1959.

MESCHOWSKI, HERBERT, *Unsolved and Unsolvable Problems in Geometry*, tr. by J. A. C. Burlak. Edinburgh: Oliver and Boyd, 1968.

MILLER, L. H., *College Geometry*. New York: Appleton-Century-Crofts, Inc., 1957.

MOISE, E. E., *Elementary Geometry from an Advanced Standpoint*. Reading, Mass.: Addison-Wesley Publishing Company, Inc., 1963.

PETERSEN, JULIUS, *Methods and Theories for the Solution of Problems of Geometrical Construction Applied to 410 Problems*, tr. by Sophus Haagensen. New York: G. E. Stechert and Company, 1927. Reprinted in *String Figures, and Other Monographs*, Chelsea Publishing Company, 1960. Original Danish volume published in 1879.

RANSOM, W. R., *Can and Can't in Geometry*. Portland, Me.: J. Weston Walsh, 1960.

ROW, SUNDARA, *Geometric Exercises in Paper Folding*, tr. by W. W. Beman and D. E. Smith. Chicago: The Open Court Publishing Company, 1901. Original volume published in Madras in 1893.

RUSSELL, J. W., *An Elementary Treatise on Pure Geometry*. New York: Oxford University Press, 1905.

SHIVELY, L. S., *An Introduction to Modern Geometry*. New York: John Wiley and Sons, Inc., 1939.

SMOGORZHEVSKII, A. S., *The Ruler in Geometrical Constructions*, tr. by Halina Moss. New York: Blaisdell Publishing Company, 1961. Original Russian volume published in 1957.

STEINER, JACOB, *Geometrical Constructions with a Ruler Given a Fixed Circle with Its Center*, tr. by M. E. Stark. New York: Scripta Mathematica, 1950. First German edition published in 1833.

SYKES, MABEL, *A Source Book of Problems for Geometry Based Upon Industrial Design and Architectural Ornament*. Boston: Allyn and Bacon, Inc., 1912.

YATES, R. C., *Geometrical Tools, a Mathematical Sketch and Model Book*. Saint Louis: Educational Publishers, Inc., 1949.

———, *The Trisection Problem*. Ann Arbor, Mich.: Edwards Brothers, Inc., 1947.

YOUNG, J. W. A., ed., *Monographs on Topics of Modern Mathematics Relevant to the Elementary Field*. New York: Longmans, Green and Company, 1911. Reprinted by Dover Publications, Inc., 1955.

5

Dissection Theory

This chapter is devoted to a subject of great theoretical and recreational interest, for dissection theory plays an important role in a rigorous development of area and volume and also furnishes a seemingly endless variety of attractive and challenging puzzles.

When two (planar) polygons are such that we may cut one of them into a finite number of polygonal pieces which can be rearranged to form the second polygon, we say that the two given polygons are *congruent by addition*. The two polygons can thus be considered as made up of sets of corresponding pieces which are congruent in pairs but perhaps fitted together differently. The large number of polygonal dissection puzzles and recreations arises from the remarkable fact that any two polygons having equal areas are congruent by addition. This fact is known as the *fundamental theorem of polygonal dissection*.

The fundamental theorem of polygonal dissection has been discussed by a number of writers from at least the beginning of the nineteenth century up to present times, and instances of the theorem appeared in geometry as early as the days of Greek antiquity. In Section 5.3 we shall present a constructive proof of this fundamental theorem. That is, we shall devise a process by which any two polygons of equal areas may be respectively dissected into polygonal pieces which are congruent in pairs. It will be interesting to notice that the process is of more theoretical than practical interest inasmuch as it will generally lead to a dissection involving more pieces than are necessary. There is at present no general theory for *minimal dissections.*

After establishing the fundamental theorem of polygonal dissection we shall, in Sections 5.4, 5.5, and 5.6, look at corresponding matters in space, and shall learn that not always can a given polyhedral solid be dissected into a finite number of polyhedral pieces which can be fitted together to form a second given polyhedral solid of the same volume. This interesting fact was proved in 1902 by Max Dehn (1878–195?) and is of far-reaching importance in the theory of volumes. It implies that although a theory of areas of polygons can be constructed without continuity considerations, such is not possible for a complete theory of volumes of polyhedra. Thus the ordinary treatment of volumes, such as we find in Euclid's *Elements*, for example, involves infinitesimals, which do not occur in the theory of polygonal areas. The theory of polyhedral volumes, if it is to be complete, as well as that of volumes of curved solids, must rest on a theory of integration.

Other dissection matters in space differ from the corresponding situations in the plane. Thus, though it is always possible to dissect a polygon into a finite number of triangles having their vertices only at the vertices of the polygon, there exist polyhedra, which we shall call *Lennes polyhedra*, which cannot be dissected into finite numbers of tetrahedra having their vertices only at the vertices of the polyhedra. Indeed, much of the theory of the dissection of polyhedra will be found to be largely of the nature of a makeshift.

The extension of the idea of congruency by addition to embrace general point sets is very intriguing, and Section 5.7 is directed to this matter. Here we encounter some of the most famous and most startling of the paradoxes of set theory.

In Section 5.8 we shall offer a small collection of planar and spatial dissection curiosities. Though this collection can easily be considerably extended, it will incorporate some of the more interesting examples of this fascinating and highly entertaining subject.

A final word of apology for our treatment is perhaps in order. In an effort to make the treatment elementary and at the same time relatively brief, we shall here avoid both the more subtle topological questions surrounding dissection matters and the delicate questions lying at the base of a rigorous theory of geometrical measure. In the language of topology, all our polygons will be homeomorphic to circles, and all our polyhedra will be homeomorphic to spheres. This means that the polygons, for example, cannot be criss-cross, nor can they have "holes" in them, or be in more than one piece; on the other hand, they need not be convex, but may be reëntrant, perhaps of spiral form, and so on. We make no effort to define "area of a polygon" or "volume of a polyhedron," but rely on intuitive notions of these ideas. A careful and sophisticated treatment of these ideas, developed from some suitable postulate base, is an important matter in a discussion of the foundations of geometry.

5.1 PRELIMINARIES

Throughout this section, upper case letters represent either polygons or polyhedra, so that the notation, definitions, and theorems apply equally to either class.

5.1.1 Notation and definitions.

(a) If P is dissected into the pieces $P_1, \ldots, P_n$, we write

$$P = P_1 + \cdots + P_n.$$

(b) If P and Q have the same measure (area or volume, as the case may be), we say that P and Q are *equivalent* and we write

$$P \simeq Q.$$

(c) If P and Q are congruent, either directly or oppositely, we write

$$P \cong Q.$$

(d) If

$$P = P_1 + \cdots + P_n,$$
$$Q = Q_1 + \cdots + Q_n,$$
$$P_i \cong Q_i \qquad\qquad (i = 1, \ldots, n),$$

we say that P and Q are *congruent by addition* and we write

$$P \cong Q\,(+).$$

(e) If there exist $R, S, U_1 \ldots, U_n, V_1, \ldots, V_n$ such that

$$R = P + U_1 + \cdots + U_n,$$
$$S = Q + V_1 + \cdots + V_n,$$
$$U_i \cong V_i \qquad\qquad (i = 1, \ldots, n),$$
$$R \cong S\,(+),$$

then we say that P and Q are *congruent by subtraction* and we write

$$P \cong Q\,(-).$$

Thus P and Q are congruent by subtraction if we can adjoin to P and Q pairs of congruent pieces so that the final results are congruent by addition. We also agree to write $P \cong Q\,(-)$ if

$$P = \sum_{i=1}^{n} P_i, \quad Q = \sum_{i=1}^{n} Q_i, \quad P_i \cong Q_i\,(-) \qquad \text{for } i = 1, \ldots, n.$$

It is apparent that congruency implies congruency by addition; congruency by addition implies congruency by subtraction; and congruency or congruency by addition or congruency by subtraction implies equivalence. We state these facts, for convenience of reference, in the form of a theorem.

5.1.2 THEOREM. (1) *If* $P \cong Q$, *then* $P \cong Q$ (+); (2) *if* $P \cong Q$ (+), *then* $P \cong Q$ (−); (3) *if* $P \cong Q$ *or* $P \cong Q$ (+) *or* $P \cong Q$ (−), *then* $P \simeq Q$.

The idea of establishing the equivalence of two planar polygons by showing that the polygons are either congruent by addition or congruent by subtraction is found in even ancient geometric studies. For example, in Proposition 35 of Book I of Euclid's *Elements*, we find essentially the following proof of the fact that two parallelograms having the same base and equal altitudes are equivalent to one another. Calling the two parallelograms P and Q we clearly have, in the case of Figure 5.1a, $P \cong Q$ (+), and, in the

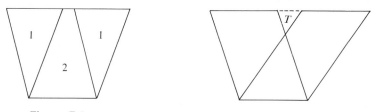

Figure 5.1a   Figure 5.1b

case of Figure 5.1b, $P + T \cong Q + T$ (+), whence $P \cong Q$ (−). There are many proofs of the Pythagorean Theorem illustrating the same procedures. For example, H. Perigal, in 1873, gave the dissection proof of the Pythagorean Theorem that is indicated in Figure 5.1c. Here the sum of the two squares

Figure 5.1c

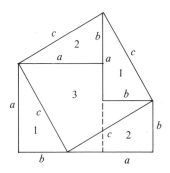

on the legs a and b of the right triangle is cut into three pieces which, by rearrangement, can be formed into the square on the hypotenuse c. A dissection proof of the Pythagorean Theorem which uses the idea of congruency by subtraction rather than congruency by addition is indicated in Figure 5.1d.

This is a very old dissection proof of the famous theorem and might well date back to Pythagorean times.

Figure 5.1d

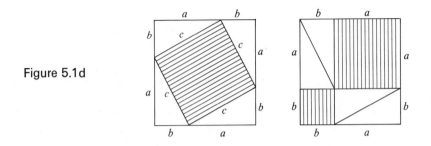

It is obvious that both congruency by addition and congruency by subtraction are reflexive and symmetric relations.* That each of these relations is also transitive† is not so obvious; we accordingly supply proofs of these two facts.

5.1.3 THEOREM. *If* $P \cong R \ (+)$ *and* $R \cong Q \ (+)$, *then* $P \cong Q \ (+)$.

We are given (see suggestive Figure 5.1e)

$$P = \sum_1^n P_i, \ R = \sum_1^n R_i, \ P_i \cong R_i \qquad (i = 1, \ldots, n),$$

$$Q = \sum_1^m Q_j, \ R = \sum_1^m R_j', \ Q_j \cong R_j' \qquad (j = 1, \ldots, m).$$

Figure 5.1e

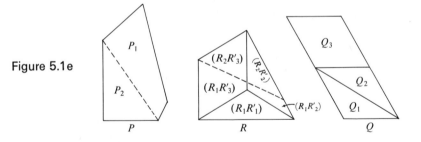

Now let $(R_i R_j')$ denote the common part(s) of R_i and R_j when the two dissections of R are superimposed on one another. Then we may dissect P_i into pieces congruent to

---

* That is, for any polygon or polyhedron P we have $P \cong P \ (+)$ and $P \cong P \ (-)$, and for any two polygons or polyhedra P and Q we have that $P \cong Q \ (+)$ implies $Q \cong P$ $(+)$ and $P \cong Q \ (-)$ implies $Q \cong P \ (-)$.
† Congruency by addition is transitive if, for any three polygons or polyhedra P, R, Q such that $P \cong R \ (+)$ and $R \cong Q \ (+)$, we have $P \cong Q \ (+)$. A similar definition holds for congruency by subtraction.

$$(R_i R_j'), \quad j = 1, \ldots, m,$$

and we may dissect Q_j into pieces congruent to

$$(R_i R_j'), \quad i = 1, \ldots, n.$$

Therefore both P and Q can be dissected into pieces congruent to

$$(R_i R_j'), \quad i = 1, \ldots, n, \quad j = 1, \ldots, m.$$

That is,

$$P \cong Q \,(+).$$

5.1.4   THEOREM.   *If* $P \cong R \,(-)$ *and* $R \cong Q \,(-)$, *then* $P \cong Q \,(-)$.

We are given (see suggestive Figure 5.1f)

$$P + \sum U_i \cong R + \sum W_i \,(+), \qquad U_i \cong W_i,$$
$$Q + \sum V_j \cong R + \sum W_j' \,(+), \qquad V_j \cong W_j'.$$

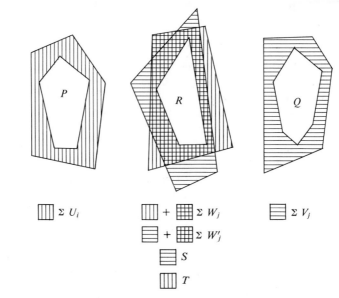

Figure 5.1f

$$\boxed{\,} \; \Sigma\, U_i \qquad \boxed{\,} + \boxed{\,} \; \Sigma\, W_i \qquad \boxed{\,} \; \Sigma\, V_j$$
$$\boxed{\,} + \boxed{\,} \; \Sigma\, W_j'$$
$$\boxed{\,} \; S$$
$$\boxed{\,} \; T$$

Now, in the superimposed figures for R, let S denote the sum of all those parts of $\sum W_j'$ which are not parts of $\sum W_i$ and let T denote the sum of all those parts of $\sum W_i$ which are not parts of $\sum W_j'$. Then

$$P + \sum U_i + S \cong R + \sum W_i + S \,(+)$$

and

$$Q + \sum V_j + T \cong R + \sum W_j' + T \,(+).$$

But

$$R + \sum W_i + S \cong R + \sum W_j' + T.$$

Hence, by Theorem 5.1.2 (1) and Theorem 5.1.3,

(1) $$P + \sum U_i + S \cong Q + \sum V_j + T\,(+).$$

Also

$$\sum U_i + S \cong \sum W_i + S\,(+), \qquad \sum V_j + T \cong \sum W_j' + T\,(+),$$

and

$$\sum W_i + S \cong \sum W_j' + T.$$

Therefore, again by Theorem 5.1.2 (1) and Theorem 5.1.3,

(2) $$\sum U_i + S \cong \sum V_j + T\,(+).$$

Relations (1) and (2) now guarantee that

$$P \cong Q\,(-).$$

Since each of the relations "congruent by addition" and "congruent by subtraction" is reflexive, symmetric, and transitive, it follows that each of the relations is an *equivalence relation* in the sense of abstract algebra. The necessity of proving Theorem 5.1.3 seems first to have occurred to W. H. Jackson in 1912.

PROBLEMS

1. Assuming the familiar formula for the area of a rectangle, give a congruency-by-addition proof of the theorem: The area of a parallelogram is equal to the product of one of its longer sides and the perpendicular distance between the two longer sides.

2. Give a complete proof, by dissection methods, of Proposition I 35 of Euclid's *Elements*.

3. Give a congruency-by-addition proof of the Pythagorean Theorem suggested by the dissection, given by H. E. Dudeney in 1917, pictured in Figure 5.1g. (Note that the square on the shorter leg is not cut.)

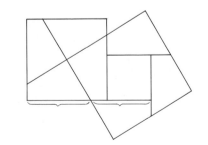

Figure 5.1g

4. Give a congruency-by-subtraction proof of the Pythagorean Theorem suggested by Figure 5.1h, said to have been devised by Leonardo da Vinci (1452–1519).

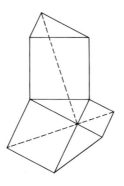

Figure 5.1h

5.2 DISSECTION OF POLYGONS INTO TRIANGLES

In this section we consider some matters connected with the dissection of a polygon into triangles. These matters become important in the program of building up a rigorous theory of polygonal area. Curiously enough, much of the material of this section cannot be extended to the analogous situations in space.

5.2.1 DEFINITION. A vertex V of a polygon P will be called a *projecting vertex* if it is possible to draw a line in the plane of P which separates vertex V from all the other vertices of P.

5.2.2 THEOREM. *A polygon possesses at least two projecting vertices.*

Referring to Figure 5.2a, denote the consecutive vertices of polygon P

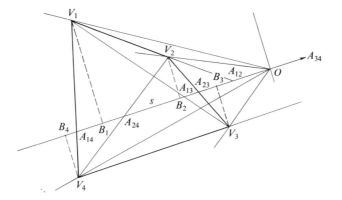

Figure 5.2a

by $V_1, \ldots, V_n$ and let s be any line in the plane of P other than a line $V_i V_j$ $(i \neq j)$. Let $V_i V_j$, produced if necessary, cut s in the (ordinary or ideal) point A_{ij}. Let B_k be the foot of the perpendicular from V_k on s. Choose on s an ordinary point O, distinct from each A_{ij}, such that all the points B_k lie on a common side of O. Draw the rays OV_k, $k = 1, \ldots, n$. These rays are distinct and all lie on one side of the perpendicular to s at O. It follows that two of the rays, say OV_a and OV_b, determine an angle within which all the other rays lie. Vertices V_a and V_b are clearly projecting vertices of P.

We now prove a stronger theorem.

5.2.3 THEOREM. *A polygon possesses at least three noncollinear projecting vertices.*

Let V_a, V_b be any two projecting vertices, guaranteed to exist by Theorem 5.2.2. There exist vertices of the polygon P on at least one side of the line $V_a V_b$; call it the left side. Draw a line t parallel to $V_a V_b$, on the left of $V_a V_b$, not on any line $V_i V_j$, and such that there are vertices of P on the left of t. Proceeding as in the proof of Theorem 5.2.2, using line t in place of line s, we find projecting vertices of P on the right and left of t. But a projecting vertex on the left of t is different from V_a, V_b and is not collinear with V_a, V_b.

5.2.4 THEOREM. *Any polygon of n sides can be dissected into n − 2 triangles whose vertices lie only at the vertices of the polygon.*

The theorem is trivially true for $n = 3$.

Suppose the theorem is true for all polygons having less than k sides and let P denote any k-sided polygon with consecutive vertices $V_1, \ldots, V_k$. By Theorem 5.2.2, there exists a projecting vertex V_a of P. Draw the segment $V_{a-1} V_{a+1}$. Two cases arise, according as all other vertices of P lie outside triangle $V_{a-1} V_a V_{a+1}$ or at least one vertex of P lies within or on this triangle.

Figure 5.2b

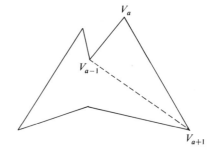

In the first case (see Figure 5.2b), segment $V_{a-1} V_{a+1}$ dissects P into triangle $V_{a-1} V_a V_{a+1}$ and a polygon Q having $k - 1$ sides and whose vertices are vertices of P. By our supposition, Q can be dissected into $(k - 1) - 2$ triangles whose vertices lie only at the vertices of Q. It follows that P can be dissected into $1 + [(k - 1) - 2] = k - 2$ triangles whose vertices lie only at the vertices of P.

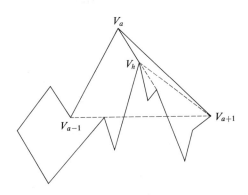

Figure 5.2c

In the second case (see Figure 5.2c), there is a vertex V_h inside or on triangle $V_{a-1}V_aV_{a+1}$ such that no other vertex is inside or on triangle $D = V_aV_{a+1}V_h$. If $h = a + 2$ we have polygons D and $R = V_hV_{h+1} \cdots V_a$; if $h \neq a + 2$ we have D, $S = V_{a+1}V_{a+2} \cdots V_h$, $T = V_hV_{h+1} \cdots V_a$. If R, S, T have r, s, t sides, then $r = k - 1$ and $s + t = k + 1$. By supposition, R, S, T can be dissected into $r - 2$, $s - 2$, $t - 2$ triangles with vertices only at those of R, S, T respectively. It follows that P can be dissected into $(r - 2) + 1 = k - 2$ or $(s - 2) + (t - 2) + 1 = k - 2$ triangles with vertices only at those of P.

Thus, in either case, it follows, from our supposition, that the theorem is true for any polygon having k sides.

The theorem is then true, by mathematical induction, for a polygon having any number of sides.

5.2.5 COROLLARY. *The sum of the interior angles of any polygon of* n *sides is equal to* n − 2 *straight angles.*

This is an immediate consequence of Theorem 5.2.4. (It should be noted that the customary proofs found in elementary geometry texts fail to hold if the polygon is nonconvex.)

5.2.6 THEOREM. *A convex polygon* can be dissected into triangles having vertices only at the vertices of the polygon and also all sharing a common vertex.*

Take any vertex of the polygon and connect it with each of the remaining

* A polygon is *convex* if it lies entirely on one side of each of its side lines.

vertices. Since the polygon is convex, these joins all lie within the polygon, and we have a dissection of the desired kind.

5.2.7 DEFINITIONS. Dissecting a triangle into two others by drawing a cevian line is called a *cevian operation*. A dissection of a triangle into triangles by a finite number of cevian operations applied to the given triangle or any resulting subtriangle is called a *cevian dissection* of the original triangle.

The following theorem is quite apparent.

5.2.8 THEOREM. *A dissection of a triangle into subtriangles in which the interior and at least one side, exclusive of the endpoints of that side, of the given triangle are free of vertices of subtriangles is a cevian dissection of the given triangle.*

5.2.9 THEOREM. *Any dissection of a triangle into subtriangles can, by additional cuts if necessary, be converted into a cevian dissection.*

Figure 5.2d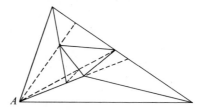

Let the given triangle T (see Figure 5.2d) be dissected into subtriangles T_k. From a vertex A of T draw cevian lines of T through all the vertices of the subtriangles T_k. These cevian lines dissect T into certain subtriangles T_t, each of which is further broken up by the original dissection into subtriangles and, perhaps, subquadrilaterals. If in each of the subquadrilaterals we draw a diagonal, then each subtriangle T_t is dissected into further subtriangles T_{tk}. By Theorem 5.2.8, the dissection of each subtriangle T_t into the subtriangles T_{tk} is cevian. It now follows that the dissection of T into the subtriangles T_{tk} is cevian.

5.2.10 THEOREM. *The additional cuts made in Theorem 5.2.9 effect a cevian dissection of each subtriangle* T_k.

Consider a subtriangle T_k. Among the cevian lines of T emanating from vertex A, there is always a definite one to be found which either coincides with a side of T_k, or which divides T_k into two subtriangles. In the first case, the side in question remains free of vertices of subtriangles T_{tk}. In the second case, the segment of the cevian line contained within T_k is a side of the two triangles arising from the division, and this side certainly remains free from further vertices of subtriangles T_{tk}. We may now apply Theorem 5.2.8.

PROBLEMS

1. Give an alternative proof of Theorem 5.2.2 by considering a line s not parallel to any $V_i V_j$ and lying entirely to the left of the polygon P. Now move s continuously to the right, always keeping it parallel to its original position.

2. Show how the customary proofs of Corollary 5.2.5 found in elementary geometry texts may fail to hold if the polygon is not convex.

3. Give a definition of a projecting vertex of a polyhedron P.

4. Prove that a polyhedron possesses at least two projecting vertices.

5.3 THE FUNDAMENTAL THEOREM OF POLYGONAL DISSECTION

We now develop the theorem that largely accounts for the great number of polygonal dissection puzzles. Some preliminary theorems will be established first.

5.3.1 THEOREM. *Any triangle is congruent by addition to the equivalent rectangle having for length a longest side of the triangle.**

Figure 5.3a

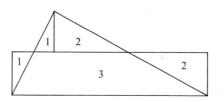

The dissection is apparent from Figure 5.3a.

5.3.2 THEOREM. *Any rectangle is congruent by addition to the equivalent square.*

Let the lengths of the sides of the rectangle be a and b, $a \geq b$. We shall consider three cases.

Case 1. $a = b$. The rectangle is congruent to the square.
Case 2. $b < a \leq 4b$. Let $s = (ab)^{1/2}$ be the side of the square equivalent to the rectangle. Place the square, $AEFH$, on the rectangle, $ABCD$, as shown in Figure 5.3b. Draw ED to cut BC in R and HF in K. Let BC cut

---

* A *longest side* of a triangle is any side which is not shorter than any other side.

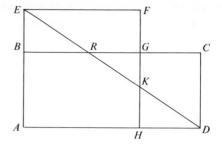

Figure 5.3b

HF in G. From the similar triangles KDH and EDA we have $HK/AE = HD/AD$, or

$$HK = (AE)(HD)/AD = s(a - s)/a = s - s^2/a = s - b.$$

But, by hypothesis, $4b \geq a$. Therefore $4b^2 \geq ab = s^2$, or $2b \geq s$, whence $s - b \leq b$. That is, K lies on the segment HG, and we have

$$\triangle EFK \cong \triangle RCD, \qquad \triangle EBR \cong \triangle KHD.$$

Case 3. $a > 4b$. Keep halving the base of the rectangle and doubling the altitude until an equivalent rectangle is obtained which satisfies case 1 or case 2. This new rectangle is clearly congruent by addition to the original rectangle. Hence, by Theorem 5.1.3, the equivalent square is congruent by addition to the original rectangle.

5.3.3 THEOREM. *Any two squares are jointly congruent by addition to the square equivalent to their sum.*

To prove this theorem is equivalent to giving a congruency-by-addition proof of the Pythagorean Theorem; this we have already indicated by Perigal's dissection in Figure 5.1c.

5.3.4 THEOREM. *Any finite number of squares are jointly congruent by addition to the square equivalent to their sum.*

Let the given squares be $S_1, \ldots, S_n$ and let $S_{12}, S_{123}, \ldots, S_{123 \ldots n}$ denote the squares equivalent to $S_1 + S_2, S_{12} + S_3, \ldots, S_{123 \ldots (n-1)} + S_n$ respectively. Then, by Theorem 5.3.3, $S_1 + S_2 \cong S_{12} (+)$, $S_{12} + S_3 \cong S_{123} (+), \ldots, S_{123 \ldots (n-1)} + S_n \cong S_{123 \ldots n} (+)$. It follows, by Theorem 5.1.3, that

$$S_1 + S_2 + \cdots + S_n \cong S_{12} + S_3 + \cdots + S_n (+)$$
$$\cong S_{123} + S_4 + \cdots + S_n (+)$$
$$\cdot \quad \cdot \quad \cdot \quad \cdot \quad \cdot$$
$$\cong S_{123 \ldots n} (+).$$

5.3.5 THEOREM. *Any polygon is congruent by addition to the equivalent square.*

Let P be the polygon. Now, by Theorem 5.2.4, we can dissect P into triangles $T_1, \ldots, T_k$. That is

(1) $$P = T_1 + \cdots + T_k.$$

By Theorem 5.3.1, there exist rectangles $R_1, \ldots, R_k$ such that

(2) $$T_i \cong R_i (+), \qquad\qquad i = 1, \ldots, k.$$

By Theorem 5.3.2, there exist squares $S_1, \ldots, S_k$ such that

(3) $$R_i \cong S_i (+), \qquad\qquad i = 1, \ldots, k.$$

Let S be the square such that $S \simeq S_1 + \cdots + S_k$. From (1), (2), (3) we have

$$P \cong \sum T_i (+), \qquad \sum T_i \cong \sum R_i (+), \qquad \sum R_i \cong \sum S_i (+),$$

and by Theorem 5.3.4 we have $\sum S_i \cong S (+)$. Therefore, by Theorem 5.1.3, $P \cong S (+)$.

5.3.6 THE FUNDAMENTAL THEOREM OF POLYGONAL DISSECTION. *Any two equivalent polygons are congruent by addition, and the dissection may be accomplished with Euclidean tools.*

Let P and Q be the polygons and let S be the square such that $S \simeq P \simeq Q$. Then, by Theorem 5.3.5, $P \cong S (+)$ and $Q \cong S (+)$. Hence, by Theorem 5.1.3, $P \cong Q (+)$.

That the dissection may be performed with Euclidean tools follows from the fact that with these tools we may decompose a polygon into triangles, construct for each triangle the corresponding rectangle of Theorem 5.3.1, next the corresponding square of Theorem 5.3.2, and then successively combine the squares as in Theorem 5.3.3.

5.3.7 THEOREM. *If two polygons are congruent by subtraction, then they are congruent by addition.*

For if $P \cong Q (-)$, then $P \simeq Q$. Hence, by Theorem 5.3.6, $P \cong Q (+)$.

The theorems of this section can be easily extended to more general polygons than those we have considered. Thus, by a simple device, we may allow our polygons to have "holes" in them, or to be in more than one piece. The device (see Figure 5.3c) is to convert such extended polygons into essentially those of the type already considered by drawing appropriate *cut lines*. Thus any polygonal line lying in the interior of a polygon containing a "hole" and connecting the boundary of the "hole" with the boundary of the surrounding polygon will convert the figure into a polygon of the kind already considered if certain points and lines are counted twice as vertices and edges. Similarly, an exterior polygonal line connecting the boundaries of two pieces joins the two pieces into essentially a polygon of the type already considered.

It might be interesting to close this section by observing an application

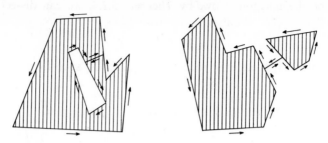

Figure 5.3c

of the foregoing theory to a specific problem. If we successively apply the dissections of Theorems 5.3.1 and 5.3.2 to the problem of cutting up an equilateral triangle into pieces which can be reassembled to form a square, we obtain the five-piece dissection pictured in Figure 5.3d. In Section 5.8 a four-piece solution of this problem is given.

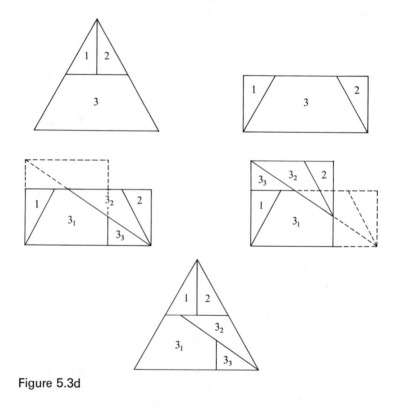

Figure 5.3d

PROBLEMS

Dissection puzzles are generally quite difficult and require considerable ingenuity on the part of the solver. Problems 1 through 17 are selected from the Problem

Department of *The American Mathematical Monthly*, and a reader may find the solutions to the problems at the indicated references. Problems 18 through 26 are selected from H. E. Dudeney's *Amusements in Mathematics* (New York: Dover Publications, Inc., 1958); solutions appear in Dudeney's book on the indicated pages.

1. A newspaper gave this problem: Cut a regular six-pointed star into the fewest number of pieces which will fit together and make a square. The newspaper gave a solution in seven pieces. First cut off two opposite points of the star. Divide each into two parts, and fit to the remaining portion of the star so as to make a rectangle. Find the mean proportional between the length and breadth of this rectangle; this is the side of the required square. Using this dimension on the two long sides of the rectangle, divide the latter into three pieces, which make the square. Total seven pieces.
   (a) Exhibit the newspaper dissection.
   (b) Obtain a five-piece solution. (Problem 2799, Apr. 1921.)

2. Cut two equilateral triangles of any relative proportions into not more than five pieces which can be assembled to form a single equilateral triangle. (Problem 3048, Mar. 1930, and Problem E 1210, Nov. 1956.)

3. Cut a 1×2 rectangle into three pieces which will fit into a Maltese cross. (Problem 3142, Aug.-Sept. 1926.)

4. Divide a triangle by two straight lines into three parts, which when properly arranged shall form a parallelogram whose angles are of given magnitudes. (Problem 3244, Feb. 1928.)

5. A rectangular board has the length l (with the grain) and the width w (across the grain). From it a square table top is to be made with the grain running all one way. The board may be sawed both parallel and perpendicular to its length; but the table top may have joints only with the grain, and *not* across the grain. What is the largest table top that can be so made? (Problem E 80, Aug.-Sept. 1934).

6. From an arbitrary point within a triangle, draw three lines to the sides of the triangle which shall trisect its area. (Problem E 199, Nov. 1936.)

7. (a) Dissect a regular hexagon by straight cuts into 6 pieces which can be reassembled to form an equilateral triangle. (b) Does there exist a five-piece solution? (Problem E 400, Aug.-Sept. 1940.)

8. (a) Given $n \geq 6$, show that it is possible to fit together n isosceles right triangles, all of different sizes, so as to make a single isosceles right triangle. (b) Can the problem be solved for $n = 5$? (c) Can a right isosceles triangle be dissected into a finite number of right isosceles triangles, no two having a common side? (Problem E 476, Mar. 1942.)

9. Given a regular polygon of n sides, $n > 4$, design a quadrilateral, Q, such that: (1) it shall be possible to fit $2n$ of the Q's to the polygon to form a new regular polygon of n sides, and (2) it shall be possible to fit $2n$ additional Q's to the new polygon to form a still larger third regular polygon of n sides. (Problem E 541, June-July 1943.)

10. Give two equal regular dodecagons. Show how to dissect one of them into twelve congruent pieces which can be fitted to the other to form another larger regular dodecagon. (Problem E 721, Jan. 1947.)

11. Show that any given triangle can be dissected by straight cuts into four pieces which can be arranged to form two triangles similar to the given triangle. (Problem E 922, Feb. 1951.)

12. Dissect a regular pentagon, by straight cuts, into six pieces which can be put together to form an equilateral triangle. (Problem E 972, Feb. 1952.)

13. Draw a straight line which will bisect both the area and the perimeter of a given quadrilateral. (Problem E 992, June-July 1952 and Jan. 1953.)

14. Given n equal unit squares, dissect each of these in exactly the same way with straight cuts into $p(n)$ parts such that the $n\,p(n)$ pieces may be assembled to form a square of edge $n^{1/2}$. (Problem E 1010, Dec. 1952.)

15. Find six-piece dissections of a regular dodecagon into a square and into a Greek cross. (Problem E 1240, May 1957.)

16. Dissect a regular five-pointed star into no more than eight pieces which can be reassembled, without turning over any of the pieces, to form a square. (Problem E 1309, Nov. 1958.)

17. The dissection of a unit square into one $\frac{1}{8} \times 1$ and two $\frac{1}{2} \times \frac{7}{8}$ rectangular pieces shows that it is possible to dissect the unit square into three polygonal pieces each having diameter $d = \sqrt{65/8}$. Prove that the unit square cannot be dissected into three polygonal pieces all of which have diameters less than d. (Problem E 1311, Dec. 1958.)

18. Dissect a Greek cross into four congruent pieces that can be reassembled to form a square. (p. 29.)

19. Dissect a square into four pieces that can be reassembled to form:
(a) two Greek crosses of different sizes,
(b) two Greek crosses of the same size. (p. 30.)

20. Dissect a right isosceles triangle into four pieces that can be reassembled to form a Greek cross. (p. 32.)

21. Dissect a Greek cross into five pieces that will form two separate squares, one of which shall contain half the area of one of the arms of the cross. (p. 34.)

22. Dissect a Greek cross into five pieces that will form two equal Greek crosses. (Problem 143, p. 168.)

23. Dissect a Greek cross into six pieces that will form an equilateral triangle. (Problem 144, p. 169.)

24. Cut out of paper a Greek cross; then so fold it that with a single straight cut of the scissors, the four pieces produced will form a square. (Problem 145, p. 169.)

25. Dissect a 1×5 rectangle into four pieces which can be reassembled to form a square. (Problem 153, p. 172.)

26. Dissect a regular pentagon into six pieces that can be reassembled to form a square. (Problem 155, p. 172.)

27. Let P and Q (not necessarily polygons) be two equivalent planar areas such that for each of P and Q all the curved boundary arcs fall into pairs of congruent portions differing only as to the side on which the interior of the figure lies. Show that P can be dissected into pieces which may be rearranged to form Q.

28. Show that in the dissection guaranteed by the fundamental theorem of polygonal dissection no piece is turned over.

5.4 LENNES POLYHEDRA AND CAUCHY'S THEOREM

In this and the next two sections we consider some matters connected with the dissection of (solid) polyhedra into subpolyhedra. We first prove the existence of polyhedra which cannot be dissected into tetrahedra having their vertices only at the vertices of the polyhedron. This interesting fact was first published in 1911 by N. J. Lennes, who constructed a polyhedron having the property that the join of any two vertices not the endpoints of a common edge lies either wholly or partly outside the polyhedron; it is obvious that such a polyhedron cannot be dissected into tetrahedra in the desired fashion.

5.4.1 DEFINITION. A polyhedron which cannot be dissected into tetrahedra having their vertices only at the vertices of the polyhedron will be called a *Lennes polyhedron*.

The polyhedron constructed by Lennes possessed seven vertices. In 1928, E. Schönhardt gave a simpler example of a Lennes polyhedron having only six vertices, and he proved that there is no Lennes polyhedron with less than six vertices. In 1948, F. Bagemihl showed the existence of Lennes polyhedra having any number of vertices not less than six. We shall here establish the existence of Lennes polyhedra by describing Schönhardt's example of 1928.

5.4.2 THEOREM. *Lennes polyhedra exist.*

Let ABC be an equilateral triangle with unit sides, and let $A'B'C'$ be the triangle obtained from ABC by first rotating ABC, in its plane, about its center through 30° in the direction ABC, and then translating it one unit in the direction perpendicular to the plane ABC. Consider the polyhedron P consisting of (see Figure 5.4a):

Figure 5.4a

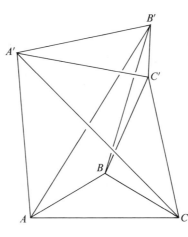

the 6 vertices: A, B, C; A', B', C';

the 12 edges: AB, BC, CA; $A'B'$, $B'C'$, $C'A'$; AA', BB', CC'; AB', BC', CA';

the 8 triangular faces: ABC, $A'B'C'$; $AA'C$, $CC'B$, $BB'A$; $AA'B'$, $BB'C'$, $CC'A'$.

The only joins of pairs of vertices which are not the endpoints of a common edge are AC', CB', BA', and these are clearly wholly outside of P. It follows that P is a Lennes polyhedron.

In contrast to Theorem 5.4.2 we have the following theorem.

5.4.3 CAUCHY'S THEOREM. *Every convex polyhedron* can be dissected into tetrahedra having their vertices only at the vertices of the polyhedron and also all sharing a common vertex.*

Consider the faces of a convex polyhedron P which do not contain a particular vertex A. Each of these faces can, by Theorem 5.2.4, be dissected into triangles having their vertices only at vertices of P. Now form the tetrahedra having these triangles as bases and A as opposite vertex. Since P is convex, these tetrahedra all lie within P, and we have P dissected in the desired manner.

Theorem 5.4.3 now enables us to prove the following fundamental theorem.

5.4.4 THEOREM. *Every polyhedron can be dissected into tetrahedra.*

Let $p_1, \ldots, p_n$ be the face planes of a polyhedron P. Then $p_1, \ldots, p_n$ divide space into a set of convex regions, a finite number of which make up P. The theorem now follows from Theorem 5.4.3.

PROBLEMS

1. Figure 5.4b gives a pattern for constructing a Lennes polyhedron of the type first discovered by Lennes. Reproduce Figure 5.4b on a convenient scale, cut it out, and bend the paper along the inner lines upwards or downwards according as these lines are full or broken. All flaps for gluing the surface together are to be bent upwards. The side of the paper upon which the pattern is drawn will end up inside of the model. The resulting polyhedron has a plane of symmetry.

2. Draw a pattern leading to a Lennes polyhedron of the Schönhardt type.

3. In Figure 5.4a, connect A and A' with a circular arc of radius so large that the open arc is on the same side of plane $CA'C'$ as A and on the same side of plane BAB' as A', and choose k distinct points $D_1, D_2, \ldots, D_k$, in this order, on the open arc. Consider the polyhedron P consisting of:

* A polyhedron is *convex* if it lies entirely on one side of each of its face planes.

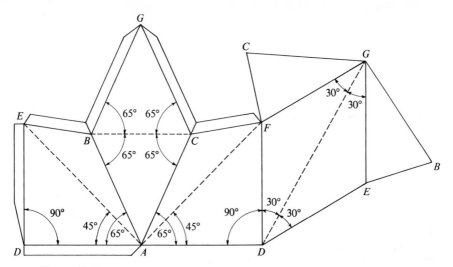

Figure 5.4b

the $k + 6$ vertices: $A, B, C, A', B', C', D_1, \ldots, D_k$;

the $3k + 12$ edges: AB, BC, CA; $A'B', B'C', C'A'$; BB', CC'; $AB', BC',$ CA'; $AD_1, D_1D_2, \ldots, D_{k-1}D_k, D_kA'$; $CD_1, CD_2, \ldots, CD_{k-1}, CD_k$; $B'D_1, B'D_2, \ldots, B'D_{k-1}, B'D_k$;

the $2k + 8$ triangular faces: $ABC, A'B'C'$; $BC'B', CBC', AB'B, A'CC'$; $CAD_1, CD_1D_2, CD_2D_3, \ldots, CD_{k-1}D_k, CD_kA'$; $B'AD_1, B'D_1D_2,$ $B'D_2D_3, \ldots, B'D_{k-1}D_k, B'D_kA'$.

Show that P is a Lennes polyhedron. (This is Bagemihl's example of a Lennes polyhedron having $k + 6$ vertices. See *The American Mathematical Monthly*, August–September 1948, pp. 411–413.)

4. Give a dissection proof of the theorem: An oblique prism is equivalent to a right prism whose base is equal to a right section of the oblique prism, and whose altitude is equal to a lateral edge of the oblique prism.

5.5 DEHN'S THEOREM

We next establish a theorem concerning the dihedral angles of any two polyhedra which are congruent by addition or by subtraction; the proof of the theorem will be elementary but somewhat long.

5.5.1 THEOREM. *If P and Q are polyhedra such that* $P \cong Q \, (+)$ *or* $P \cong Q$ $(-)$, *and if P and Q have dihedral angles* $\alpha_1, \ldots, \alpha_n$ *and* $\beta_1, \ldots, \beta_m$ *respectively, then there exist positive integers* p_i, q_i, p, p', q, q' *such that*

$$\sum_1^n p_i \alpha_i + (p + 2p')\pi = \sum_1^m q_i \beta_i + (q + 2q')\pi.$$

Case 1. $P \cong Q \, (+)$.

We have

$$P = P_1 + P_2 + \cdots + P_k,$$
$$Q = Q_1 + Q_2 + \cdots + Q_k,$$
$$P_i \cong Q_i, \quad i = 1, \ldots, k.$$

Consider an edge of a subpolyhedron of P. The vertices of other subpolyhedra of P perhaps divide this edge into intervals. Each of these intervals (or lacking such, the edge itself) will be known as an a-interval of the edge. Thus the edges of all the subpolyhedra of P are divided into a-intervals, and every a-interval belongs to edges of one or more subpolyhedra of P.

Now concentrate on some particular a-interval and let e_{P_i} denote the edge of one subpolyhedron P_i of P to which it belongs. Mark this a-interval on the corresponding edge e_{Q_i} of the corresponding subpolyhedron Q_i of Q, and temporarily designate the marked interval by a_i. Now a_i may be further divided by vertices of other subpolyhedra of Q. This division of a_i, when transferred to the original a-interval of P, will be known as the division of that a-interval into c-intervals belonging to the edge e_{P_i}. The original a-interval will thus be divided into sets of c-intervals, a set arising from each subpolyhedron of P on an edge of which the a-interval is to be found. We suppose all the a-intervals of P to be divided in this manner into their sets of c-intervals.

Let us designate as b-intervals of Q those intervals on the edges of the subpolyhedra of Q which are analogous to the a-intervals of P. In finding the sets of c-intervals of the a-intervals of P, as outlined in the preceding paragraph, the b-intervals of Q have been divided into analogous sets which we shall refer to as sets of d-intervals. It is important to notice that to any particular c-interval of P belonging, say, to edge e_{P_i}, there is a corresponding equal d-interval of Q belonging to the corresponding edge of e_{Q_i}, and conversely. Moreover, in the final division of the a and b-intervals, corresponding c and d-intervals are divided in the same way into the same number of parts.

We now assign *weights* and *arguments* to our intervals. The weight of an interval (a, b, c, or d) will be the number of parts into which it is finally divided. The argument of a c or d-interval will be the measure of the dihedral angle of the subpolyhedral edge to which it belongs. The argument of an a-interval will be π if it lies in the face of P or lies inside P but in a face of some P_i, 2π if it lies inside P but not in a face of some P_i, or α_h if it lies on that edge of P whose dihedral angle is α_h. Thus the argument of an a-interval is the sum of the dihedral angles of the subpolyhedra along that a-interval. The argument of a b-interval is similarly defined with reference to Q.

If an a-interval of weight w and argument t be divided in several ways by c-intervals with weights

$$w_1^{(1)}, w_1^{(2)}, w_1^{(3)}, \ldots,$$
$$w_2^{(1)}, w_2^{(2)}, w_2^{(3)}, \ldots,$$
$$\cdot \quad \cdot \quad \cdot \cdots,$$

and arguments

$$t_1^{(1)}, t_1^{(2)}, t_1^{(3)}, \ldots,$$
$$t_2^{(1)}, t_2^{(2)}, t_2^{(3)}, \ldots,$$
$$\cdot \quad \cdot \quad \cdot \quad \cdots,$$

we have, by the definitions given above,

$$t_1^{(1)} = t_1^{(2)} = t_1^{(3)} = \ldots,$$
$$t_2^{(1)} = t_2^{(2)} = t_2^{(3)} = \ldots,$$
$$\cdot \quad \cdot \quad \cdot \quad \cdots,$$

and

$$\sum w_1^{(i)} = \sum w_2^{(i)} = \cdots = w,$$

and

$$t_1^{(1)} + t_2^{(1)} + \cdots = t.$$

Therefore

$$\sum w_1^{(i)} t_1^{(i)} + \sum w_2^{(i)} t_2^{(i)} + \cdots$$
$$= t_1^{(1)} \sum w_1^{(i)} + t_2^{(1)} \sum w_2^{(i)} + \cdots$$
$$= w(t_1^{(1)} + t_2^{(1)} + \cdots)$$
$$= wt.$$

Hence

(1) $$\sum_c wt = \sum_a wt,$$

where $\sum_c$ and $\sum_a$ are sums over all c-intervals of P and a-intervals of P. Let $p_1, \ldots, p_n, p, p'$ be the sums of the weights of the a-intervals with arguments $\alpha_1, \ldots, \alpha_n, \pi, 2\pi$ respectively. Then (1) becomes

$$\sum_c wt = p_1\alpha_1 + \cdots + p_n\alpha_n + (p + 2p')\pi.$$

Similarly treating Q we have

$$\sum_d wt = q_1\beta_1 + \cdots + q_m\beta_m + (q + 2q')\pi.$$

But

$$\sum_c wt = \sum_d wt,$$

and p_i, q_i, p, q, p', q' are positive integers. Hence the theorem for case 1.

Case 2. $P \cong Q\ (-)$.
We have

$$P + P_1 + \cdots + P_k \cong Q + Q_1 + \cdots + Q_k\ (+),$$

where

$$P_j \cong Q_j, \quad j = 1, \ldots, k.$$

Let the dihedral angles of P_j (and Q_j) be $\alpha_{j1}, \ldots, \alpha_{ja_j}$. Then we have, by case 1,

$$\sum_{1}^{n} p_i \alpha_i + \sum_{1}^{\alpha_1} p_{1i} \alpha_{1i} + \cdots + \sum_{1}^{\alpha_k} p_{ki} \alpha_{ki} + (p + 2p')\pi$$

$$= \sum_{1}^{m} q_i \beta_i + \sum_{1}^{\alpha_1} q_{1i} \alpha_{1i} + \cdots + \sum_{1}^{\alpha_L} q_{ki} \alpha_{ki} + (q + 2q')\pi.$$

But $p_{ji} = q_{ji}$, since either one is equal to the total number of parts of that edge of P_j whose dihedral angle is α_{ji}. Hence the theorem for case 2.

Theorem 5.5.1 gives us a *necessary* condition for two polyhedra to be congruent by addition or by subtraction. It follows that if we can produce two equivalent polyhedra which do not satisfy the condition of Theorem 5.5.1, then the two polyhedra cannot be congruent either by addition or by subtraction, and the analogue in space of the fundamental theorem of planar dissection does not hold. We shall show that a regular tetrahedron and an equivalent cube do not satisfy the condition of Theorem 5.5.1. In order to do this we need the following purely arithmetical fact, which we shall establish with the aid of a little trigonometry.

5.5.2 THEOREM. *If $\theta = \arccos \frac{1}{3}$, then there do not exist integers* r *and* s *such that* $r\theta = s\pi$.

Suppose there are integers r and s such that

(1) $$r\theta = s\pi,$$

where, without loss of generality, we may assume r and s to be relatively prime. Consider the binomial expansion

$$(\cos \theta + i \sin \theta)^r = \sum_{n=0}^{r} \binom{r}{n} i^n \cos^{r-n} \theta \sin^n \theta.$$

Taking the imaginary parts of both members, and remembering (1), we find*

$$0 = \sin r\theta = \binom{r}{1} \cos^{r-1} \theta \sin \theta - \binom{r}{3} \cos^{r-3} \theta \sin^3 \theta + \cdots.$$

Substituting $\cos \theta = \frac{1}{3}$, $\sin \theta = \sqrt{8}/3$, we find

$$0 = (\sqrt{8}/3^r)\left\{ r - \binom{r}{3} 8 + \binom{r}{5} 8^2 - \cdots \right\}.$$

Therefore the quantity in braces must vanish, and r must be even (divisible by 8 in fact). Since r and s were chosen as relatively prime, s must be odd. Thus $\cos(r\theta/2) = \cos(s\pi/2) = 0$. But

$$(\cos \theta + i \sin \theta)^{r/2} = \sum_{n=0}^{r/2} \binom{r/2}{n} i^n \cos^{r/2-n} \theta \sin^n \theta.$$

* By De Moivre's theorem, $(\cos \theta + i \sin \theta)^r = \cos r\theta + i \sin r\theta$.

Taking the real parts of both members we have

$$0 = \cos(r\theta/2)$$

$$= \cos^{r/2}\theta - \binom{r/2}{2}\cos^{r/2-2}\theta \sin^2\theta + \binom{r/2}{4}\cos^{r/2-4}\theta \sin^4\theta - \cdots.$$

Substituting $\cos\theta = \frac{1}{3}$, $\sin\theta = \sqrt{8}/3$, we find

$$0 = \frac{1 - \binom{r/2}{2}8 + \binom{r/2}{4}8^2 - \cdots}{3^{r/2}}$$

$$= \frac{1 + \text{even number}}{3^{r/2}}.$$

But this is impossible since the last numerator is odd and hence cannot vanish. Therefore the original assumption (1) is untenable, and the theorem is proved.

5.5.3 DEHN'S THEOREM (1902). *If P and Q are equivalent polyhedra, it does not necessarily follow that* $P \cong Q\ (+)$ *or* $P \cong Q\ (-)$.

For consider a regular tetrahedron T and an equivalent cube C. Suppose $T \cong C\ (+)$ or $T \cong C\ (-)$. Then the condition of Theorem 5.5.1 applies. But the dihedral angles of a regular tetrahedron are all equal to $\theta = $ arc cos $\frac{1}{3}$, and the dihedral angles of a cube are all equal to $\pi/2$. Theorem 5.5.1 then states that

(1)
$$\sum_1^6 p_i\theta + (p + 2p')\pi = \sum_1^{12} q_i(\pi/2) + (q + 2q')\pi.$$

Doubling both sides of (1) and doing some transposing we finally have a relation of the form

$$r\theta = s\pi,$$

where r and s are integers. But such a relation is impossible by Theorem 5.5.2. Hence the theorem.

PROBLEMS

1. Prove that the dihedral angles of a regular tetrahedron are all equal to arc cos $\frac{1}{3}$.

2. Prove that a frustum of a triangular pyramid is equivalent to the sum of three triangular pyramids whose common altitude is the altitude of the frustum and whose bases are the lower base, the upper base, and the mean proportional between the two bases of the frustum.

3. Prove: (a) A truncated triangular prism is equivalent to the sum of three triangular pyramids whose common base is the base of the prism and whose vertices are the three vertices of the inclined section.

(b) The volume of a truncated right triangular prism is equal to the product of its base and one-third the sum of its lateral edges.

(c) The volume of any truncated triangular prism is equal to the product of its right section and one-third the sum of its lateral edges.

4. How may the total surface of a sphere be divided into the largest possible number of congruent pieces, if each side of each piece is an arc of a great circle less than a quadrant? (See Problem E 5, *The American Mathematical Monthly*, Feb. 1933.)

5. (a) Can any sealed rectangular envelope, after a single straight cut, be folded into two congruent tetrahedra? Will the position of the cut affect the size of the tetrahedra?

(b) How should the cut be made to make the total number of folds and unfolds a minimum?

(c) What should be the relative dimensions of the envelope in order that the tetrahedra be regular? (See Problem E 841, *The American Mathematical Monthly*, June-July, 1949.)

6. In the February 1957 issue of *Scientific American* the following problem appeared: "A carpenter, working with a buzz saw, wishes to cut a wooden cube, three inches on a side, into 27 one-inch cubes. He can do this easily by making six cuts through the cube, keeping the pieces together in the cube shape. Can he reduce the number of necessary cuts by rearranging the pieces after each cut?" Generalize this to find the number of cuts necessary to dissect an $n \times n \times n$ cube into n^3 one-inch cubes. (See Problem E 1279, *The American Mathematical Monthly*, March 1958.)

5.6 CONGRUENCY (T) AND SUSS' THEOREM

So far we have considered whether two equivalent polyhedra can be dissected into subpolyhedra which are congruent in pairs, and we have found that in general they cannot. We now examine the question whether they can be dissected into corresponding tetrahedra such that in each pair of corresponding tetrahedra we have a face of one equivalent to a face of the other and the altitudes on these faces equal. This matter was first investigated by W. Süss in 1920.

5.6.1 NOTATION AND DEFINITIONS

(a) If two tetrahedra T_1 and T_2 are such that a face of one is equivalent to a face of the other, and the altitudes on these faces are equal, we say that T_1 and T_2 are *congruent* (T) and we write

$$T_1 \cong T_2 \ (T).$$

(b) If polyhedron P can be dissected into tetrahedra $T_1, \ldots, T_n$ and polyhedron Q into tetrahedra $T'_1, \ldots, T'_n$, where

$$T_i \cong T'_i \ (T), \qquad\qquad i = 1, \ldots, n,$$

we say that P and Q are *congruent* $(T+)$ and we write

$$P \cong Q \ (T+).$$

(c) If there exist polyhedra R, S, $P_1, \ldots, P_n, Q_1, \ldots, Q_n$ such that

$$R = P + P_1 + \cdots + P_n,$$
$$S = Q + Q_1 + \cdots + Q_n,$$
$$P_i \cong Q_i \ (T+), \qquad\qquad i = 1, \ldots, n,$$
$$R \cong S \ (T+),$$

then we say that P and Q are *congruent* $(T-)$ and we write

$$P \cong Q \ (T-).$$

Thus P and Q are congruent $(T-)$ if we can adjoin to P and Q pieces which are congruent $(T+)$ so that the results are congruent $(T+)$. We also say $P \cong Q \ (T-)$ if

$$P = \sum_{i=1}^{n} P_i, \ Q = \sum_{i=1}^{n} Q_i, \ P_i \cong Q_i \ (T-) \text{ for } i = 1, \ldots, n.$$

The following theorem is apparent from the preceding definitions, and the next one is almost apparent.

5.6.2   THEOREM.   *Let* P *and* Q *be polyhedra.* (1) *If* P $\cong$ Q, *then* P $\cong$ Q $(T+)$; (2) *if* P $\cong$ Q $(T+)$, *then* P $\cong$ Q $(T-)$.

5.6.3   THEOREM.   *If* T_1, T_2, T_3 *are three tetrahedra such that* $T_1 \cong T_3$ (T) *and* $T_2 \cong T_3$ (T), *then* $T_1 \cong T_2$ $(T-)$.

For we have $T_1 + T_3 \cong T_2 + T_3$ $(T+)$.

5.6.4   THEOREM.   *If* T_1 *and* T_2 *are any two equivalent tetrahedra, then* $T_1 \cong T_2$ $(T-)$.

Let the vertices of T_1 be a_1, a_2, a_3, a_4 and those of T_2 be b_1, b_2, b_3, b_4.

Case 1. $a_1 a_2 a_3 \simeq b_1 b_2 b_3$.
In this case the altitudes of T_1 and T_2 through a_4 and b_4 are equal and $T_1 \cong T_2$ (T), whence, of course, $T_1 \cong T_2$ $(T-)$.

Case 2. $a_1 a_2 a_3 > b_1 b_2 b_3$.
Take b_1' (see Figure 5.6a) on $b_2 b_1$ produced such that $b_1' b_2 b_3 \simeq a_1 a_2 a_3$. Through b_1 pass a plane parallel to $b_1' b_4 b_3$ to cut $b_2 b_4$ in b_4'. Set $T_3 \equiv b_1' b_2 b_3 b_4'$. Then $T_2 = T_{21} + T_{22}$, $T_3 = T_{31} + T_{32}$, where

$$T_{21} \equiv T_{31} \equiv b_1 b_2 b_3 b_4', \ T_{22} \equiv b_1 b_3 b_4 b_4', \ T_{32} \equiv b_1 b_1' b_3 b_4'.$$

Since $b_1 b_4'$ is parallel to $b_1' b_4$, we have $b_4 b_4' b_1 \simeq b_1' b_4' b_1$, and $T_{22} \cong T_{32}$ (T). It follows that $T_2 \cong T_3$ $(T+)$, or $T_2 \simeq T_3$. Therefore $T_1 \simeq T_3$, and, since face $a_1 a_2 a_3$ of T_1 is equivalent to face $b_1' b_2 b_3$ of T_3, we can pass a

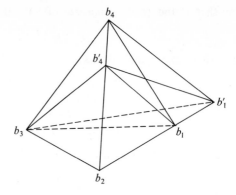

Figure 5.6a

plane through edge $a_3 a_4$ which dissects T_1 into subtetrahedra T_{11} and T_{12} where $T_{11} \cong T_{31}\ (T)$ and $T_{12} \cong T_{32}\ (T)$. We now have

$$T_{11} \cong T_{31}\ (T) \quad \text{and} \quad T_{21} \cong T_{31}\ (T),$$
$$T_{12} \cong T_{32}\ (T) \quad \text{and} \quad T_{22} \cong T_{32}\ (T),$$

whence (by Theorem 5.6.3)

$$T_{11} \cong T_{21}\ (T-) \text{ and } T_{12} \cong T_{22}\ (T-).$$

It follows that $T_1 \cong T_2\ (T-)$.

5.6.5 SÜSS' THEOREM (1920). *If* P *and* Q *are any two equivalent polyhedra, then* P $\cong$ Q (T−).

By Theorem 5.4.4, we have

$$P = \sum_{i=1}^{n} T_i, \qquad Q = \sum_{j=1}^{m} T_j',$$

where the T_i and the T_j' are tetrahedra. Consider a tetrahedron $T \simeq P \simeq Q$, and let e (see Figure 5.6b) denote some particular edge of T. There exists

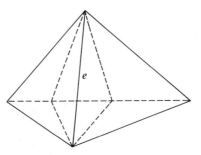

Figure 5.6b

a set of planes through e which dissect T into adjacent tetrahedra $S_1, \ldots, S_n$ such that $S_i \simeq T_i$, and a second set of planes through e which dissect T into adjacent tetrahedra $S_1', \ldots, S_m'$ such that $S_j' \simeq T_j'$. Each tetrahedron

S_i is possibly subdivided by the second set of planes into a set of subtetrahedra S_{ij}, and each tetrahedron S'_j is possibly subdivided by the first set of planes into a set of subtetrahedra S'_{ji}. Now divide each T_i (by planes through an edge of T_i) into subtetrahedra T_{ij} equivalent to the S_{ij}, and divide each T'_j (by planes through an edge of T'_j) into subtetrahedra T'_{ji} equivalent to the S'_{ji}. But the S_{ij} are the S'_{ji} rearranged. It follows that the T_{ij} are equivalent to the T'_{ji} in some arrangement, whence (by Theorem 5.6.4)

$$\sum T_{ij} \cong \sum T'_{ji} \,(T-).$$

That is, $P \cong Q \; (T-)$.

It is now apparent that the whole difficulty in constructing a theory of polyhedral volume lies in the consideration of tetrahedra which are congruent (T). Such tetrahedra are not in general congruent either by addition or by subtraction, and two such tetrahedra can be proven equivalent only by introducing infinitesimals in connection with some sort of integration process. This state of affairs is reflected in Book XII of Euclid's *Elements* and in the American high school texts on solid geometry. On the other hand, infinitesimals are not needed in the construction of a theory of polygonal area.

PROBLEMS

1. Show that there are tetrahedra T_1, T_2, T_3 such that $T_1 \cong T_3 \,(T)$, $T_2 \cong T_3 \,(T)$, but not $T_1 \cong T_2 \,(T)$.

2. Prove that any two equivalent polyhedra can be dissected into tetrahedra which are equivalent in pairs.

3. If P and Q are two equivalent polyhedra, show that there exists a polyhedron R such that $P \cong R \,(T+)$ and $Q \cong R \,(T+)$, and that, moreover, the same dissection of R can serve for both congruences.

4. If P, Q, R are three polyhedra and $P \cong R \,(T-)$ and $Q \cong R \,(T-)$, show that $P \cong Q \,(T-)$.

5. Look up the proof, in a high school text of solid geometry, that two tetrahedra which are congruent (T) are equivalent.

6. Assuming (1) two tetrahedra which are congruent (T) are equivalent, and (2) the volume of a prism is the product of its base and its altitude, prove, by dissection methods, that the volume of a tetrahedron is one-third the product of its base and its altitude.

5.7 CONGRUENCY BY DECOMPOSITION*

An outstanding characteristic of much of twentieth-century mathematics is its pronounced tendency toward generalization. Many concepts and many

* See L. M. Blumenthal, "A paradox, a paradox, a most ingenious paradox," *The American Mathematical Monthly*, vol. 47 (June–July 1940), pp. 346–353. Here also appears a discussion of the part played by the axiom of choice in the paradoxes of congruence theory.

theorems of mathematics have been refined and generalized to a surprising degree. Such geometrical notions as those of curve, surface, space, parallelism, dimension, distance, area and volume, to name only a few, have undergone remarkable generalization and abstraction. From one point of view this is excellent, for it is good to know the extreme limits to which a theorem or a concept can be stretched, and there is an aesthetically pleasing and very practical economy of effort in developing an abstract theory which can cover hosts of specific situations. This tendency toward abstraction and generalization in geometry will become very apparent in the second volume of this work. In this section we illustrate the modern generalizing tendency of mathematics by briefly considering the extension of the idea of congruency by addition so as to embrace not just polygons and polyhedra but perfectly arbitrary sets of points. The section will be largely descriptive since so many of the proofs would require a level of mathematical development and sophistication that we are not assuming in this book. The terminology and notation of elementary set theory will be employed. We commence by formulating a few definitions.

5.7.1 DEFINITIONS AND NOTATION

(a) Two sets of points S and T (on a line, in a plane, or in space) are said to be *congruent*, and we write $S \cong T$, if there exists an isometry which makes one set coincide with the other.

(b) Two sets of points S and T are said to be *congruent by finite decomposition*, and we write $S \cong T$ (fd), if

$$S = \bigcup_{i=1}^{n} S_i, \qquad S_i \cap S_j = \emptyset \text{ for } i \neq j,$$

$$T = \bigcup_{i=1}^{n} T_i, \qquad T_i \cap T_j = \emptyset \text{ for } i \neq j,$$

$$S_i \cong T_i \qquad\qquad (i = 1, \ldots, n).$$

That is, S and T are unions of a finite number n of disjoint subsets which are congruent in pairs.

(c) Two sets of points S and T are said to be *congruent by denumerable decomposition*, and we write $S \cong T$ (dd), if

$$S = \bigcup_{i=1}^{\infty} S_i, \qquad S_i \cap S_j = \emptyset \text{ for } i \neq j,$$

$$T = \bigcup_{i=1}^{\infty} T_i, \qquad T_i \cap T_j = \emptyset \text{ for } i \neq j,$$

$$S_i \cong T_i \qquad\qquad (i = 1, 2, \ldots).$$

5.7.2 *Point sets congruent to a proper part of themselves.* In the sense that *congruency* has been used in connection with polygons and polyhedra, no polygon or polyhedron is congruent to a proper part of itself. In contrast to this, it is easy to exhibit a point set which, according to Definition 5.7.1 (a), *is* congruent to a proper part of itself. Consider, for example, the set S_1 of all points on a given horizontal line m and lying to the right of a selected point A on m (see Figure 5.7a), and the set S_2 of all points on m lying to

Figure 5.7a

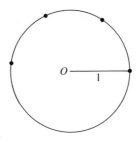

the right of a selected point B on m which is to the right of point A. Clearly S_2 is a proper part of S_1. But $S_1 \cong S_2$, since S_1 can be made to coincide with S_2 by subjecting it to the translation $T(AB)$. It is easy to construct planar and spatial analogues of this example.

But the set S_1 in the above example is an *unbounded* set; that is, there does not exist any sphere completely containing set S_1 in its interior. One might think that surely there is no bounded point set which is congruent to a proper part of itself. It is true that there are no bounded *linear* points sets which are congruent to a proper part of themselves, but there do exist bounded *planar* point sets of this sort. Consider, for example, the set S_1 consisting of the points having polar coordinates $(1, n^r)$, where $n = 0, 1, \ldots$ and n^r denotes n radians. These points (see Figure 5.7b) lie on the unit

Figure 5.7b

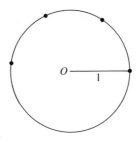

circle with center at the origin O of coordinates and having 1 radian of arc between points corresponding to consecutive values of n. Since π is irrational, the points $(1, i^r)$ and $(1, j^r)$ of S_1 are distinct if $i \neq j$. Clearly set S_1 is a bounded point set. Let S_2 be the point set consisting of all the points of S_1 except the point $(1, 0^r)$. Then S_2 is a proper part of S_1. But $S_1 \cong S_2$, since S_1 can be made to coincide with S_2 by subjecting it to the rotation $R(O, 1^r)$.

5.7.3 *von Neumann's decomposition.* Although a bounded linear set cannot be congruent to a proper part of itself, there is an unexpected property possessed by a line segment (which, of course, is a bounded linear set).

Whether an end point A (or B) of a line segment AB is to be considered as belonging or not belonging to the segment will be indicated by using a bracket or parenthesis, respectively, about the letter A (or B). Using this notation we then have the four related segments $[AB]$, $[AB)$, $(AB]$, (AB). The first segment is called a *closed* segment, the next two are said to be *half-closed* (or *half-open*), and the last segment is called an *open* segment. Now it is easy to decompose a half-closed segment $[AB)$ into a finite number of pairwise disjoint subsets S_1, S_2, ..., S_n which are all congruent to one another (merely divide the segment AB into n equal parts by the points C_1, C_2, ..., C_{n-1} and take $S_1 = [AC_1)$, $S_2 = [C_1 C_2)$, ..., $S_n = [C_{n-1}B)$). J. von Neumann showed, in 1928, that it is also possible to decompose the segment $[AB)$ into a denumerably infinite number of pairwise disjoint subsets S_1, S_2, ... which are all congruent to one another. Indeed, this can be done for a closed, half-closed, or open segment. Of course the subsets S_i are no longer themselves segments.

5.7.4 *Noncongruent segments which are congruent by denumerable decomposition.* Of the four segments $[AB]$, $[AB)$, $(AB]$, (AB), considered as sets of points, it is clear that only the two half-closed ones are congruent to one another. Nevertheless, we can prove that all four segments are congruent to one another by denumerable decomposition. To accomplish this let M_1 be the midpoint of AB, M_2 the midpoint of $M_1 B$, M_3 the midpoint of $M_2 B$, and so on, and let E denote the set of all points of $[AB]$ with the exception of points A, B, M_1, M_2, M_3, ... Then $[AB]$ is the union of the pairwise disjoint sets E, A, B, M_1, M_2, ...; $[AB)$ is the union of the pairwise disjoint sets E, A, M_1, M_2, M_3, ...; $(AB]$ is the union of the pairwise disjoint sets E, B, M_1, M_2, M_3, ...; (AB) is the union of the pairwise disjoint sets E, M_1, M_2, M_3, M_4, ... Since $E \cong E$, $A \cong A$, $B \cong M_1$, $M_1 \cong M_2$, $M_2 \cong M_3$, ..., it follows that $[AB] \cong [AB)$ (*dd*). Similarly we can show that any two of the four segments are congruent to one another by denumerable decomposition.

5.7.5 *The Sierpiński-Mazurkiewicz paradox.* In Article 5.7.2 we saw how certain sets E can be decomposed into two nonempty disjoint subsets E_1 and E_2 such that $E \cong E_1$. Beyond the fact that E_2 is not empty (thus making E_1 a proper part of E), this second part of E was disregarded. The Polish mathematician W. Sierpiński proposed the question: Does there exist a set E which may be decomposed into two nonempty disjoint subsets E_1, E_2 such that $E_1 \cong E_2 \cong E$? The question was answered in the affirmative in 1914 by S. Mazurkiewicz, another Polish mathematician, in the following way:

Let E consist of a point O and all points obtainable from O by a finite number of rotations $R(O, 1^r)$ and translations $T(AB)$, where AB is a fixed unit segment. A point P of E will belong to subset E_1 of E if and only if the final transformation in the sequence of transformations yielding the point P is a rotation $R(O, 1')$; otherwise P will belong to subset E_2 of E. Clearly $E = E_1 \cup E_2$ and E_1 and E_2 are nonempty. It can be shown that also $E_1 \cap E_2 = \emptyset$. But $E \cong E_1$, for E can be made to coincide with E_1 by subjecting it to the rotation $R(O, 1')$. Also $E \cong E_2$, for E can be made to coincide with E_2 by subjecting it to the translation $T(AB)$. It follows that $E_1 \cong E_2 \cong E$.

The affirmative answer to Sierpiński's question is so contrary to the dictates of common sense that the word "paradox" has been attached to it.

5.7.6 *The Hausdorff paradox.* If the Sierpiński-Mazurkiewicz paradox taxes one's credulity, even more so does a discovery made by F. Hausdorff and published by him in 1914 in his famous book on set theory. Hausdorff's paradox, as slightly improved in 1924 by S. Banach and A. Tarski, may be stated as follows:

The surface S *of a sphere may be decomposed into three pairwise disjoint and congruent sets, each of which is congruent to the union of the other two.*

Banach has shown that a similar situation cannot exist for a planar set S.

5.7.7 *The Banach-Tarski paradox.* The most astonishing paradox (in the sense of being a true statement that surely seems false) concerning the notion of congruence was obtained in 1924 by Banach and Tarski. The paradox may be stated as follows:

In any Euclidean space of dimension n > 2, *any two arbitrary sets containing interior points are congruent by finite decomposition.*

Without explaining the meanings of all the technical terms in this statement, we shall merely illustrate by an example. Consider two solid spheres, P and S, where P is the size of a pea and S is the size of the sun. Then, according to the statement, it is possible to decompose the set of points making up P into a finite number of pairwise disjoint subsets such that, by ordinary rigid motions (translations and rotations), these subsets may be made to fill out the entire solid sphere S. That is, in the notation of set theory, we have subsets $P_1, \ldots, P_n$ of P and subsets $S_1, \ldots, S_n$ of S such that

$$P = P_1 \cup P_2 \cup \cdots \cup P_n, \quad P_i \cap P_j = \emptyset \qquad \text{for } i \neq j,$$
$$S = S_1 \cup S_2 \cup \cdots \cup S_n, \quad S_i \cap S_j = \emptyset \qquad \text{for } i \neq j,$$

and P_i is congruent to S_i for each i. Surely such a state of affairs must appear astonishing to even the most sophisticated mathematician.

It is to be noted that the Banach-Tarski paradox holds only in Euclidean spaces of dimension greater than 2. It can be shown that two polygons are congruent by finite decomposition if and only if they have the same area. On the other hand, any two subsets (bounded or not) of *any* Euclidean space are congruent by denumerable decomposition, provided they contain interior points.

A decomposition has been found of a solid sphere into nine pairwise disjoint sets such that the first five of these sets, or the last four of them, can, after appropriate rotations, completely fill out the solid sphere.

PROBLEMS

1. Construct planar and spatial analogues of the linear example illustrated in Figure 5.7a.

2. If i and j are distinct nonnegative integers, show that the points with polar coordinates $(1, i^r)$ and $(1, j^r)$ are distinct.

3. Do you think that $[AB]$ or (AB) can be decomposed into two disjoint congruent subsets?

4. Show that for every positive integer n, a plane can be decomposed into n congruent connected parts. (See Problem E 1515, *The American Mathematical Monthly*, Jan. 1963.)

5. Two sets S and T are said to be *equivalent* if there exists a one-to-one correspondence between the elements of S and the elements of T. Show that the segments $[AB]$, $[AB)$, $(AB]$, (AB), considered as sets of points, are equivalent to one another.

6. In 5.7.5, choose a Cartesian frame of reference with origin at O and positive x axis parallel to AB, and represent a point (x, y) in the plane by the complex number $z = x + iy$.
   (a) Show that $R(O, 1^r)$ carries a point z into the point ze^i, and that $T(AB)$ carries a point z into the point $z + 1$.
   (b) Show that if z is a point obtained from O by a finite number of rotations $R(O, 1^r)$ and translations $T(AB)$, then z can be expressed as a polynomial in e^i having integral coefficients.
   (c) Show that the assumption that $E_1 \cap E_2 \neq \emptyset$ implies that e^i satisfies a polynomial equation with integral coefficients.
   (d) Show that $E_1 \cap E_2 = \emptyset$.

7. Are two tetrahedra of different volumes congruent by finite decomposition?

5.8 A BRIEF BUDGET OF DISSECTION CURIOSITIES

We conclude the chapter with a small collection of planar and spatial dissection curiosities. It is not to be regarded as a proper part of our general textual development, but as a browsing ground for any interested reader. The examples, while illustrating some of the fascination of this odd geometrical art, may appeal to some reader's recreational instincts. A number of unsolved problems are stated.

5.8.1 *Dudeney's dissection of an equilateral triangle into a square.* Henry Ernest Dudeney (1857–1931), unquestionably England's foremost inventor of puzzles, was spectacularly adept in the art of geometrical dissection. His best-known discovery in this field is his four-piece dissection of an equilateral triangle into a square. A five-piece solution of this problem (see the end of Section 5.3) had been known for some time, and it was generally believed that this could not be bettered. Dudeney's discovery was made about 1902, and has since become famous in dissection literature. Figure 5.8a explains

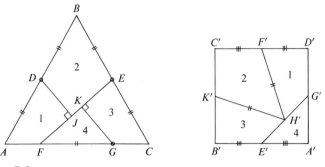

Figure 5.8a

the dissection; a proof of its correctness is left to the reader. The segments *AD, DB, BE, EC, FG* are all equal to half the side of the triangle; *EF* is equal to the side of the equivalent square; *DJ* and *GK* are each perpendicular to *EF*.

If the four pieces 1, 2, 3, 4 are successively hinged to one another at the points *D, E, G*, then, holding piece 1 fixed and swinging the connected set of pieces 4–3–2 counterclockwise, the equilateral triangle is neatly carried into the square. A set of four connected tables has been built based upon this fact; swinging the tables in one direction causes the tops to fit together into a single equilateral triangular table, and swinging them in the other direction causes the tops to fit together into a single square table.

L. V. Lyons has used Dudeney's construction to obtain a dissection of the plane into a mosaic of interlocking equilateral triangles and squares, as pictured in Figure 5.8b.

5.8.2 *Dissection of a regular octagon into a square.* Dudeney bettered several long-established records in polygonal dissection. He was the first, for example, to dissect a regular pentagon into a square with only six pieces (see Problem 26, section 5.3), and to dissect a given square into three equal squares with only six pieces. Very curious along this line is a statement made by Alice Dudeney (Mrs. H. E. Dudeney) in the preface to *Puzzles and Curious Problems*, by H. E. Dudeney, published in 1932, after the author had died. The book was reprinted in 1936, 1941, and 1948, revised by James Travers. The statement is: "It is remarkable that the regular octagon can be cut into as few as four pieces to form a corresponding square." The implication is that such a four-piece dissection was another of Dudeney's triumphs. But no one has since been able to find such a solution. A beautiful five-piece

Figure 5.8b

dissection, illustrated in Figure 5.8c, was published by Travers in 1933. It is strange that Travers, who revised the book in question, made no correction of or comment on Alice Dudeney's unverifiable statement. It is difficult to believe that a four-piece solution of the problem is possible.

Figure 5.8c

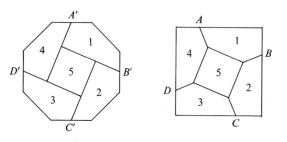

5.8.3 *The minimal dissection problem.* The minimal dissection problem is that of determining the least number of pieces needed to dissect a given polygon into another given equivalent polygon. Only very small inroads have been made into this problem. W. B. Carver and Alfred Tarski showed that if n is the minimum number of pieces needed to dissect a rectangle of dimensions a and b, $a \geq b$, into a square, then

$$n \le E(\sqrt{x^2 - 1}) + 2,$$

where $x = (a/b)^{1/2}$, and $E(y)$ means the least integer not less than y. This result was improved upon when E. E. Moise showed that $n \le E(x) + 1$ and Tarski, in a later paper, showed that $n \ge E(x)$. It follows, then, that $n = E(x)$ or $E(x) + 1$. In particular, the reader might like to try to answer the following two questions:

(a) Is there a three-piece dissection of a 2×15 rectangle into a square? (Here $E(x) = 3$.)

(b) Is there a five-piece dissection of a 3×64 rectangle into a square? (Here $E(x) = 5$.)

5.8.4 *"Squaring" the square.* In 1925, Z. Morón noted (see Figure 5.8d)

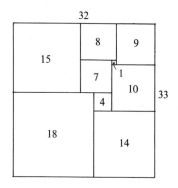

Figure 5.8d

that a 32×33 rectangle can be dissected into nine squares, no two of which are equal. This raised the question of whether a square can be dissected into a finite number of squares, no two of which are equal. It was felt that this latter problem was impossible, but such turned out not to be the case. The first published example of a square dissected into unequal squares appeared in 1939; the dissection was given by R. Sprague of Berlin, and contained 55 subsquares. In 1940, R. C. Brooks, C. A. B. Smith, A. H. Stone, and W. T. Tutte, in a joint paper, published a dissection containing only 26 pieces. These men ingeniously established a connection between the problem of dissection and certain properties of currents in electrical networks. In 1948, T. H. Willcocks published the 24-piece dissection of a square into unequal squares pictured in Figure 5.8e, and this dissection is today the record so far as least number of pieces is concerned.

5.8.5 *Perfect dissections.* A dissection is said to be *perfect* if all the pieces are similar but unequal; the number of pieces is called the *order* of the dissection. A dissection is called *finite* or *infinite* according as the order of the dissection is finite or infinite. The problem of "squaring" the square is that of finding a finite perfect dissection of a square into squares. A great

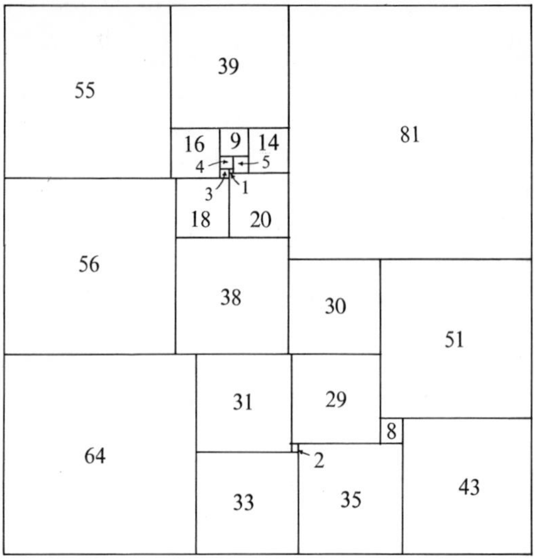

Figure 5.8e

deal of work has been done in recent years in connection with perfect dissections. We here list some of the results found, and conclude with a couple of questions that still remain unanswered.

(a) If a rectangle can be finitely dissected perfectly into squares, then the sides of the rectangle are commensurable.

(b) If the sides of a rectangle are commensurable, then the rectangle can in an infinity of ways be finitely dissected perfectly into squares.

(c) There is an infinite perfect dissection of a rectangle into squares.

(d) There is no perfect dissection of a rectangle into squares of order less than 9, and exactly two of order 9. (Proved by H. Reichardt and H. Toepken in 1940. One of the two possible dissections is pictured in Figure 5.8d.)

(e) There is no finite perfect dissection of an equilateral triangle into equilateral triangles. (Proved by W. T. Tutte in 1948.)

(f) There is no finite perfect dissection of a rectangular parallelepiped into cubes.

There is a proof of this that is so pretty and simple that we sketch it here. Suppose there is a finite perfect dissection of a rectangular parallelepiped P into cubes. Then the bottom face of P is a "squared" rectangle. Within this "squared" rectangle there is a smallest square, which clearly cannot be along an edge of the rectangle. This means that the smallest cube A resting on the bottom base of P is surrounded by larger cubes. On top of cube A, smaller cubes must rest, forming a "squared" square on the top face of A. Within this "squared" square there is a smallest square, giving rise to a cube B that is the smallest cube resting on the top of cube A. Continuing the argument, there must be a smallest cube C resting on top of cube B, and so on ad infinitum. But this is impossible.

A perfect dissection of a rectangle into squares is said to be *simple* if the dissection does not contain within it a perfect dissection of a smaller rectangle. Note that the perfect dissection illustrated in Figure 5.8d is simple, whereas that illustrated in Figure 5.8e is not simple.

(g) The least known order to-date of a simple perfect dissection of a square into squares is 37. (Found by T. H. Willcocks in 1959.)

(h) There is a simple perfect dissection (of order 69) of a square into squares in which no four subsquares share a common vertex and no square other than the four corner squares is bisected by a diagonal of the complete figure. (Found by W. T. Tutte.)

(i) What is the smallest possible order for a perfect dissection of a square into squares?

(j) Does there exist a simple perfect dissection of a 1 × 2 rectangle into squares?

5.8.6 *Dissecting a checkerboard.* The problem of enumerating the number of ways in which an $n \times n$ checkerboard can be cut, with cuts made only along edges of squares of the board, into two congruent connected parts, seems not to be easy. A solution is known for a 6 × 6 board.

5.8.7 *Dissection into dominoes.* We call a 1 × 2 rectangle a *domino*. One can ask a number of interesting questions, such as:

(a) In how many ways can a 2 × *n* rectangle be dissected into dominoes? (See Problem E 1470, *The American Mathematical Monthly*, Jan. 1962.)

(b) In how many ways can 2*k* squares be cut from an 8 × 8 checkerboard so that the remaining part can be dissected into dominoes?

5.8.8 *Regular tessellations.* A *planar regular tessellation* is a dissection of the Euclidean plane into regular polygons. If the polygons are all congruent, then they must have 3, 4, or 6 sides. More interesting is the situation where two or more sizes of the same kind of polygon are admitted, or where two or more different kinds of polygons are admitted. The literature on regular tessellations is very large, and the subject becomes particularly interesting in the Lobachevskian non-Euclidean plane, where one can form tessellations of congruent regular *n*-gons with *p* of them about each vertex for any *p* such that $1/n + 1/p < 1/2$. Solid regular tessellations have also been extensively discussed.

5.8.9 *The translation restriction.* H. Lindgren, of Australia, has considered dissections wherein one polygon is cut into pieces which can, by translations alone, be rearranged to form a second polygon. Such a dissection may be called a *translation dissection*. Examples of Lindgren's achievements along this line are:

(a) A 4-piece translation dissection of a quadrilateral into any equivalent quadrilateral having the same angles.

(b) A 10-piece translation dissection of a regular nine-sided polygon into an equilateral triangle.

5.8.10 *Parallelogramic dissection of a parpolygon.* A *parpolygon* is a polygon of an even number of sides in which opposite sides are equal and parallel. It is a nice geometric application of mathematical induction to prove that: Every parpolygon of $2n$ sides can be dissected into $n(n-1)/2$ parallelograms.

The theorem is certainly true for $n = 2$. Suppose it is true for $n = k$, and let P_{k+1} be any parpolygon of $2(k+1)$ sides. Then P_{k+1} can be dissected (as indicated in Figure 5.8f) into k parallelograms and a parpolygon P_k of

Figure 5.8f

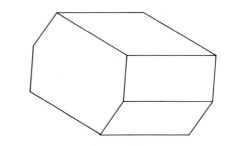

$2k$ sides. By our supposition, P_k can be dissected into $k(k-1)/2$ parallelograms. It follows that P_{k+1} can be dissected into

$$k + k(k-1)/2 = (k+1)k/2$$

parallelograms. The theorem now follows by mathematical induction.

It is interesting to note that if the parpolygon is equilateral, then the parallelograms become rhombuses of the same length side. A regular $2n$-gon is an example of an equilateral parpolygon.

5.8.11 *The Fibonacci numbers and a dissection puzzle.* There is a familiar geometrical paradox, pictured in Figure 5.8g, where a square, subdivided like a checkerboard into 64 small unit squares, is cut into four pieces which can apparently be reassembled to form a rectangle containing 65 unit

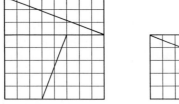

Figure 5.8g

squares. The explanation of where the additional unit square comes from lies in the fact that the edges of the four pieces which seem to lie along a diagonal of the rectangle really do not do so, but form a very flat parallelogram of exactly one square unit of area. A number of mathematicians, among them Lewis Carroll of *Alice in Wonderland* fame, have considered the problem of generalizing this paradox. A neat generalization is furnished by the so-called Fibonacci numbers, defined by $f_1 = f_2 = 1, f_{i+1} = f_{i-1} + f_i$ ($i = 2, 3, \ldots$). It can be shown that

$$f_{n+1} f_{n-1} = f_n^2 + (-1)^n,$$

whence a square of side f_n can be dissected, as in Figure 5.8h, into four

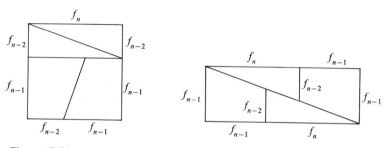

Figure 5.8h

pieces which can be reassembled to *almost* form an $f_{n-1} \times f_{n+1}$ rectangle. If we take $n = 6$, we obtain the original dissection puzzle.

5.8.12 *The law of cosines.* Figure 5.8i indicates a simple proof, utilizing

Figure 5.8i

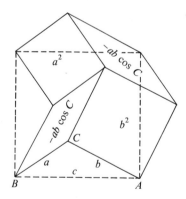

dissection notions, of the law of cosines:

$$c^2 = a^2 + b^2 - 2ab \cos C,$$

where a, b, c are the sides of any triangle and C is the angle opposite side c.

5.8.13 *An equilateral triangle problem.* Some geometrical problems can be neatly solved with dissection methods. This is illustrated here, and in the next two items. Consider the problem: Find the area of an equilateral triangle if a point P within the triangle is at distances 3, 4, 5 from the three vertices.

Referring to Figure 5.8j, cut the triangle into three pieces along the solid

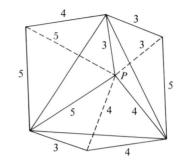

Figure 5.8j

lines from P and then arrange these pieces around the triangle as indicated to form a hexagon of twice the area of the triangle. Cut this hexagon along the solid and dashed lines from P to form three equilateral triangles and three right triangles, all of known sides. The desired area is now readily computed to be

$$[\sqrt{3}(3^2 + 4^2 + 5^2)/4 + 18]/2.$$

5.8.14 *An area problem.* If A', B', C' (see Figure $5.8k_1$) are the first

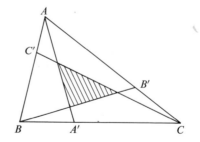

Figure $5.8k_1$

trisection points of the sides BC, CA, AB of a given triangle ABC, show that the cevians AA', BB', CC' are the side lines of a triangle whose area is one-seventh the area of the given triangle.

A very simple dissection proof is indicated in Figure $5.8k_2$.

5.8.15 *Another area problem.* The following problem appeared on the

Figure 5.8k $_2$

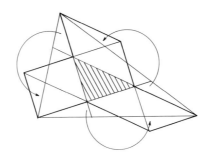

1958 Kürschák Prize Competition, an annual high school mathematics contest held in Hungary: If in the convex hexagon *ABCDEF* every pair of opposite sides are parallel, prove that triangles *ACE* and *BDF* have equal areas.

Dissect the hexagon in two ways as indicated by the solid and the dashed lines of Figure 5.8l. In each dissection we see that the area of the hexagon

Figure 5.8l

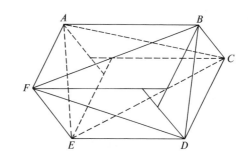

is twice the area of one of the concerned triangles decreased by the area of a little triangle whose sides are equal to the differences of the pairs of opposite sides of the hexagon. The desired result is now evident.

5.8.16 *Cutting the cheese.* Two interesting related questions are: (1) Into how many regions can *n* straight lines divide a plane? (2) Into how many regions can *n* planes divide space? We shall prove that the answers to the two questions are $(n^2 + n + 2)/2$ and $(n^3 + 5n + 6)/6$ respectively.

THEOREM. n *straight lines can divide a plane into at most* $(n^2 + n + 2)/2$ *regions.*

The theorem is clearly true for $n = 1$. Suppose it true for $n = k$, and suppose we have k lines so drawn that the plane is divided into $(k^2 + k + 2)/2$ regions. Draw a $(k + 1)$st line not parallel to any of the k given lines and not passing through a point of intersection of any pair of the k given lines. Then this $(k + 1)$st line will cut into $k + 1$ regions, adding this many regions to the number we already have, and this is the best we can do. We now have

$$(k^2 + k + 2)/2 + (k + 1) = (k^2 + 3k + 4)/2$$
$$= [(k + 1)^2 + (k + 1) + 2]/2$$

regions, and the theorem follows by mathematical induction.

THEOREM. n *planes can divide space into at most* $(n^3 + 5n + 6)/6$ *regions.*

The theorem is clearly true for $n = 1$. Suppose it true for $n = k$, and suppose we have k planes so drawn that space is divided into $(k^3 + 5k + 6)/6$ regions. Draw a $(k + 1)$st plane not parallel to any of the k given planes, not passing through the line of intersection of any two of the k given planes, and not passing through a point of intersection of any three of the k given planes. Then this $(k + 1)$st plane will be intersected by the k given planes in k straight lines dividing the $(k + 1)$st plane into at most $(k^2 + k + 2)/2$ areas. For each of these areas, the $(k + 1)$st plane divides a region of space already formed by the k given planes into two parts, and this increases the total number of regions by at most $(k^2 + k + 2)/2$. But

$$(k^3 + 5k + 6)/6 + (k^2 + k + 2)/2 = [(k + 1)^3 + 5(k + 1) + 6]/6,$$

and the theorem follows by mathematical induction.

It is interesting to note that:

1. n points can divide a line into $1 + n$ parts,

2. n lines can divide a plane into $1 + n + \binom{n}{2}$ parts,

3. n planes can divide space into $1 + n + \binom{n}{2} + \binom{n}{3}$ parts.

It has been shown that n $(m - 1)$-spaces can divide an m-space into

$$1 + n + \binom{n}{2} + \binom{n}{3} + \cdots + \binom{n}{m}$$

m-dimensional regions.

5.8.17 *Duplicating the cube.*' In Oct. 1932, W. F. Cheney, Jr., proposed the problem of dissecting a $1 \times 1 \times 2$ wooden block into as few polyhedral pieces as possible which can be reassembled into a cube. In Feb. 1933, W. R. Ransom offered an eight-piece solution. In Oct. 1935, a seven-piece solution by A. H. Wheeler was published. All these references can be found in the appropriate issues of *The American Mathematical Monthly* in connection with Problem E 4. The seven-piece dissection of Wheeler makes a splendid three-dimensional puzzle. Given the seven pieces, one is hard pressed to assemble them into either the cube or the $1 \times 1 \times 2$ block.

5.8.18 *Cutting a cube into cubes.* It is easy to cut a cube into 8 subcubes (by planes parallel to pairs of opposite faces and midway between them), and it is quite apparent that it is impossible to cut a cube into only 2 sub-cubes. For a given positive integer k, then, it may or may not be possible to cut a cube into k subcubes. William Scott, in 1946, showed that a cube can be cut into any $k > 54$ subcubes. It was also shown that a cube can be cut into k subcubes where k is

$$1, 8, 15, 20, 22, 27, 29, 34, 36, 38, 39,$$
$$41, 43, 45, 46, 48, 49, 50, 51, 52, 53.$$

It was natural, then, to wonder if $k = 54$ is the largest number of subcubes into which a cube cannot be cut. This question was still unanswered when the fifth printing of the first edition of the present book appeared in 1968. Reading this, and challenged by the question, Von Christian Thiel, of Germany, attacked the problem and in 1969 managed to show that a cube can be cut into 54 subcubes. The question now is: Is $k = 47$ the largest number of subcubes into which a given cube cannot be cut? For a consideration of the extension of this problem to n-dimensional space, see Problem E 724, *The American Mathematical Monthly*, Jan. 1947.

5.8.19 *A volume dissection.* Leo Moser has shown that if each face of a polyhedron has central symmetry, then the polyhedron can be dissected into subpolyhedra which can be reassembled to form a cube. His proof may be found in Problem E 860, *The American Mathematical Monthly*, Dec. 1949.

5.8.20 *Passing a cube through a cube of the same size.* A well-known and entertaining problem is that of cutting a hole through a given solid cube (the hole being completely surrounded by material of the cube) through which a second cube of the same size as the given one can be passed. For a solution to the problem see Problem E 888, *The American Mathematical Monthly*, May 1950.

5.8.21 *Ham sandwich problems.* One of the most surprising dissection theorems states that if A, B, C are any three volumes in space, there exists a plane which bisects all three volumes simultaneously. The theorem becomes very striking when phrased in connection with a ham sandwich as follows: It is possible by a planar cut to divide a ham sandwich into two parts so that the parts contain equal amounts of bread, of butter, and of ham. The two-dimensional version of the theorem is also true, namely, if A and B are any two areas in the plane, there exists a straight line in the plane which bisects A and B simultaneously. An allied theorem is: Given a single area in the plane, there exists a pair of perpendicular lines in the plane which divide the area into four equivalent parts. Proofs of these theorems require a knowledge of limits and continuity, and will be taken up in a later chapter.

PROBLEMS

1. Establish the correctness of Dudeney's four-piece dissection of an equilateral triangle into a square.

2. Using the Dudeney dissection, describe how a square can be cut into four pieces by three straight cuts of equal length so that the pieces can be reassembled to form an equilateral triangle.

3. Describe the Travers five-piece dissection of a regular octagon into a square.

4. Show how any of the three dissections of a regular octagon, given by H. Lindgren and pictured in Figure 5.8m, lead to a four-piece dissection of a regular octagon into a rectangle.

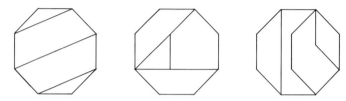

Figure 5.8m

5. (a) Show that $E(y) = -[-y]$, where $[z]$ is the "integral part of z."
   (b) Show that for a 2×15 rectangle $E(x) = 3$, and for a 3×64 rectangle $E(x) = 5$.

6. Show that if a rectangle can be finitely dissected perfectly into squares in one way, then it can be finitely dissected perfectly into squares in an infinity of ways.

7. Assemble nine squares of sides 2, 5, 7, 9, 16, 25, 28, 33, 36 into a 61×69 rectangle.

8. Show that if two diagonally opposite corner squares of an 8×8 checkerboard are removed, then the remaining part of the board cannot be dissected into dominoes.

9. A very interesting problem of tessellations is to fill the plane with congruent regular polygons.
   (a) If we do not permit a vertex of one polygon to lie on a side of another, show that the number of polygons at each vertex is given by $2 + 4/(n - 2)$, and hence that we must have $n = 3$, 4, or 6. Construct illustrative tessellations.
   (b) If we insist that a vertex of one polygon lie on a side of another, show that the number of polygons clustered at such a vertex is given by $1 + 2/(n - 2)$, whence we must have $n = 3$ or 4. Construct illustrative tessellations.
   (c) Construct tessellations containing (1) two sizes of equilateral triangles, the larger having a side twice that of the smaller, (2) two sizes of squares, the larger having a side twice that of the smaller, (3) congruent equilateral triangles and congruent regular dodecagons, (4) congruent equilateral triangles and congruent regular hexagons, (5) congruent squares and congruent regular octagons.
   (d) Suppose we have a tessellation composed of regular polygons of three different kinds at each vertex. If the three kinds of polygons have p, q, r sides, respectively, show that

$$1/p + 1/q + 1/r = 1/2.$$

One integral solution of this equation is $p = 4$, $q = 6$, $r = 12$. Construct a tessellation of the type under consideration and composed of congruent squares, congruent regular hexagons, and congruent regular dodecagons.

10. Dissect a regular dodecagon into 15 rhombuses having the same length side.

11. If f_n is the nth term of the Fibonacci sequence $1, 1, 2, 3, \ldots, x, y, x + y, \ldots$, show that $f_{n+1}f_{n-1} = f_n{}^2 + (-1)^n$.

12. Draw Figure 5.8i for the case where (a) angle C is acute, (b) angle C is a right angle.

13. Generalize Problem 5.8.13 to the situation where P is at distances a, b, c from the vertices of the equilateral triangle, where $a^2 + b^2 = c^2$.

14. (a) Dissect an obtuse triangle into the least number of acute triangles.
    (b) Dissect an obtuse triangle into the least number of isosceles acute triangles.

15. Divide a circular disc into 22 pieces by 6 straight cuts.

16. (a) Show that if a cube can be cut into k subcubes, then it can be cut into $k + 7q$ subcubes, where q is any nonnegative integer.
    (b) Assuming that a cube can be cut into 1, 20, 38, 39, 49, 51, 61 subcubes, show that a cube can be cut into any number $k > 54$ subcubes.

BIBLIOGRAPHY

BALL, W. W. ROUSE, *Mathematical Recreations and Essays*, revised by H. S. M. Coxeter. New York: The Macmillan Company, 1939.

BOLTYANSKII, *Equivalent and Equidecomposable Figures*, tr. by A. K. Henn and C. E. Watts. Boston: D. C. Heath and Company, 1963.

DUDENEY, H. E., *Amusements in Mathematics*. New York: Dover Publications, Inc., 1958.

———, *The Canterbury Puzzles*. New York: Dover Publications, Inc., 1958.

FORDER, H. G., *The Foundations of Euclidean Geometry*. New York: Cambridge University Press, 1927.

GARDNER, MARTIN, *The Scientific American Book of Mathematical Puzzles and Diversions*. New York: Simon and Schuster, 1959.

———, *The 2nd Scientific American Book of Mathematical Puzzles and Diversions*. New York: Simon and Schuster, 1961.

HILBERT, DAVID, *Foundations of Geometry*, 10th ed., tr. by Leo Unger. La Salle, Ill.: The Open Court Publishing Company, 1971.

KRAITCHIK, MAURICE, *Mathematical Recreations*. New York: W. W. Norton and Company, Inc., 1942.

LINDGREEN, HARRY, *Geometric Dissections*. Princeton, N.J.: D. Van Nostrand Company, Inc., 1964.

MESCHKOWSKI, HERBERT, *Unsolved and Unsolvable Problems in Geometry*, tr. by J. A. C. Burlak. Edinburgh: Oliver and Boyd, 1966.

STEINHAUS, H., *Mathematical Snapshots*. New York: Oxford University Press, 1950.

YATES, R. C., *Geometrical Tools, a Mathematical Sketch and Model Book*. Saint Louis: Educational Publishers, Inc., 1949.

6

Projective Geometry

Consider, for a moment, the problem an artist faces in attempting to paint a true picture of some object. As the artist looks at the object, rays of light from the object enter his eye. If a transparent screen should be placed between the artist's eye and the object, these rays of light would intersect the screen in a collection of points. It is this collection of points, which may be called the *image*, or *projection*, of the object on the screen, that the artist must draw on his paper or canvas if a viewer of the picture is to receive the same impression of the form of the object as he would receive were he to view the object itself. Since the artist's paper or canvas is not a transparent screen, the task of accurately drawing the desired projection presents a real problem to the artist. In an effort to produce more realistic pictures, many of the Renaissance artists and architects became deeply interested in discovering the formal laws controlling the construction of the projections of objects on a screen, and, in the fifteenth century, a number of these men created the elements of an underlying geometrical theory of perspective.

The theory of perspective was considerably extended in the early seventeenth century by a small group of French mathematicians, the motivator of whom was Gérard Desargues, an engineer and architect who was born in Lyons in 1593 and who died in the same city about 1662. Influenced by the growing needs of artists and architects for a deeper theory of perspective,

240

Desargues published, in Paris in 1639, a remarkably original treatise on the conic sections which exploited the idea of projection. But this work was so neglected by most other mathematicians of the time that it was soon forgotten and all copies of the publication disappeared. Two centuries later, when the French geometer Michel Chasles (1793–1880) wrote a history of geometry, there was no way to estimate the value of Desargues' work. Six years later, however, in 1845, Chasles happened upon a manuscript copy of Desargues' treatise, made by one of Desargues' few followers, and since that time the work has been recognized as one of the classics in the early development of projective geometry.

There are several reasons for the initial neglect of Desargues' little volume. It was overshadowed by the more supple analytic geometry introduced by Descartes two years earlier. Geometers were generally either developing this new powerful tool or trying to apply infinitesimals to geometry. Also, Desargues unfortunately adopted a style and a terminology that were so eccentric that they beclouded his work and discouraged others from attempting properly to evaluate his accomplishments.

The reintroduction of projective considerations into geometry did not occur until the late eighteenth century, when the great French geometer Gaspard Monge (1746–1818) created his descriptive geometry. This subject, which concerns a way of representing and analyzing three dimensional objects by means of their projections on certain planes, had its origin in the design of fortifications. Monge was a very inspiring teacher, and there gathered about him a group of brilliant students of geometry, among whom were L. N. Carnot (1753–1823), Charles J. Brianchon (1785–1864), and Jean Victor Poncelet (1788–1867).

The real revival of projective geometry was launched by Poncelet. As a Russian prisoner of war, taken during Napoleon's retreat from Moscow, and with no books at hand, Poncelet planned his great work on projective geometry, which, after his release and return to France, he published in Paris in 1822.* This work gave tremendous impetus to the study of the subject and inaugurated the so-called "great period" in the history of projective geometry. There followed into the field a host of mathematicians, among whom were Gergonne, Brianchon, Chasles, Plücker, Steiner, Staudt, Reye, and Cremona—great names in the history of geometry, and in the history of projective geometry in particular.

The work of Desargues and of Poncelet, and their followers, led geometers to classify geometric properties into two categories, the *metric properties*, in which the measure of distances and of angles intervenes, and the *descriptive properties*, in which only the relative positional connection of the geometric elements with respect to one another is concerned. The Pythagorean Theorem, that *the square on the hypotenuse of a right triangle is equal to the sum of the squares on the two legs*, is a metric property. As an example of a

---

* *Traité des propriétés projectives des figures.*

descriptive property we might mention the remarkable "mystic hexagram" theorem of Blaise Pascal, which we have earlier considered for the case of a circle and which was inspired by the work of Desargues: *If a hexagon be inscribed in a conic, then the points of intersection of the three pairs of opposite sides are collinear, and, conversely, if the points of intersection of the three pairs of opposite sides of a hexagon are collinear, then the hexagon is inscribed in a conic.*

The distinction between the two types of geometric properties, at least in the case of plane figures, becomes clearer when viewed from the fact that the descriptive properties are unaltered when the figure is subjected to a projection, whereas the metric properties may no longer hold when the figure is projected. Thus, under a projection from one plane to another, a right triangle does not necessarily remain a right triangle, and so the Pythagorean relation does not necessarily hold for the projected figure; the Pythagorean Theorem is a metric theorem. In the case of Pascal's Theorem, however, a hexagon inscribed in a conic projects into a hexagon inscribed in a conic and collinear points project into collinear points, and hence the theorem is preserved; Pascal's Theorem is a descriptive theorem.

Many descriptive properties present themselves in the seeming form of metric properties. For example we have seen (in Theorem 2.5.7) that the cross ratio (AB,CD) of four points A, B, C, D on a straight line is unaltered when the line containing the four points is projected into another line and the four points A', B', C', D'. In other words, though the lengths of the various corresponding segments on the two lines are not necessarily equal to one another, nevertheless the two compound ratios

$$(\overline{A'C'}/\overline{C'B})/(\overline{A'D'}/\overline{D'B'}) \quad \text{and} \quad (\overline{AC}/\overline{CB})/(\overline{AD}/\overline{DB}),$$

that is, the two cross ratios (AB,CD) and $(A'B',C'D')$, are equal in value, and thus the cross ratio of four collinear points is a descriptive property of those points.

The study of the descriptive properties of geometric figures is known as *projective geometry.*

Projective geometry has grown into a vast and singularly beautifully developed branch of geometry, and has become basic for many geometrical studies. Some of its more elementary aspects will be examined in this chapter; some of its deeper aspects will appear in subsequent chapters. A sharper definition of the subject will be given later.

6.1 PERSPECTIVITIES AND PROJECTIVITIES

Let π and π' (see Figure 6.1a) be two given fixed nonideal planes of *extended* space, and let V be a given fixed point not lying on either π or π'. Since the space is extended, π and π' are extended planes, and point V may be either an ordinary or an ideal point. Let P be any point, ordinary or ideal, of

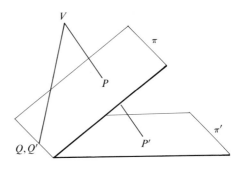

Figure 6.1a

plane π. Then line VP will intersect plane π' in a unique ordinary or ideal point P' of π'. In this way the extended plane π is mapped onto the extended plane π'. Indeed, since distinct points of π have distinct images in π', the mapping is actually a transformation of the set of all points of the extended plane π onto the set of all points of the extended plane π'. The points on the line of intersection of planes π and π' are invariant points of the transformation.

6.1.1 DEFINITIONS. A transformation such as described above is called a *perspectivity*, or a *perspective transformation,* and the point V is called the *center* of the perspectivity. If V is an ordinary point of space, the perspectivity is called a *central perspectivity*; if V is an ideal point of space, the perspectivity is called a *parallel perspectivity*. The line of intersection of π and π' is called the *axis of perspectivity*. The line in π (π') which maps into the line at infinity in π' (π) is called the *vanishing line* of π (π'). The point in which a line of π (π') meets the vanishing line of π (π') is called the *vanishing point* of the line.

A perspectivity or a product of two or more perspectivities is called a *projectivity*, or a *projective transformation.* For example, a perspectivity of center V of plane π onto plane π', followed by a perspectivity of center W of plane π' onto plane π'', is a projectivity of plane π onto plane π''.

It is clear that the two planes π and π' of a perspectivity must be taken as *extended* planes, since otherwise the correspondence between the points of the two planes might not be one-to-one. It is for this reason that an extended plane is often called a *projective plane*.

It is easily seen that a projectivity carries a straight line into a straight line. Very useful is the following special situation.

6.1.2 THEOREM. *If π is a given plane, l a given line in π, and V a given point not on π, then there exists a plane π' such that the perspectivity of center V carries line l of π into the line at infinity of π'.*

Choose for π' (see Figure 6.1b) any plane parallel to (but not coincident

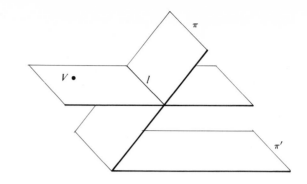

Figure 6.1b

with) the plane determined by V and l. Then it is clear that the line joining V to any point P on l will be parallel to plane π', that is, will meet π' at infinity.

6.1.3 DEFINITION. The operation of selecting a suitable center of perspectivity V and a plane π' so that a given line l of a given plane π shall be mapped into the line at infinity of π' is called *projecting the given line to infinity*.

The operation of projecting a given line to infinity can often greatly simplify the proof of a theorem. We give some examples; the reader should note the application of the *transform-solve-invert* procedure. We first establish one of the harmonic properties of a complete quadrangle.

6.1.4 THEOREM. *Let* PQRS *(see Figure 6.1c$_1$) be a complete quadrangle and let* PQ *and* SR *intersect in* A, PR *and* SQ *intersect in* B, PS *and* QR *intersect in* C, AB *and* PS *intersect in* D. *Then* (PS,DC) $= -1$.

Project line AC to infinity. Then $P'Q'R'S'$ (see Figure 6.1c$_2$) is a parallelogram and D' is the midpoint of $P'S'$. Since C' is at infinity, we have $(P'S',D'C') = -1$. The theorem now follows.

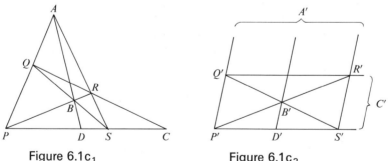

Figure 6.1c$_1$ Figure 6.1c$_2$

6.1.5 THEOREM. *If the three lines* UP_1P_2, UQ_1Q_2, UR_1R_2 *(see Figure* $6.1d_1$) *intersect the two lines* OX_1 *and* OX_2 *in* P_1, Q_1, R_1 *and* P_2, Q_2, R_2 *respectively, then the points of intersection of* Q_1R_2 *and* Q_2R_1, R_1P_2 *and* R_2P_1, P_1Q_2 *and* P_2Q_1 *are collinear on a line that passes through point* O.

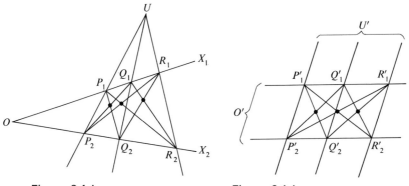

Figure $6.1d_1$ Figure $6.1d_2$

Project line OU to infinity. The projected figure appears as in Figure $6.1d_2$, where $P_1'P_2'$, $Q_1'Q_2'$, $R_1'R_2'$ are all parallel, and $P_1'Q_1'R_1'$ and $P_2'Q_2'R_2'$ are parallel. It is clear that the points of intersection of $Q_1'R_2'$ and $Q_2'R_1'$, $R_1'P_2'$ and $R_2'P_1'$, $P_1'Q_2'$ and $P_2'Q_1'$ are collinear on a line parallel to lines $P_1'Q_1'R_1'$ and $P_2'Q_2'R_2'$ (they lie on the line midway between lines $P_1'Q_1'R_1'$ and $P_2'Q_2'R_2'$). It follows that the corresponding points in the original figure are collinear on a line passing through point O.

6.1.6 DESARGUES' TWO-TRIANGLE THEOREM. *Copolar triangles in a plane are coaxial, and conversely.*

Let the two triangles (see Figure 6.1e) be $A_1B_1C_1$ and $A_2B_2C_2$, and let B_1C_1 and B_2C_2 intersect in L, C_1A_1 and C_2A_2 intersect in M, A_1B_1 and A_2B_2 intersect in N.

Suppose A_1A_2, B_1B_2, C_1C_2 are concurrent in a point O. Project line MN to infinity. Then $A_1'B_1'$ and $A_2'B_2'$ are parallel, and $A_1'C_1'$ and $A_2'C_2'$ are parallel. It follows that $O'B_1'/O'B_2' = O'A_1'/O'A_2' = O'C_1'/O'C_2'$, whence $B_1'C_1'$ and $B_2'C_2'$ are also parallel. That is, the intersections of corresponding sides of triangles $A_1'B_1'C_1'$ and $A_2'B_2'C_2'$ are collinear (on the line at infinity). It follows that the intersections of corresponding sides of triangles $A_1B_1C_1$ and $A_2B_2C_2$ are collinear. That is, copolar triangles in a plane are coaxial.

Now suppose L, M, N are collinear. Project line LMN to infinity. Then the two triangles $A_1'B_1'C_1'$ and $A_2'B_2'C_2'$ have their corresponding sides parallel, and hence are homothetic to one another, whence $A_1'A_2'$, $B_1'B_2'$, $C_1'C_2'$ are concurrent in a point O'. It follows that A_1A_2, B_1B_2, C_1C_2 are concurrent in a point O. That is, coaxial triangles in a plane are copolar.

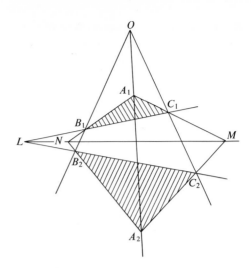

Figure 6.1e

6.1.7 THEOREM. *There is a perspectivity that carries a given triangle* ABC *and a given point* G *in its plane (but not on a side line of the triangle) into a triangle* A′B′C′ *and its centroid* G′.

Let *AG, BG, CG* (see Figure 6.1f) intersect the opposite sides *BC, CA, AB* in points *D, E, F*. Then triangles *DEF* and *ABC* are copolar, and hence also coaxial. That is, the points *L, M, N* of intersection of *EF* and *BC, FD* and *CA, DE* and *AB* are collinear. By Theorem 6.1.4, $(BC,DL) = -1$. Project line *LMN* to infinity. Then *D′* is the midpoint of *B′C′*. Similarly *E′* and *F′* are the midpoints of *C′A′* and *A′B′* respectively. It follows that *G′* is the centroid of triangle *A′B′C′*.

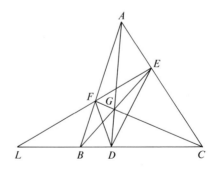

Figure 6.1f

PROBLEMS

1. (a) Prove that a projectivity carries a straight line into a straight line.
   (b) Prove that under a perspectivity a straight line and its image intersect each other on the axis of perspectivity.

2. (a) Prove that under a perspectivity the angle between the images m' and n' of any two lines m and n is equal to the angle which the vanishing points of m and n subtend at the center V of perspectivity.

(b) Prove that under a perspectivity all angles whose sides have the same vanishing points map into equal angles.

3. (a) Given a plane π containing a line l and two angles ABC and DEF, where A, C, D, F are on l in the order A, D, C, F. Show that there exists a perspectivity which projects l to infinity and angles ABC and DEF into angles of given sizes α and β respectively.

(b) Must the segments AC and DF in part (a) separate each other?

4. Show that there exists a perspectivity which projects a given quadrilateral $ABCD$ into a square.

5. Prove that the cross ratio of a pencil of four distinct coplanar lines is preserved by projection.

6. Show that in a perspectivity carrying a plane π into a nonparallel plane π' there are in each plane exactly two points such that every angle at either of them projects into an equal angle. (These points are called the *isocenters* of the perspectivity.)

7. Show that in a perspectivity carrying a plane π into a nonparallel plane π' there is in each plane, besides the axis of perspectivity, a line whose segments are projected into equal segments. (These lines are called the *isolines* of the perspectivity.)

6.2 FURTHER APPLICATIONS

If AD, BE, CF are concurrent cevian lines for a triangle ABC, Ceva's Theorem states that

$$\frac{(\overline{BD})(\overline{CE})(\overline{AF})}{(\overline{DC})(\overline{EA})(\overline{FB})} = +1.$$

The left member of this equation is a ratio of a product of some directed segments to a product of some other directed segments, and this ratio of products of directed segments has the two following interesting properties:

(1) If we replace $\overline{BD}$ by $B \times D$, and similarly treat all other segments appearing, and then regard the resulting expression as an algebraic one in the letters B, D, etc., these letters can all be cancelled out.

(2) If we replace $\overline{BD}$ by a letter, say a, representing the line on which the segment is found, and similarly treat all other segments appearing, and then regard the resulting expression as an algebraic one in the letters a, etc., these letters can all be cancelled out.

6.2.1 DEFINITION. A ratio of a product of directed segments to another product of directed segments, where all the segments lie in one plane, is called an h-*expression* if it has the properties (1) and (2) described above.

6.2.2 THEOREM. *The value of an h-expression is invariant under any projectivity.*

It is sufficient to show that a given *h*-expression has a value which is invariant under an arbitrary perspectivity. Let V be the center of a perspectivity and let $\overline{AB}$ be any one of the segments appearing in the *h*-expression. Let p denote the perpendicular distance from V to AB. Then

$$p\,\overline{AB} = (VA)(VB) \sin \overline{AVB},$$

since each side is twice the area of $\triangle \overline{VAB}$. It follows that

$$\overline{AB} = [(VA)(VB) \sin \overline{AVB}]/p.$$

Replace each segment $\overline{AB}$ in the *h*-expression by $[(VA)(VB) \sin \overline{ABV}]/p$. Since property (1) above holds, VA, VB, etc. cancel out; since property (2) above holds, p, etc. cancel out. We are left with an expression containing only the sines $\sin \overline{AVB}$, etc. Now, since $\angle AVB = \angle A'VB'$, etc., the same relation that holds among the sines $\sin \overline{AVB}$, etc. holds among the sines $\sin \overline{A'VB'}$, etc. Introducing factors VA', VB', p', etc., by reversing the earlier cancellations, leads to the same relation among segments $\overline{A'B'}$, etc. as was given among segments $\overline{AB}$, etc.

The reader will note that the procedure employed in the above proof is essentially the way we proved, in Section 2.5, that the cross ratio of four collinear points is invariant under a perspectivity.

6.2.3 CEVA'S THEOREM. *If* AD, BE, CF *are concurrent cevian lines for a triangle* ABC, *then*

(1)
$$\frac{(\overline{BD})(\overline{CE})(\overline{AF})}{(\overline{DC})(\overline{EA})(\overline{FB})} = +1.$$

The expression on the left of (1) is an *h*-expression, and is therefore (by Theorem 6.2.2) invariant in value under projection. By Theorem 6.1.7, there is a perspectivity which carries triangle ABC and the point G of concurrence of the three cevian lines into a triangle $A'B'C'$ and its centroid G'. Then $\overline{B'D'}/\overline{D'C'} = \overline{C'E'}/\overline{E'A'} = \overline{A'F'}/\overline{F'B'} = 1$, and clearly

$$\frac{(\overline{B'D'})(\overline{C'E'})(\overline{A'F'})}{(\overline{D'C'})(\overline{E'A'})(\overline{F'B'})} = +1.$$

The theorem now follows.

6.2.4 MENELAUS' THEOREM. *If* D, E, F *are collinear menelaus points on the sides* BC, CA, AB *of a triangle* ABC, *then*

(1)
$$\frac{(\overline{BD})(\overline{CE})(\overline{AF})}{(\overline{DC})(\overline{EA})(\overline{FB})} = -1.$$

The expression on the left of (1) is an *h*-expression, and is therefore (by Theorem 6.2.2) invariant in value under projection. Project line DEF to infinity. Then $\overline{B'D'}/\overline{D'C'} = \overline{C'E'}/\overline{E'A'} = \overline{A'F'}/\overline{F'B'} = -1$, and clearly

$$\frac{(\overline{B'D'})(\overline{C'E'})(\overline{A'F'})}{(\overline{D'C'})(\overline{E'A'})(\overline{F'B'})} = -1.$$

The theorem now follows.

We conclude the section by giving another projective proof of Desargues' Two-Triangle Theorem and a proof of a theorem due to Pappus. The latter theorem is a generalization of Theorem 6.1.5, and is an instance of a descriptive theorem that was known to the ancient Greeks. In Chapter 8 we shall see that Pappus' Theorem and Desargues' Two-Triangle Theorem are very important in a study of the foundations of projective geometry.

6.2.5 Desargues' two-triangle theorem. *Copolar triangles (in space or in a plane) are coaxial, and conversely.*

(a) Let the two triangles be ABC and $A'B'C'$ and suppose they lie in different planes π and π' respectively (see Figure 6.2a). Also suppose AA',

Figure 6.2a

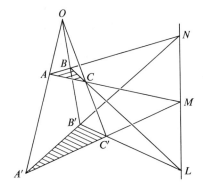

BB', CC' are concurrent in a point O. Then BC and $B'C'$ are coplanar and thus intersect in a point L. Similarly, CA and $C'A'$ are coplanar and thus intersect in a point M, and AB and $A'B'$ are coplanar and thus intersect in a point N. The points L, M, N then lie in both planes π and π', and hence on the line of intersection of these two planes. That is, copolar triangles in different planes are coaxial. The converse follows by reversing the above argument.

(b) Now suppose the two planes π and π' coincide, and (see Figure 6.2b) that the points L, M, N of intersection of BC and $B'C'$, CA and $C'A'$, AB and $A'B'$ are collinear on a line l in π. Let π_1 be a plane distinct from π and passing through l, let P be a point not on π or π_1, and let PA', PB', PC' cut π_1 in A_1, B_1, C_1, respectively. Then B_1, C_1, B', C' are coplanar, as also

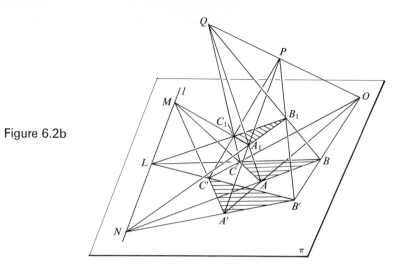

Figure 6.2b

are C_1, A_1, C', A' and A_1, B_1, A', B', and we see that BC, $B'C'$, B_1C_1 meet in L; CA, $C'A'$, C_1A_1 meet in M; AB, $A'B'$, A_1B_1 meet in N. Thus the two triangles $A_1B_1C_1$ and ABC are coaxial and therefore, by part (a), are copolar. That is, AA_1, BB_1, CC_1 are concurrent in a point Q. Let QP cut π in point O. Then A, A', O all lie in the plane determined by PA_1A' and QA_1A. It follows that AA' passes through O. Similarly, BB' and CC' pass through O, and triangles ABC and $A'B'C'$ are copolar. The converse is obtained by applying the above to triangles $AA'M$ and $BB'L$.

6.2.6 PAPPUS' THEOREM.   *If the vertices* 1, 2, 3, 4, 5, 6 *of a hexagon* 123456 *lie alternately on a pair of lines, then the three intersections* P, Q, R *of the opposite sides* 23 *and* 56, 45 *and* 12, 61 *and* 34 *of the hexagon are collinear* (see Figure 6.2c$_1$).

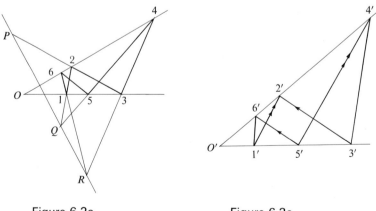

Figure 6.2c$_1$ Figure 6.2c$_2$

Project line PQ to infinity and suppose the intersection O of lines 135 and 246 does not lie on PQ. Then $1'$, $3'$, $5'$ are collinear, $2'$, $4'$, $6'$ are collinear, O' is a finite point, $2'3'$ is parallel to $5'6'$, and $4'5'$ is parallel to $1'2'$. We must prove that $6'1'$ is parallel to $3'4'$. Now (see Figure 6.2c$_2$), since $2'3'$ is parallel to $5'6'$ and $4'5'$ is parallel to $1'2'$,

$$\overline{O'6'}/\overline{O'2'} = \overline{O'5'}/\overline{O'3'} \text{ and } \overline{O'1'}/\overline{O'5'} = \overline{O'2'}/\overline{O'4'}.$$

It follows that $\overline{O'6'}/\overline{O'1'} = \overline{O'4'}/\overline{O'3'}$ and $6'1'$ is parallel to $3'4'$.

If O lies on PQ, then lines $1'3'5'$ and $2'4'6'$ are parallel. But then $\overline{1'5'} = \overline{2'4'}$ and $\overline{5'3'} = \overline{6'2'}$. It follows that $\overline{1'3'} = \overline{6'4'}$ and $6'1'$ is parallel to $3'4'$.

PROBLEMS

1. Lines VAA', VBB', VCC' cut two lines OX, OY in A and A', B and B', C and C' respectively. AB' and $A'C$ intersect in P; $A'B$ and AC' intersect in Q. Prove that line PQ passes through V.

2. Philippe de la Hire (1640–1718), a pupil of Desargues, invented the following interesting mapping of a plane onto itself. Let a and b be two given parallel lines and O a given point in their plane. Let P be an arbitrary point of the plane and through P draw a line cutting a in A and b in B. Then the image P' of P is taken as the intersection with PO of the parallel to AO through B.
   (a) Show that P' is independent of the particular line PAB through P used to determine it.
   (b) Into what does line a map?
   (c) Into what does line b map?
   (d) Into what does a line parallel to lines a and b map?
   (e) Into what does a line through O parallel to lines a and b map?
   (f) Into what does the line at infinity map?
   (g) Show that all straight lines map into straight lines.
   (h) Where must we take the image of O?
   (i) Generalize la Hire's mapping to the situation where lines a and b need not be parallel.

3. If the sides of a hexagon 123456 pass alternately through two fixed points P and Q, prove that the three diagonals joining opposite pairs of vertices of the hexagon are concurrent.

4. If P, Q, R are any three points on the sides BC, CA, AB respectively of a triangle ABC, show that $(\overline{BP}/\overline{PC})(\overline{CQ}/\overline{QA})(\overline{AR}/\overline{RB})$ is invariant in value under projection.

5. In the complete quadrangle $ABCD$, a transversal cuts sides AB and CD in M and M', sides BC and AD in N and N', and sides AC and BD in P and P'. Show that

$$(\overline{MN})(\overline{MN'})/(\overline{MP})(\overline{MP'}) = (\overline{M'N})(\overline{M'N'})/(\overline{M'P})(\overline{M'P'}).$$

6.3 PROPER CONICS

We commence by informally stating a number of definitions which are very likely already familiar to the reader.

A *circular cone* (see Figure 6.3a) is a surface generated by a straight line

Figure 6.3a

which moves so that it always intersects a given circle *c* and passes through a fixed point *V* not in the plane of the circle. The generating line, in each of its positions, is called an *element* of the cone; the fixed point is called the *vertex* of the cone. The vertex divides each element into two half-lines and divides the cone into two *nappes*, each of which is generated by one of the half-lines. If the line joining the vertex of the cone to the center of the given circle is perpendicular to the plane of the circle, the cone is called a *right circular cone*; otherwise it is called an *oblique circular cone*.

The curves called parabolas, ellipses, and hyperbolas were named by Apollonius, and were investigated by him as certain planar sections of right and oblique circular cones. Since these curves are intersections of circular cones by planes, they are examples of *conic sections*, or more briefly, *conics*. If the cutting plane does not pass through the vertex of the cone, the conic is called a *proper* conic. A *parabola* is a proper conic whose plane of section is parallel to one and only one element of the cone; an *ellipse* (including a circle as a special case) is a proper conic whose plane of section cuts all the elements of one nappe of the cone; a *hyperbola* is a proper conic whose plane of section cuts into both nappes of the cone. Figure 6.3b shows a right circular cone sectioned to yield a parabola *p*, and ellipse *e*, and a hyperbola *h*.

In ordinary space, a parabola is clearly a one-piece nonclosed curve; an ellipse is a one-piece closed curve; a hyperbola is a two-piece nonclosed curve. In extended space, each curve is one-piece and closed.

Since a proper conic is a section of a circular cone, and the section does not pass through the vertex of the cone, it follows that a proper conic is the image of a circle under a perspectivity. Therefore any property of a circle which is descriptive, that is, is unaltered by projection, can be transferred at once to the conic. In this way we obtain the following sequence of theorems about proper conics.

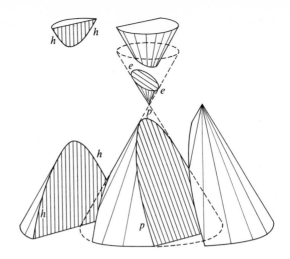

Figure 6.3b

6.3.1 THEOREM. *A straight line in the plane of a proper conic cuts the conic in two points, is tangent to the conic, or fails to cut the conic.*

6.3.2 THEOREM. *There is a unique tangent line to a proper conic at each point on the conic.*

6.3.3 THEOREM. *A proper conic divides its plane, exclusive of the conic itself, into two regions such that from any point in one of the regions two lines can be drawn tangent to the conic, and from any point in the other region no line can be drawn tangent to the conic.*

6.3.4 DEFINITIONS. The two regions mentioned in Theorem 6.3.3 are called, respectively, the *outside* and the *inside* of the proper conic.

6.3.5 THEOREM. *If a variable line through a given point* P *in the plane of a proper conic* c *intersects the conic, the harmonic conjugates of the point with respect to the intersections of the line and conic all lie on a straight line* p. (See Corollary 3.9.8.)

6.3.6 DEFINITIONS. The straight line *p* of Theorem 6.3.5 is called the *polar* of the point *P* for the conic *c*, and point *P* is called the *pole* of line *p* for the conic *c*.

6.3.7 THEOREM. (1) *The polar of a point for a proper conic intersects the conic, is tangent to the conic at the point, or does not intersect the conic, according as the point is outside, on, or inside the conic.* (2) *If point* P *is outside a proper conic, then its polar for the conic passes through the points of contact of the tangents to the conic from* P. (See Theorem 3.9.2.)

6.3.8 THEOREM. (1) *If, for a given proper conic, the polar of* P *passes through* Q, *then the polar of* Q *passes through* P. (2) *If, for a given proper conic, the pole of line* p *lies on line* q, *then the pole of* q *lies on* p. (3) *If, for a given proper conic,* P *and* Q *are the poles of* p *and* q, *then the pole of line* PQ *is the point of intersection of* p *and* q. (See Theorem 3.9.3.)

6.3.9 DEFINITIONS. Two points such that each lies on the polar of the other, for a given proper conic, are called *conjugate points* for the conic; two lines such that each passes through the pole of the other, for a given proper conic, are called *conjugate lines* for the conic.

6.3.10 THEOREM. *For a given proper conic:* (1) *Each point of a line has a conjugate point on that line.* (2) *Each line through a point has a conjugate line through that point.* (3) *Of two distinct conjugate points on a line that cuts the conic, one is inside and the other outside the conic.* (4) *Of two distinct conjugate lines that intersect outside the conic, one cuts the conic and the other does not.* (5) *Any point on the conic is conjugate to all the points on the tangent to the conic at the point.* (6) *Any tangent to the conic is conjugate to all the lines through its point of contact with the conic.* (See Theorem 3.9.6.)

6.3.11 THEOREM. *If, for a given proper conic, two conjugate points lie on a line which intersects the conic, they are harmonically separated by the points of intersection.* (See Theorem 3.9.7.)

6.3.12 THEOREM. *If, for a given proper conic, two conjugate lines intersect outside the conic, they are harmonically separated by the tangents to the conic from their point of intersection.* (See Theorem 3.9.9.)

6.3.13 THEOREM. *If* A, B, C, D *are four distinct collinear points, and* a, b, c, d *are their polars for a given proper conic, then* (AB,CD) = (ab,cd). (See Theorem 3.9.10.)

6.3.14 THEOREM. *If* ABCD *is a complete quadrangle inscribed in a proper conic, then each diagonal point of the quadrangle is the pole, for the conic, of the line determined by the other two diagonal points.* (See Theorem 3.10.2.)

PROBLEMS

1. (a) Show that a parabola is tangent to the line at infinity.
   (b) Show that no two tangents to a parabola can be parallel.
   (c) *TP* and *TQ* touch a parabola at *P* and *Q*; *TV* bisects *PQ* in *V* and intersects the parabola in *R*. Show that *R* is the midpoint of *TV*.
   (d) Prove that the line half way between a point and its polar for a parabola is tangent to the parabola.
   (e) Prove that the lines joining the midpoints of the sides of a triangle which is self-conjugate for a parabola are tangent to the parabola.

2. We define the *center* of a proper conic to be the pole for the conic of the line at infinity. Prove the following:

   (a) The center of a parabola is at infinity.

   (b) The center of a hyperbola is an ordinary point lying outside the curve.

   (c) The center of an ellipse is an ordinary point lying inside the curve.

   Hyperbolas and ellipses are called *central conics*.

   (d) The center of a central conic bisects every chord of the conic through it, and is thus a center of symmetry of the conic.

   (e) All proper conics circumscribing a parallelogram have their centers at the center of the parallelogram.

   (f) If C is the center of a central conic, TP and TQ touch the conic at P and Q, and CT cuts PQ in V and the conic in U, then $(\overline{CV})(\overline{CT}) = (CU)^2$.

3. The locus of the midpoints of a family of parallel chords of a proper conic is called a *diameter* of the conic. Prove the following:

   (a) Diameters of a proper conic are straight lines.

   (b) All diameters of a central conic pass through the center.

   (c) All diameters of a parabola are parallel.

   (d) The tangents at the ends of a diameter of a central conic are parallel to the chords which the diameter bisects.

   (e) If the tangents at the ends of a chord of a central conic are parallel, the chord is a diameter of the conic.

   (f) Two chords of a central conic which bisect each other are diameters of the conic.

4. Two diameters of a central conic which are conjugate lines for the conic are called a pair of *conjugate diameters* of the conic. Prove the following:

   (a) Each of two conjugate diameters of a central conic bisects the chords parallel to the other.

   (b) The diagonals of a parallelogram circumscribing a central conic are conjugate diameters of the conic; the points of contact are the vertices of a parallelogram whose sides are parallel to the above diagonals.

   (c) If each diameter of a central conic is perpendicular to its conjugate diameter, the conic is a circle.

5. The tangents to a hyperbola from its center are called the *asymptotes* of the hyperbola. Prove the following:

   (a) The asymptotes of a hyperbola harmonically separate every pair of conjugate diameters of the hyperbola.

   (b) Let a line cut a hyperbola in Q and Q', and cut its asymptotes in R and R'. Then $RQ = Q'R'$.

   (c) The intercept between the asymptotes of a hyperbola on any tangent to the hyperbola is bisected by the point of contact of the tangent.

6.4 APPLICATIONS

In this section we apply projection to extend to the general proper conic some descriptive properties first established for the circle.

6.4.1 THE GENERALIZED BUTTERFLY THEOREM. *Let* O (see Figure 6.4a) *be*

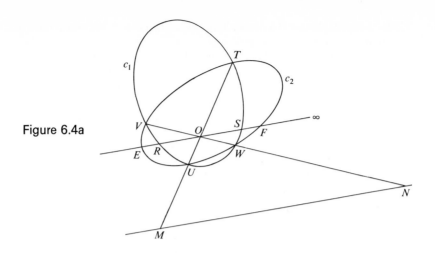

Figure 6.4a

the midpoint of a given chord RS of a proper conic c_1, let two other chords TU and VW be drawn through O, and let a conic c_2 through U, V, T, W cut the given chord in E and F. Then O is the midpoint of EF.

Let MN, the polar of O for conic c_1, cut TU in M and VW in N. Then, since $(VW,ON) = (TU,OM) = -1$, MN is also the polar of O for conic c_2. Moreover, RS is parallel to MN since $(RS,O\infty) = -1$. Therefore $(EF,O\infty) = -1$ and O is the midpoint of EF. (See Theorem 3.10.3.)

6.4.2 PASCAL'S THEOREM. *If a hexagon (not necessarily convex) is inscribed in a proper conic, the three points of intersection of pairs of opposite sides of the hexagon are collinear.* (See Theorem 2.4.4.)

6.4.3 BRIANCHON'S THEOREM. *If a hexagon (not necessarily convex) is circumscribed about a proper conic, the three lines joining pairs of opposite vertices of the hexagon are concurrent.* (See Theorem 3.10.1.)

6.4.4 CHASLES' THEOREM. *Let A, B, C, D be four distinct points on a proper conic and let the tangents to the conic at A, B, C, D meet a fixed tangent t to the conic in the points A', B', C', D' respectively. Then, if O is any point on the conic, O(AB,CD) = (A'B',C'D').*

Since the theorem is a projective one, it suffices to establish it for the case where the proper conic is a circle. Let K be the center of the circle (see Figure 6.4b), and T the point of contact of the tangent t. Then $\angle A'KT = \frac{1}{2}\angle AKT = \angle AOT$. It follows that pencils O(ABCD) and K(A'B'C'D') are congruent, and $O(AB,CD) = K(A'B',C'D') = (A'B',C'D')$.

6.4.5 CARNOT'S THEOREM. *If (see Figure 6.4c) the sides BC, CA, AB of a triangle ABC cut a proper conic in the points A_1 and A_2, B_1 and B_2, C_1 and C_2 respectively, then*

(1) $$(\overline{AC}_1)(\overline{AC}_2)(\overline{BA}_1)(\overline{BA}_2)(\overline{CB}_1)(\overline{CB}_2)$$
$$=(\overline{AB}_1)(\overline{AB}_2)(\overline{BC}_1)(\overline{BC}_2)(\overline{CA}_1)(\overline{CA}_2).$$

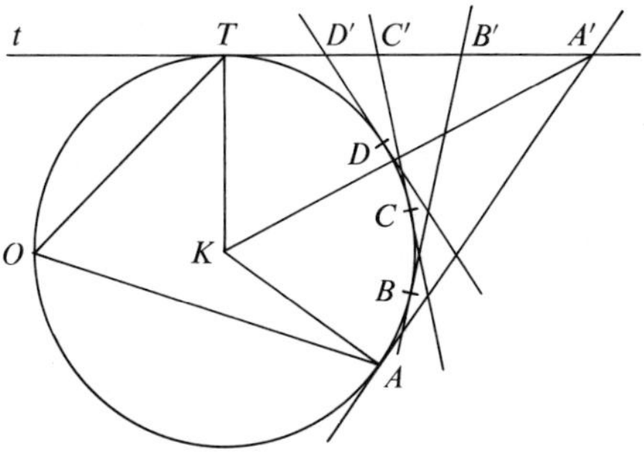

Figure 6.4b

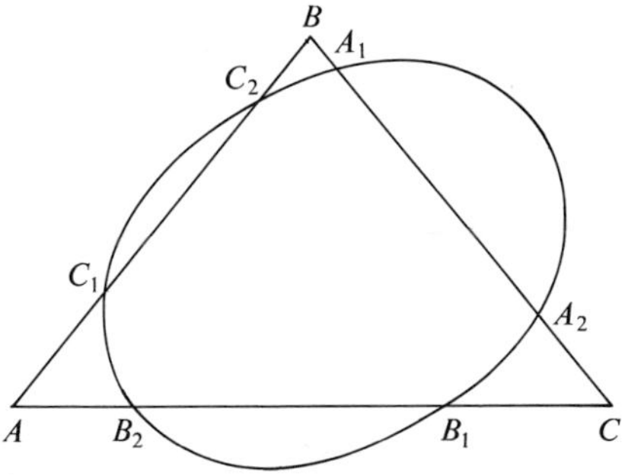

Figure 6.4c

We shall assume that no one of the points of intersection is either an ideal point or a vertex of the triangle. Then the left member of (1) divided by the right member of (1) is an h-expression, and is therefore (by Theorem 6.2.2) invariant in value under projection. By the definition of a proper conic, there is a perspectivity which carries the conic into a circle. Now relation (1) holds for a circle, since $(\overline{A'C_1'})(\overline{A'C_2'}) = (\overline{A'B_1'})(\overline{A'B_2'})$, etc. Hence the h-expression has the value $+1$, and the theorem follows.

6.4.6   THEOREM.   *Let* PQR *be a triangle and let* P', Q', R' *be the poles of* QR, RP, PQ *for a proper conic* K. *Then* P, Q, R *are the poles of* Q'R', R'P', P'Q' *for the conic* K. (See Theorem 3.10.4.)

6.4.7   DEFINITIONS.   Two triangles are said to be *conjugate,* or *polar,* for a proper conic if each vertex of one triangle is the pole of a side of the other triangle. If a triangle is conjugate to itself—that is, each vertex is the pole of the opposite side—the triangle is said to be *self-conjugate,* or *self-polar,* for the conic.

6.4.8   THEOREM.   *Two triangles which are conjugate for a proper conic are copolar to one another.* (See Theorem 3.10.6.)

6.4.9 HESSE'S THEOREM (1840). *If two pairs of opposite vertices of a complete quadrilateral are conjugate points for a proper conic, then the third pair of opposite vertices are also conjugate points for the conic.* (See Theorem 3.10.7.)

PROBLEMS

1. If PQ is a chord of a proper conic, U any point on line PQ, and V the point where the polar of U for the conic cuts line PQ, show that $1/\overline{PU} + 1/\overline{PV}$ = constant.

2. TP and TQ touch a proper conic at P and Q, and the bisector of angle PTQ cuts the chord PQ in U. RUR' is any chord through U. Show that TU bisects angle RTR'.

3. Let A, B, C be three points on a proper conic, CT the tangent to the conic at C, and CD such that $C(TD,AB) = -1$. Show that CD passes through the pole of AB.

4. TP and TQ touch a proper conic at P and Q. The tangent at a point R on the conic cuts TP, TQ, PQ in L, M, N respectively. Show that $(LM,RN) = -1$.

5. A chord PQ is drawn through the midpoint U of a chord AB of a proper conic. If the tangents at P and Q cut AB in L and M, prove that U is the midpoint of LM.

6. TP and TQ touch a proper conic at P and Q. A tangent to the conic parallel to PQ cuts TP and TQ in M and N respectively, and touches the conic at R. Show that R is the midpoint of MN.

7. A, B, C, D are four points on a proper conic; AB and CD intersect in R; AC and BD intersect in S; the tangents to the conic at A and D intersect in T. Prove that R, S, T are collinear.

8. With straightedge alone: (a) Draw the tangents to a proper conic from a given point outside the conic. (b) Draw the tangent to a proper conic at a given point on the conic.

9. (a) If a pentagon 12345 is inscribed in a proper conic, show that the pairs of lines 12, 45; 23, 51; 34 and the tangent at 1, intersect in three collinear points.
(b) Show that the pairs of opposite sides of a quadrangle inscribed in a proper conic, together with the pairs of tangents at opposite vertices, intersect in four collinear points.
(c) Show that if a triangle is inscribed in a proper conic, then the tangents at the vertices intersect the opposite sides in three collinear points.

10. (a) If a pentagon $ABCDE$ is circumscribed about a proper conic, and F is the point of contact of side AB, show that BE, CA, DF are concurrent.
(b) Show that the lines joining the points of contact of pairs of opposite sides of a quadrilateral circumscribed about a proper conic, and the two diagonals of the quadrilateral, are concurrent.
(c) Show that if a triangle is circumscribed about a proper conic, then the lines joining the vertices and the points of contact of the opposite sides are concurrent.

11. The sides AB, BC, CD, ... of a polygon intersect a proper conic in A_1 and A_2, B_1 and B_2, C_1 and C_2, ... Show that

$$(\overline{AA_1})(\overline{AA_2})(\overline{BB_1})(\overline{BB_2})(\overline{CC_1})(\overline{CC_2}) \cdots$$
$$= (\overline{BA_1})(\overline{BA_2})(\overline{CB_1})(\overline{CB_2})(\overline{DC_1})(\overline{DC_2}) \ldots.$$

(This is a generalization of Carnot's Theorem.)

12. B_1B_2 and A_1A_2 are two variable chords of a proper conic drawn through two fixed points A and B respectively. If A_1A_2, B_1B_2 intersect in C, show that

$$(BA_1)(BA_2)(CB_1)(CB_2)/(AB_1)(AB_2)(CA_1)(CA_2)$$

is constant.

13. A proper conic intersects the sides BC, CA, AB of a triangle ABC in P_1 and P_2, Q_1 and Q_2, R_1 and R_2 respectively. BQ_2 and CR_2 intersect in X, AP_1 and CR_1 intersect in Y, AP_2 and BQ_1 intersect in Z. Show that AX, BY, CZ are concurrent.

14. Triangle $A'B'C'$ is circumscribed about a proper conic; points A, B, C are the points of contact of the sides $B'C'$, $C'A'$, $A'B'$ respectively. If D, E, F are taken on BC, CA, AB such that AD, BE, CF are concurrent, show that $A'D$, $B'E$, $C'F$ are also concurrent.

15. O is a variable point on a proper conic and M and N are a fixed pair of conjugate points for the conic; OM and ON cut the conic again in P and Q. Show that PQ passes through a fixed point.

16. Show that the proofs of Theorems 3.10.6 and 3.10.7 are descriptive, and can therefore be applied directly to prove Theorems 6.4.8 and 6.4.9.

6.5 THE CHASLES-STEINER DEFINITION OF A PROPER CONIC

In this section we introduce an alternative definition of a proper conic that was obtained and utilized by both Michel Chasles and Jacob Steiner in certain of their works. The definition is a singularly convenient one for the development of the projective properties of conics. We commence by stating a projective generalization of Theorem 2.6.5.

6.5.1 THEOREM. *If* A, B, C, D *are any four distinct points on a proper conic, and if* U *and* V *are any two points on the conic, then* U(AB,CD) = V(AB,CD), *where* U(A), *for example, is taken as the tangent to the conic at* A *if* U *should coincide with* A. (See Theorem 2.6.5.)

An easy consequence of Theorem 6.5.1 is the following.

6.5.2 THEOREM. *If* P *is a variable point on a proper conic, and if* U *and* V *are any two distinct points on the conic, then the complete pencils* U(P) *and* V(P) *are homographic and such that line* U(V) *does not correspond to line* V(U).

For if P_1, P_2, P_3, P_4 are any four distinct positions of P we have, by

Theorem 6.5.1, $U(P_1P_2,P_3 P_4) = V(P_1P_2,P_3 P_4)$. Hence the complete pencils $U(P)$ and $V(P)$ are homographic. Also, line $U(V)$ cannot correspond to line $V(U)$, for if it did the locus of P would (by Theorem 2.7.3(2)) be a straight line, which is impossible since the locus of P is a proper conic.

This last theorem may now be restated in the following form.

6.5.3   THEOREM. *A proper conic passing through two distinct points* U *and* V *may be regarded as the locus of the points of intersection of corresponding lines of two homographic pencils on* U *and* V, *where line* U(V) *does not correspond to line* V(U).

It is natural to wonder if the converse of Theorem 6.5.3 is also true. We prove that it is.

6.5.4   THEOREM. *The locus of the points of intersection of corresponding lines of two homographic pencils on distinct vertices* U *and* V, *where line* U(V) *does not correspond to line* V(U), *is a proper conic passing through* U *and* V.

Let P (see Figure 6.5a) be a variable point on the locus; then pencil

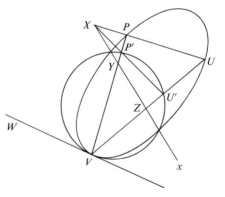

Figure 6.5a

$U(P)$ and $V(P)$ are homographic. Let line $V(W)$ correspond to line $U(V)$. By hypothesis $V(W)$ does not coincide with $U(V)$. Draw any circle tangent to VW at V, and let this circle cut VU in U' and the variable line VP in P'.

Since pencils $U(V,P)$ and $V(W,P)$ are homographic, and (by Theorem 6.5.2) pencils $V(W,P)$ and $U'(V,P')$ are homographic, it follows that pencils $U(V,P)$ and $U'(V,P')$ are homographic. But in these last two pencils, the corresponding lines $U(V)$ and $U'(V)$ are coincident. It follows (by Theorem 2.7.3(2)) that the variable intersection X of corresponding variable lines $U(P)$ and $U'(P')$ traces a line x. Let PP' and UU' cut line x in points Y and Z.

Now rotate the figure of the circle about line x out of its original plane, and let V'', U'', P'' be the new positions of V, U', P'. Then $U''P''$, $V''P''$, $V''U''$ cut line x in points X, Y, Z respectively. Thus triangles VUP and $V''U''P''$ are coaxial (along x), and must therefore be copolar. That is, VV'', UU'', PP'' are concurrent in a point O, and the figure VPU is the projection from O of the figure $V''P''U''$. Since P'' describes a circle passing through U'' and V'', it follows that P describes a proper conic passing through U and V.

Theorems 6.5.3 and 6.5.4 give us a necessary and sufficient condition for a locus of points to be a proper conic. The condition can then be taken as an alternative definition of a proper conic, as follows.

6.5.5   THE CHASLES-STEINER DEFINITION OF A PROPER CONIC.   A *proper conic* is the locus of the points of intersection of corresponding lines of two homographic pencils on distinct vertices U and V, where line $U(V)$ does not correspond to line $V(U)$.

We now prove a couple of important theorems with the assistance of the Chasles-Steiner definition of a proper conic.

6.5.6   THEOREM.   *Any projectivity carries a proper conic into a proper conic.*

It is sufficient to prove that any perspectivity carries a proper conic into a proper conic. Let c' be the image, under a perspectivity, of a proper conic c. By the Chasles-Steiner definition, c is the locus of the points of intersection of corresponding lines of two homographic pencils $U(P)$ and $V(P)$ on distinct vertices U and V, where line $U(V)$ does not correspond to line $V(U)$. Since homographic pencils of this sort project into homographic pencils of the same sort, it follows that c' is the locus of the points of intersection of corresponding lines of two homographic pencils $U'(P')$ and $V'(P')$, where the vertices U' and V' are distinct and line $U'(V')$ does not correspond to line $V'(U')$. By the Chasles-Steiner definition, c' is seen to be a proper conic.

6.5.7   THEOREM.   *There is one, and only one, proper conic passing through five distinct coplanar points no three of which are collinear.*

Let the five points be U, V, A, B, C. Take U and V as vertices of pencils. Through U draw any line $U(X)$ and let $V(Y)$ be such that $U(AB,CX) = V(AB,CY)$. Then the lines $U(X)$ and $V(Y)$ generate homographic pencils of which $U(A)$ and $V(A)$, $U(B)$ and $V(B)$, $U(C)$ and $V(C)$ are corresponding lines. Since no three of the five points are collinear, line $U(V)$ does not correspond to line $V(U)$. It follows, by the Chasles-Steiner definition, that the locus of the point P of intersection of $U(X)$ and $V(Y)$ is a proper conic passing through the five given points U, V, A, B, C. Now there can be only the one proper conic passing through U, V, A, B, C, for the other point P in which an arbitrary line $U(X)$ cuts a proper conic through U, V, A, B, C

is uniquely determined by the relation $U(AB,CP) = V(AB,CP)$. Hence an arbitrary line through U cuts all proper conics through U, V, A, B, C in the same point; that is, all the proper conics coincide.

6.5.8 COROLLARY. *Two distinct proper conics can intersect in at most four points.*

PROBLEMS

1. If the circle and conic of Figure 6.5a intersect in two points other than V, prove that line x passes through the two points.

2. If the three points of intersection of pairs of opposite sides of a (not necessarily convex) hexagon are collinear, and if no three of the vertices of the hexagon are collinear, prove that the six vertices of the hexagon lie on a proper conic. (This is the converse of Pascal's Theorem.)

3. If on the sides BC, CA, AB of a triangle, pairs of points A_1 and A_2, B_1 and B_2, C_1 and C_2 are taken such that

$$(\overline{AC_1})(\overline{AC_2})(\overline{BA_1})(\overline{BA_2})(\overline{CB_1})(\overline{CB_2})$$
$$= (\overline{AB_1})(\overline{AB_2})(\overline{BC_1})(\overline{BC_2})(\overline{CA_1})(\overline{CA_2}),$$

and if no three of the six points A_1, A_2, B_1, B_2, C_1, C_2 are collinear, prove that the six points lie on a proper conic. (This is the converse of Carnot's Theorem.)

4. Prove that the parallels through any point to the sides of a triangle cut the sides in six points on a proper conic.

5. (a) We are given five points on a proper conic, but we are not given the conic itself. Draw the tangent to the conic at one of the five points.
   (b) We are given four points on a proper conic and the tangent at one of them, but we are not given the conic itself. Construct further points on the conic.
   (c) We are given three points on a proper conic and the tangents at two of them, but we are not given the conic itself. Construct the tangent at the third point.

6. A proper conic c intersects the sides BC, CA, AB of a triangle ABC at A_1, A_2; B_1, B_2; C_1, C_2 respectively. A_1' and A_2' are the harmonic conjugates of A_1 and A_2 for B and C; B_1' and B_2' are the harmonic conjugates of B_1 and B_2 for C and A; C_1' and C_2' are the harmonic conjugates of C_1 and C_2 for A and B. Prove that A_1', A_2', B_1', B_2', C_1', C_2' lie on a proper conic or on two straight lines.

7. On the tangent at point O of a proper conic a variable point P is taken, and PT is drawn to touch the conic at T. If S is any other point on the conic, show that the locus of the intersection of OT and SP is another proper conic touching the original conic at O and S.

8. C and C' are two fixed points on a proper conic c, r is a fixed straight line, P is a variable point on c, CP and $C'P$ cut r in R and R'. Show that the locus of the intersection Q of CR' and $C'R$ is another conic.

9. A, B, C, D, U, V are six points on a proper conic. Let J, K, L, M denote the points of intersection of UA and VC, UB and VD, UC and VA, UD and VB. Show that J, K, L, M lie on a proper conic passing through U and V.

6.6 A PROPER CONIC AS AN ENVELOPE OF LINES

A curve may be regarded either as the locus of its points or as the envelope of its tangents (see Figure 6.6a). From the first point of view, the curve is

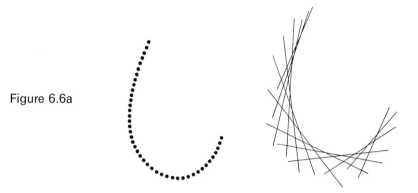

Figure 6.6a

called a *point curve*; from the second point of view, it is called a *line curve*. So far we have been considering a proper conic from the first point of view, and we now turn to the second point of view. We first introduce some convenient notation.

6.6.1 NOTATION. The cross ratio of four distinct points in which a line u is cut by four lines a, b, c, d will be denoted by $u(ab,cd)$. By $u(a)$ we mean the point in which base u is cut by line a.

6.6.2 THEOREM. *If* a, b, c, d *are any four distinct tangents of a proper conic, and if* u *and* v *are any two tangents of the conic, then* u(ab,cd) = v(ab,cd), *where* u(a), *for example, is taken as the point of contact of tangent* a *if* u *should coincide with* a.

This is an immediate consequence of Chasles' Theorem 6.4.4.

An easy consequence of Theorem 6.6.2 is the following.

6.6.3 THEOREM. *If* p *is a variable tangent of a proper conic, and if* u *and* v *are any two distinct tangents of the conic, then the complete ranges* u(p) *and* v(p) *are homographic and such that point* u(v) *does not correspond to point* v(u).

For if p_1, p_2, p_3, p_4 are any four distinct positions of p we have, by Theorem 6.6.2, $u(p_1p_2,p_3p_4) = v(p_1p_2,p_3p_4)$. Hence the complete ranges $u(p)$ and $v(p)$ are homographic. Also, point $u(v)$ cannot correspond to point $v(u)$, for if it did the envelope of p would (by Theorem 2.7.3(1)) be a point, which is impossible since the envelope of p is a proper conic.

This last theorem may now be restated in the following form.

6.6.4   THEOREM.   *A proper conic touching two distinct lines* u *and* v *may be regarded as the envelope of the lines joining corresponding points of two homographic ranges on* u *and* v, *where point* u(v) *does not correspond to point* v(u).

It is natural to wonder if the converse of Theorem 6.6.4 is also true. We prove that it is.

6.6.5   THEOREM.   *The envelope of the lines joining corresponding points of two homographic ranges on distinct bases* u *and* v, *where point* u(v) *does not correspond to point* v(u), *is a proper conic tangent to* u *and* v.

Let *p* (see Figure 6.6b) be a variable line of the envelope and let *p* cut *u*

Figure 6.6b

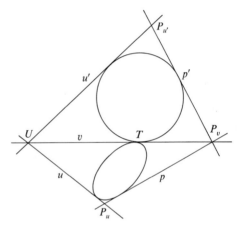

in P_u and *v* in P_v. Then ranges (P_u) and (P_v) are homographic. Let *U* be the intersection of *u* and *v*, and let *T* on *v* correspond to *U* on *u*. By hypothesis *T* does not coincide with *U*. Draw any circle tangent to *v* at *T* and let *u'* and *p'* be the other tangents to the circle from *U* and P_v respectively. Denote the intersection of *u'* and *p'* by $P_{u'}$.

Since ranges (U,P_u) and (T,P_v) are homographic, and (by Theorem 6.6.3) ranges (T,P_v) and $(U,P_{u'})$ are homographic, it follows that ranges (U,P_u) and $(U,P_{u'})$ are homographic.

Now rotate the circle and its tangents *u'* and *p'* about the line *v* out of the original plane, and let *u''*, *p''*, $P_{u''}$ denote the new positions of *u'*, *p'*, $P_{u'}$, respectively. Since ranges (U,P_u) and $(U,P_{u''})$ are homographic, and their common point *U* is self-corresponding, it follows (by Theorem 2.7.3(1)) that the variable line $P_u P_{u''}$ passes through a fixed point *O*. With *O* as center of perspectivity, we see that variable line *p''* projects into variable line *p*.

But the variable line p'' envelops a circle (the circle in its new position) touching lines v and u''. It follows that p must envelop a proper conic touching lines v and u.

We conclude the section with the following theorem and its corollary.

6.6.6 THEOREM. *There is one, and only one, proper conic tangent to five distinct coplanar lines no three of which are concurrent.*

Let the five lines (see Figure 6.6c) be u, v, a, b, c. Take u and v as bases

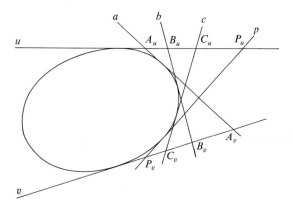

Figure 6.6c

of ranges and let a, b, c cut u and v in A_u, B_u, C_u and A_v, B_v, C_v respectively. Take any point P_u on u and let P_v on v be such that $(A_u B_u, C_u P_u) = (A_v B_v, C_v P_v)$. Then points P_u and P_v generate homographic ranges in which A_u and A_v, B_u and B_v, C_u and C_v are corresponding points. Since no three of the five lines are concurrent, point $u(v)$ does not correspond to point $v(u)$. It follows (by Theorem 6.6.5) that the envelope of $P_u P_v$ is a proper conic touching the five lines u, v, a, b, c. Now there can be only the one proper conic touching the five given lines u, v, a, b, c, for the other tangent p, to a proper conic touching u, v, a, b, c, through an arbitrary point P_u of u is uniquely determined as the line $P_u P_v$ where $(A_u B_u, C_u P_u) = (A_v B_v, C_v P_v)$. It follows that all proper conics touching u, v, a, b, c are the envelope of the same variable line p, that is, all the proper conics coincide.

6.6.7 COROLLARY. *Two distinct proper conics have at most four common tangents.*

PROBLEMS

1. If the three joins of pairs of opposite vertices of a (not necessarily convex) hexagon are concurrent, and if no three of the sides of the hexagon are concurrent, prove that the six sides of the hexagon envelop a proper conic. (This is the converse of Brianchon's Theorem.)

2. (a) We are given five tangents to a proper conic, but we are not given the conic itself. Construct the point of contact of one of the five tangents.

(b) We are given four tangents to a proper conic and the point of contact of one of them, but we are not given the conic itself. Construct further tangents to the conic.

(c) We are given three tangents to a proper conic and the points of contact of two of them, but we are not given the conic itself. Construct the point of contact of the third tangent.

3. A variable tangent PP' intersects two fixed parallel tangents to a proper conic in P and P'. If A and B are the points of contact of the two fixed tangents, prove that $(\overline{AP})(\overline{BP'})$ is constant.

4. A variable tangent intersects the asymptotes of a hyperbola at P and P'. If C is the intersection of the asymptotes, prove that $(\overline{CP})(\overline{CP'})$ is constant, and therefore that triangle CPP' has a constant area.

5. Two ellipses c and c' touch externally at O. From a variable point R on the common tangent at O lines are drawn touching c and c' at P and P'. Show that PP' passes through the point of intersection of the other two common tangents to c and c'.

6. OPQ is a fixed triangle and A and B are fixed points not collinear with O. R and S are variable points on OP and OQ such that RS is parallel to PQ. Prove that the locus of the point of intersection of AR and BS is a conic passing through A, B, O.

7. (a) The sides of a triangle pass through fixed noncollinear points and two of the vertices lie on fixed lines. What is the locus of the remaining vertex?

(b) The vertices of a triangle lie on fixed nonconcurrent lines and two of the sides pass through fixed points. What is the envelope of the remaining side?

8. Prove that if ABC and $A'B'C'$ are two triangles inscribed in a proper conic, then their six sides touch another proper conic. (This result is due to Brianchon.)

6.7 RECIPROCATION AND THE PRINCIPLE OF DUALITY

There is, in plane projective geometry, a remarkable symmetry between points and lines, such that if in a true projective proposition about "points" and "lines" we interchange these two words, and perhaps smooth out the language, we obtain another true projective proposition about "lines" and "points." The reader very likely has already noticed this symmetric pairing of the propositions of projective geometry. As a simple example, consider the following two propositions related in this way:

Any two distinct points determine one and only one line on which they both lie.

Any two distinct lines determine one and only one point through which they both pass.

In an ordinary plane, only the first of these two propositions is true without exception. In an extended, or projective, plane, however, both propositions are true without exception—regardless of whether the points and lines involved are ordinary or ideal.

We note that in passing from the statement of the first proposition to the statement of the second proposition, we not only interchanged the words "point" and "line," but we also altered the final phrase somewhat. Had we not altered the final phrase, the second proposition would have sounded a trifle awkward. This awkwardness of sound, however, results only because of a blemish in our language. Since the relation of a point P lying on a line l and that of a line l passing through a point P are symmetrical relations, a perfect language would express the two relations by a symmetrical terminology. We accordingly agree, in projective geometry, to express these two relations by the two phrases "point P is on line l" and "line l is on point P." Employing this modified language, we can pass directly from the statement of either of the propositions to that of the other by a mere mechanical interchange of the words "point" and "line."

For convenience, we call our modified language the *on-language*. The observed symmetry between points and lines in plane projective geometry leads to the so-called *principle of duality* for the plane, which may now be stated as follows:

6.7.1 THE PRINCIPLE OF DUALITY FOR THE PLANE. *If in a true projective proposition, stated in the on-language, about points and lines in a plane, we interchange the words "point" and "line," we obtain a second true projective proposition, stated in the on-language, about lines and points in a plane.*

The principle of duality, which pairs the propositions of plane projective geometry, is of far-reaching consequence, and was first explicitly stated by Joseph-Diez Gergonne in 1826, though it was led up to by the works of Poncelet and others during the first quarter of the nineteenth century. Once the principle of duality is in some way established, then the proof of one proposition of a dual pair automatically carries with it the proof of the other.

Pascal's Theorem is a projective proposition; let us dualize it. We first restate Pascal's Theorem in the on-language:

If the six points 1, 2, 3, 4, 5, 6 *lie on a proper point conic, then the points determined by the three pairs of lines* (12), (45); (23), (56); (34), (61) *lie on a line.*

Dualizing we obtain:

If the six lines 1, 2, 3, 4, 5, 6 *lie on a proper line conic, then the lines determined by the three pairs of points* (12), (45); (23), (56); (34), (61) *lie on a point.*

This dual is, of course, Brianchon's Theorem. Using less artificial language, Pascal's Theorem and its dual may now be stated as:

If a hexagon is inscribed in a proper conic, then the points of intersection of the three pairs of opposite sides are collinear.

If a hexagon is circumscribed about a proper conic, then the lines joining the three pairs of opposite vertices are concurrent.

With a little practice, one becomes adept at dualizing a proposition of projective geometry without having to resort to the artificial on-language.

Poncelet maintained that the principle of duality is a consequence of the theory of poles and polars, which, though known earlier, received its first systematic development in his hands. If Γ is a fixed proper conic, then to each point P of the extended plane can be associated the polar p' of P for the conic, and to each line l of the extended plane can be associated the pole L' of l for the conic. This correspondence, which maps the points and lines of the extended plane onto the lines and points of the extended plane, possesses some important properties. First of all, the correspondence is invariant under projection. Further, an incident point and line map into an incident line and point, collinear points map into concurrent lines, and concurrent lines map into collinear points.

Now imagine that we have established a projective property about a plane figure F composed of points and lines. Let F' be the figure which one obtains by replacing the points and lines of F by their polars and poles with respect to a given proper conic Γ lying in the plane of F. Then one will obtain from the projective property of figure F a corresponding projective property of figure F', in which the roles played by the words "point" and "line" have been interchanged. The two propositions will be the duals of each other.

It must be confessed that there seems to be something wanting in the above justification of the principle of duality. The figures of two dual theorems need not actually be related to one another by a pole-polar transformation. Indeed, since the statement of the principle of duality does not mention any proper conic Γ, it would seem desirable to justify the principle without the intervention of any such conic. One feels that the principle of duality needs a deeper and more general handling than that given by Poncelet. Such a general treatment will be discussed in Chapter 8, and still another general treatment—an analytical one—will be discussed early in the second volume of our work.

Before closing this section, we very briefly consider Poncelet's pole-polar transformation in a proper conic Γ; the transformation is a generalization of the one considered in Section 3.9, and much use is made of it in more extended treatments of projective geometry.

6.7.2 DEFINITIONS. Let Γ be a fixed proper conic. Then the transformation that maps each point P of the extended plane into its polar line p', and each line p of the extended plane into its pole P', with respect to Γ is called *reciprocation* in Γ. The conic Γ is called the *base-conic* of the reciprocation.

6.7.3 THEOREM. (1) *A point curve reciprocates into a line curve.* (2) *A line (considered as a range of collinear points) reciprocates into a point (considered as a pencil of concurrent lines).* (3) *The base-conic reciprocates into itself.* (4) *Reciprocation in a given base-conic is involutoric.*

6.7.4 THEOREM. *A proper conic reciprocates into a proper conic.*

Consider (see Theorem 6.5.3) a proper conic obtained as the locus of the point *P* of intersection of corresponding lines of two homographic pencils on distinct vertices *U* and *V*, where line *U*(*V*) does not correspond to line *V*(*U*). These homographic pencils on distinct vertices reciprocate into two homographic ranges on distinct bases, and the intersections of corresponding lines of the pencils reciprocate into the joins of corresponding points of the two ranges. It follows that the reciprocal of the proper conic is the envelope of the joins *p'* of corresponding points of two homographic ranges on distinct bases *u'* and *v'*, where point *u'*(*v'*) does not correspond to point *v'*(*u'*). It follows (by Theorem 6.6.5) that the reciprocal of the proper conic is a proper conic.

A plane passing through the vertex of a circular cone may cut the cone in a pair of straight lines. Such a section of a circular cone is called an *improper,* or *degenerate, conic of the first type.* In order to have Theorem 6.7.4 hold for both proper and improper conics, it is customary to define a pair of points (considered as the vertices of the pencils through them) as an *improper conic of the second type.* Note that Pappus' Theorem 6.2.6 is Pascal's Theorem for an improper conic of the first type.

PROBLEMS

1. Write out a description of each of the following figures, dualize the description, and sketch the dual figure.

(a)

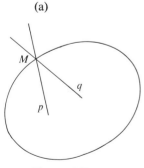

(b)

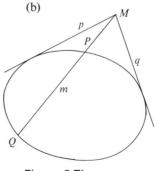

Figure 6.7a   Figure 6.7b

(c)

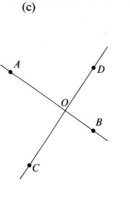

Figure 6.7c

(d)

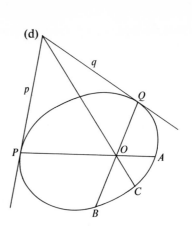

Figure 6.7d

(e)

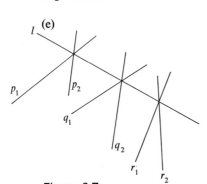

Figure 6.7e

(f)

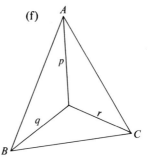

Figure 6.7f

(g)

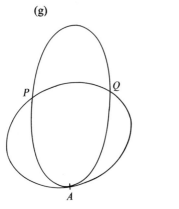

Figure 6.7g

(h)

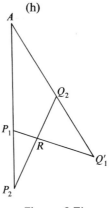

Figure 6.7h

2. What is the dual of a complete quadrangle?

3. State the dual of the following theorem: If points X, Y, U are collinear and if $X(AB,CU) = Y(AB,CU)$ then points A, B, C are collinear.

4. State the dual of the following theorem: ABC is a triangle, L a fixed point on AB, O a variable point on CL. If AO cuts CB in P and BO cuts CA in Q, then PQ intersects AB in a fixed point M.

5. State the dual of the following theorem: Copolar triangles are coaxial.

6. State the dual of Pappus' Theorem 6.2.6.

7. Dualize: (a) Theorem 6.3.1, (b) Theorem 6.3.5, (c) Theorem 6.3.8, (d) Theorem 6.3.14.

8. Show that Theorems 6.3.11 and 6.3.12 are duals of one another.

9. Prove that the reciprocal of a pole and polar for a proper conic is a polar and pole for the reciprocal conic.

10. Show that a self-conjugate triangle for a conic c reciprocates into a self-conjugate triangle for the reciprocal conic of c.

11. State Pascal's Theorem, Pappus' Theorem, and their converses as a single theorem.

12. (a) A point moves along a straight line. Show that the point of intersection of its polar lines with respect to two given proper conics traces a conic.
    (b) State the dual of the theorem of part (a).

13. (a) Given six coplanar points A, B, C, D, E, P, no three of A, B, C, D, E being collinear. Construct the polar of P with respect to the proper conic that passes through A, B, C, D, E.
    (b) Given six coplanar lines a, b, c, d, e, p, no three of a, b, c, d, e being concurrent. Construct the pole of p with respect to the proper conic touching a, b, c, d, e.

14. (a) A point moves on a proper conic. Show that the polar line of the point with respect to a second proper conic envelops a third proper conic.
    (b) State the dual of the theorem of part (a).

15. (a) The polar of a point A with respect to one conic is a, and the pole of a with respect to another conic is A'. Show that as A travels along a line, A' also travels along a line.
    (b) State the dual of the theorem of part (a).

16. (a) Prove that the lines joining the vertices of a triangle to any two points in the plane of the triangle meet the opposite sides in six points which lie on a conic.
    (b) State the dual of the theorem of part (a).

17. (a) If AL, BM, CN are three concurrent cevian lines of a triangle ABC, show that there is a proper conic tangent to BC, CA, AB at L, M, N respectively.
    (b) State the dual of the theorem of part (a).

18. A *complete plane* n-*point* is a figure consisting of n coplanar points, no three of which are collinear, and the $_nC_2$ lines determined by them. Describe a *complete plane* n-*line* if it is the dual of a complete plane n-point.

19. A planar figure consisting of points and lines is called a (plane) *configuration* if it consists of a_{11} points and a_{22} lines such that each point is on the same number (a_{12}) of lines and each line is on the same number (a_{21}) of points. A (plane) configuration is represented symbolically by the matrix

$$\begin{pmatrix} a_{11} & a_{12} \\ a_{21} & a_{22} \end{pmatrix}.$$

(a) Show that the dual of a configuration is a configuration, and give its matrix.
(b) Show that the figure of Desargues' Theorem in the plane is a self-dual configuration, and give its matrix.
(c) Show that a complete plane n-point and a complete plane n-line are configurations, and give their matrices.

20. A *complete space* n-*point* is a figure consisting of n points, no four coplanar, no three collinear, together with all the $_nC_2$ lines and $_nC_3$ planes determined by them. Show that the Desargues' configuration in a plane is a plane section of a complete space 5-point.

6.8 THE FOCUS-DIRECTRIX PROPERTY

We have defined a proper conic as any section of a right or oblique circular cone made by a plane not passing through the vertex of the cone. It is a remarkable fact, which we do not establish here, that all proper conics can be obtained as sections of only *right* circular cones.

Using the above fact that a proper conic is always a section of a right circular cone, we now derive a basic property of these curves which is customarily employed when studying these curves by analytic geometry.

6.8.1 THEOREM. *A noncircular proper conic is the locus of a point moving in a plane so that the ratio of its distance from a fixed point in the plane to its distance from a fixed line in the plane, not passing through the fixed point, is a constant.*

Consider the conic as a section made by a plane p cutting a right circular cone (see Figure 6.8a). Let s be a sphere inside the cone, touching the cone along a circle k whose plane we shall call q, and also touching plane p at point F. Let planes p and q intersect in a line d. From P, any point on the conic section, drop a perpendicular PR onto line d and a perpendicular PS onto plane q. Let the element of the cone through P cut q in point E. Finally, let α be the angle between planes p and q, and β the angle an element of the cone makes with plane q. Then, since $PF = PE$ (the segments being tangent lines from an external point to a sphere), and triangles PSE and PSR are right triangles with right angles at S, we have

$$PF/PR = PE/PR = (PS/PR)/(PS/PE) = \sin \alpha/\sin \beta = e,$$

a constant independent of the position of P on the conic. The conic may therefore be considered as the curve generated by a point P moving in the plane p such that the ratio of its distance from the point F of p to its distance from the line d of p is a constant e.

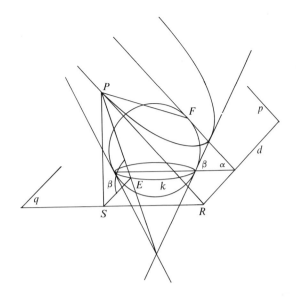

Figure 6.8a

6.8.2   Definitions.   The fixed point F of Theorem 6.8.1 is called a *focus* of the conic, the fixed line d is called a *directrix* of the conic, and the constant e is called the *eccentricity* of the conic.

6.8.3   Theorem.   *If* e *is the eccentricity of a proper conic, then the conic is a parabola, an ellipse, or a hyperbola according as* e = 1, e < 1, *or* e > 1.

If plane p in Figure 6.8a is parallel to one and only one element of the cone, then $\alpha = \beta$, and $e = 1$; if plane p cuts every element of one nappe of the cone, then $\alpha < \beta$, and $e < 1$; if p cuts both nappes of the cone, then $\alpha > \beta$, and $e > 1$.

6.8.4   Remark.   When plane p of Figure 6.8a is parallel to plane q, the section is a circle. In this case there is no finite directrix d, but one easily sees that this situation can be considered as a limiting position obtained by letting the intersection d of planes p and q move farther and farther away from the cone, the angle α, and hence also the fraction $\sin \alpha / \sin \beta$, becoming closer and closer to 0. This state of affairs is described by saying that the directrix of a circle is at infinity and the eccentricity of a circle is 0.

6.8.5   Theorem.   *An ellipse is the locus of a point moving in a plane such that the sum of its distances from two fixed points in the plane is a constant.*

Consider the ellipse as a section of a right circular cone (see Figure 6.8b). Let s_1 and s_2 be two spheres touching the plane p of the ellipse at the points

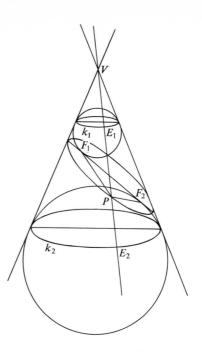

Figure 6.8b

F_1 and F_2 respectively, and touching the cone along the parallel circles k_1 and k_2 respectively. Join an arbitrary point P of the ellipse to F_1 and F_2, and let the element of the cone through P cut k_1 and k_2 in E_1 and E_2 respectively. Then $PF_1 = PE_1$ and $PF_2 = PE_2$ (each pair of equal segments being tangent lines from an external point to a sphere). It follows that

$$PF_1 + PF_2 = PE_1 + PE_2 = E_1E_2,$$

a constant independent of the position of P on the ellipse.

Note that if the ellipse is a circle, then F_1 and F_2 coincide at the center of the circle.

6.8.6   THEOREM.   *A hyperbola is the locus of a point moving in a plane such that the difference of its distances from two fixed points in the plane is a constant.*

We leave the establishment of this theorem, along the lines of the proof of Theorem 6.8.5, to the reader.

The simple and elegant approach of this section was discovered around the first quarter-mark of the nineteenth century by the two Belgian geometers Adolphe Quetelet (1796–1874) and Germinal Dandelin (1794–1847).

PROBLEMS

1. Establish Theorem 6.8.6.

2. Show that hyperbolas and noncircular ellipses have two distinct foci and two associated directrices, each directrix being perpendicular to the line joining the two foci.

3. Show that there exists an ellipse for each positive eccentricity $e < 1$, and there exists a hyperbola for each eccentricity $e > 1$.

4. Show that there are two plane sections of a right circular cone which have a focus at a given point within the cone.

5. Show that sections of a right circular cone made by parallel planes have equal eccentricities.

6. If an ellipse and a right circular cone are given, find a section of the cone congruent to the ellipse.

7. Prove that the sum of the distances of the vertex of a right circular cone from the ends of any diameter of a given elliptic section is constant.

6.9 ORTHOGONAL PROJECTION

Let π and π' be two planes not perpendicular to one another. The mapping of plane π onto plane π' which carries each point P of π into the foot P' of the perpendicular from P to π' is called the *orthogonal projection* of plane π onto plane π'.

Orthogonal projection, being a particular kind of perspectivity, has all the properties of a perspectivity, but it also has some special properties not possessed by all perspectivities. It is because of these special properties that orthogonal projection is such a highly useful transformation. We now establish some of the special properties.

6.9.1 THEOREM. *Under an orthogonal projection of a plane π onto a plane π' not perpendicular to π:*

(1) *ordinary points correspond to ordinary points,*

(2) *the lines at infinity correspond,*

(3) *parallel lines correspond to parallel lines,*

(4) *the ratio of two segments on a line is unaltered,*

(5) *the ratio of two segments on parallel lines is unaltered,*

(6) *the centroid of a triangle corresponds to the centroid of the corresponding triangle,*

(7) *if A and A' are two corresponding areas, then $A' = A \cos \theta$, where θ is the angle between planes π and π',*

(8) *the ratio of two areas is unaltered,*

(9) *circles map into homothetic ellipses.*

The first property is obvious and the next two are consequences of the first one. Properties (4) and (5) follow from the fact that the length of a line segment in π' is equal to the length of the corresponding segment in π multiplied by the cosine of the angle between the lines of the two segments. Property (6) is a consequence of property (4). Property (7) follows from the fact that any area in π can be considered as the limit of the sum of thin rectangular strips of common width parallel to the line of intersection of planes π and π', the limit being taken as the number of strips increases indefinitely; the lengths of these strips are unaltered by the projection, but their width w becomes $w \cos \theta$. Property (8) is a consequence of property (7). A circle clearly projects into an ellipse having its major axis parallel to the line of intersection of planes π and π'; also, if a and b are the semimajor and semiminor axes of the ellipse, and r is the radius of the circle, $a = r$ and $b = r \cos \theta$, and property (9) follows.

6.9.2 Theorem. *Any ellipse may be orthogonally projected into a circle.*

Let π be the plane of the ellipse and let a and b be the semimajor and semiminor axes of the ellipse. Through a line in π parallel to the minor axis of the ellipse, pass a plane π' making an angle $\theta = \cos^{-1}(b/a)$ with π. The reader may now easily show that the orthogonal projection of the ellipse to plane π' is a circle of radius b.

6.9.3 Corollary. *If an ellipse is orthogonally projected into a circle, then circles in the plane of projection correspond to those ellipses of the original plane which are homothetic to the given ellipse.*

6.9.4 Theorem. *Any triangle may be orthogonally projected into an equilateral triangle.*

Let the given triangle be ABC and let L, M, N be the midpoints of the sides BC, CA, AB respectively. By Problem 17(a), Section 6.7, there is a proper conic, which in this case clearly must be an ellipse, tangent to BC, CA, AB at L, M, N. Orthogonally project this ellipse into a circle (by Theorem 6.9.2), triangle ABC and points L, M, N thereby mapping into triangle $A'B'C'$ and points L', M', N'. Then, since $B'L' = L'C' = C'M' = M'A' = A'N' = N'B'$, it follows that triangle $A'B'C'$ is equilateral.

We now give a few applications of orthogonal projection; further applications will be found in the problems at the end of the section. The reader should note the employment of the *transform-solve-invert* procedure.

6.9.5 Theorem. *A necessary and sufficient condition for a triangle inscribed in an ellipse to have maximum area is that the centroid of the triangle coincide with the center of the ellipse.*

Orthogonally project the ellipse into a circle (Theorem 6.9.2). Triangles of maximum area in the ellipse correspond to triangles of maximum area in the circle (Theorem 6.9.1(8)). But a necessary and sufficient condition for a triangle inscribed in a circle to have maximum area is that the triangle be equilateral, and a necessary and sufficient condition for the triangle to be equilateral is that its centroid coincide with the center of the circle. The theorem now follows (Theorem 6.9.1(6)).

6.9.6 THEOREM. *The envelope of a chord of an ellipse which cuts off a segment of constant area is a concentric homothetic ellipse.*

Orthogonally project the ellipse into a circle (Theorem 6.9.2). The variable chord of the ellipse corresponds to a variable chord of the circle which cuts off a circular segment of constant area (Theorem 6.9.1(8)). But the envelope of the variable chord of the circle is a concentric circle. The theorem now follows (Corollary 6.9.3).

6.9.7 THEOREM. *The area of a triangle DEF inscribed in a given triangle ABC cannot be less than the area of each of the other three triangles formed.* (See Problem 4908, *The American Mathematical Monthly*, Apr. 1961, p. 386.)

Orthogonally project the inscribed triangle *DEF* into an equilateral triangle *D'E'F* (see Figure 6.9a). It suffices to prove the theorem for the

Figure 6.9a

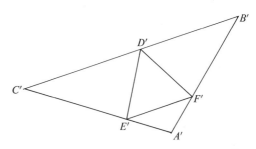

projected figure (Theorem 6.9.1(8)). Now if triangle *A'B'C'* is equilateral, then clearly the area of no subtriangle can exceed that of *D'E'F'*. Otherwise there is some angle, say *A'*, greater than 60°. But then the perpendicular from *A'* on *E'F'* is less than the perpendicular from *D'* on *E'F'*, and the area of triangle *D'E'F'* is greater than the area of triangle *A'E'F'*.

6.9.8 THEOREM. *If in the convex hexagon ABCDEF every pair of opposite sides are parallel, then triangles ACE and BDF have equal areas.* (See Article 5.8.15.)

Let *AB* and *DC* intersect in *P*. Orthogonally project triangle *BCP* into an equilateral triangle *B'C'P'* (see Figure 6.9b). Then the given hexagon *ABCDEF* projects into a hexagon *A'B'C'D'E'F'* whose opposite sides are still parallel, but all of whose angles are equal to 120°. Denoting the consecutive sides of *A'B'C'D'E'F'*, starting with side *A'B'*, by *a'*, *b'*, *c'*, *d'*, *e'*,

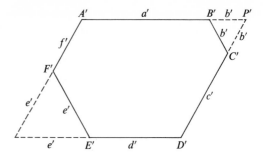

Figure 6.9b

f', it then suffices (because area $A'B'C' = (a'b' \sin 120°)/2$, etc.) to show that

(1) $$a'b' + c'd' + e'f' = f'a' + b'c' + d'e'.$$

But (see the figure) $a' = d' + e' - b'$ and $c' = e' + f' - b'$, and when these values for a' and c' are substituted in (1) we obtain an identity. Hence the theorem.

PROBLEMS

1. Show that the area of an ellipse with semiaxes a and b is πab.

2. Show that a triangle of maximum area inscribed in an ellipse has its sides parallel to the tangents to the ellipse at the opposite vertices.

3. Find the maximum area possessed by a triangle inscribed in an ellipse of semi-axes a and b.

4. Find the minimum area possessed by an ellipse circumscribed about a given triangle of area K.

5. Prove that if an ellipse is projected orthogonally into a circle, then pairs of perpendicular diameters of the circle correspond to pairs of conjugate diameters of the ellipse. (For a definition of *conjugate diameters* see Problem 4, Section 6.3.)

6. Prove that the area of any triangle, two of whose sides are conjugate radii of an ellipse, is constant.

7. Prove that a triangle of maximum area inscribed in an ellipse, and having a fixed chord for base, has its opposite vertex at an extremity of the diameter conjugate to the base.

8. Construct the maximum triangle which can be inscribed in a given ellipse and which has one of its vertices at a given point on the ellipse.

9. If PP', QQ' are a pair of variable parallel chords of an ellipse through the two fixed points P and Q on the ellipse, show that $P'Q'$ envelops a concentric homothetic ellipse.

10. AOB and $A'O'B'$, COD and $C'O'D'$ are pairs of parallel chords of an ellipse. Show that

$$(OA)(OB)/(OC)(OD) = (O'A')(O'B')/(O'C')(O'D').$$

11. We are given two concentric homothetic ellipses. Prove that all chords of the outer one which touch the inner one are bisected by the point of contact.

12. Show that two parallel tangents to an ellipse are met by any other tangent in points which lie on conjugate diameters.

13. *TP*, *TQ* are tangents, and *TRS* is a secant, of an ellipse. If *V* is the midpoint of *RS* and if *QV* cuts the ellipse again in *U*, prove that *PU* is parallel to *TS*.

14. Prove that the greatest ellipse which can be inscribed in a parallelogram touches the parallelogram at the midpoints of the four sides.

15. Prove that the chord of an ellipse which passes through a fixed interior point and cuts off the greatest or least area is bisected by the point.

16. Three congruent homothetic ellipses of semiaxes *a* and *b* are such that each touches the other two externally. Find the area of the curvilinear triangle bounded by the three ellipses.

17. Using orthogonal projection obtain a generalization of the following theorem: The feet of the perpendiculars dropped from any point of the circumcircle of a triangle on the sides of the triangle are collinear.

18. Using orthogonal projection obtain a generalization of the following theorem: The center of a circle inscribed in a quadrilateral is collinear with the midpoints of the diagonals of the quadrilateral.

19. Using orthogonal projection obtain a generalization of the following theorem: If a chord *AQ* of a circle of radius *r* cuts the diameter of the circle perpendicular to the diameter through *A* in point *R*, then $(AQ)(AR) = 2r^2$.

20. Prove, without using limits, that if *A* and *A'* are the areas of two corresponding triangles under an orthogonal projection, then $A' = A \cos \theta$, where θ is the angle between the planes of the two triangles.

BIBLIOGRAPHY

(Purely analytical treatments of projective geometry are not listed here.)

ADLER, C. F., *Modern Geometry, an Integrated First Course.* New York: McGraw-Hill Book Company, Inc., 1967.

ASKWITH, F. H., *A Course of Pure Geometry, Containing a Complete Geometrical Treatment of the Properties of the Conic Sections.* New York: Cambridge University Press, 1926.

BAKER, H. F., *An Introduction to Plane Geometry.* New York: Cambridge University Press, 1943.

COURANT, RICHARD, and HERBERT ROBBINS, *What is Mathematics?* New York: Oxford University Press, 1941.

COXETER, H. S. M., *The Real Projective Plane.* New York: McGraw-Hill Book Company, Inc., 1949.

———, *Projective Geometry.* Boston: Blaisdell Publishing Company, 1964.

CREMONA, LUIGI, *Elements of Projective Geometry*, tr. by Charles Leudesdorf. New York: Oxford University Press, 1885.

DOWLING, L. W., *Projective Geometry*. New York: McGraw-Hill Book Company, Inc., 1917.

FAULKNER, T. E., *Projective Geometry*. New York: Interscience Publishers, Inc., 1949.

FILON, L. N. G., *An Introduction to Projective Geometry*, 4th ed. London: Edward Arnold and Company, 1935.

FISHBACK, W. T., *Projective and Euclidean Geometry*, 2nd ed. New York: John Wiley and Sons, Inc., 1969.

FORDER, H. G., *Geometry*. New York: Hutchinson's University Library, 1950.

GOODSTEIN, R. L., and E. J. F. PRIMROSE, *Axiomatic Projective Geometry*. Leicester: University College, 1953.

GRAUSTEIN, W. C., *Introduction to Higher Geometry*. New York: The Macmillan Company, 1930.

HEADING, J., *An Elementary Introduction to the Methods of Pure Projective Geometry*. New York: St. Martin's Press, 1958.

HILBERT, DAVID, and S. COHN-VOSSEN, *Geometry and the Imagination*, tr. by P. Nemenyi. New York: Chelsea Publishing Company, 1952.

HOLGATE, T. F., *Projective Geometry*. New York: The Macmillan Company, 1930.

JAMES, GLENN, ed., *The Tree of Mathematics*. Pacoima, Calif.: The Digest Press, 1957.

LEHMER, D. N., *An Elementary Course in Synthetic Projective Geometry*. Boston: Ginn and Company, 1917.

LING, G. H., GEORGE WENTWORTH, and D. E. SMITH, *Elements of Projective Geometry*. Boston: Ginn and Company, 1922.

MACAULEY, F. S., *Geometrical Conics*, 2d ed. New York: Cambridge University Press, 1906.

MATHEWS, G. B., *Projective Geometry*. New York: Longmans, Green and Company, 1914.

MILNE, J. J., *An Elementary Treatise on Cross-Ratio Geometry*. New York: Cambridge University Press, 1911.

MILNE, W. P., *Projective Geometry for Use in Colleges and Schools*. New York: The Macmillan Company, 1911.

PATTERSON, B. C., *Projective Geometry*. New York: John Wiley and Sons, Inc., 1937.

PERFECT, HELEN, *Topics in Geometry*. New York: The Macmillan Company, 1963.

RAINICH, G. Y., and S. M. DOWDY, *Geometry for Teachers*. New York: John Wiley and Sons, Inc., 1968.

ROBINSON, G. DE B., *The Foundations of Geometry*. Toronto: University of Toronto Press, 1946.

RUSSELL, J. W., *An Elementary Treatise on Pure Geometry*, 2d ed. New York: Oxford University Press, 1905.

SANGER, R. G., *Synthetic Projective Geometry*. New York: McGraw-Hill Book Company, Inc., 1939.

SEIDENBERG, A., *Lectures in Projective Geometry*. Princeton, N.J.: D. Van Nostrand Company, Inc., 1962.

TULLER, ANNITA, *A Modern Introduction to Geometries*. Princeton, N.J.: D. Van Nostrand Company, Inc., 1967.

VEBLEN, OSWALD, and J. W. YOUNG, *Projective Geometry*, 2 vols. Boston: Ginn and Company, 1910, 1918.

WINGER, R. M., *An Introduction to Projective Geometry*. Boston: D. C. Heath and Company, 1923. Reprinted by Dover Publications, Inc., 1962.

YOUNG, J. W., *Projective Geometry*. Carus Mathematical Monograph No. 4. Buffalo, N.Y.: Mathematical Association of America, 1930.

YOUNG, J. W. A., ed., *Monographs on Topics of Modern Mathematics Relevant to the Elementary Field*. New York: Longmans, Green and Company, 1911. Reprinted by Dover Publications, Inc., 1955.

7

Non-Euclidean
Geometry

Shortly after the first quarter of the nineteenth century a geometrical event took place which proved to be of tremendous significance, not only for geometry, but for the whole of mathematics; a geometry was invented which differs radically from the traditional geometry of Euclid. Prior to this event it was believed that there was, and indeed could be, only one possible geometry, and that any description of space contrary to the Euclidean description must of necessity be inconsistent and contradictory.

Geometry was liberated from its traditional mold, and the postulates of geometry became, for the mathematician, mere hypotheses whose physical truth or falsity need not concern him. It became apparent that the mathematician may choose his postulates to suit his pleasure, so long as they be consistent with one another. A postulate, as employed by the mathematician, need have nothing to do with "self-evidence" or "factualness."

With the possibility of inventing such purely "artificial" geometries it became apparent that physical space must be viewed as an empirical concept derived from our external experiences, and that the postulates of a geometry designed to describe physical space are simply expressions of this experience —like the laws of a physical science. This point of view is in striking contrast to the Kantian theory of space that dominated philosophical thinking at the time of the invention of the new geometry. The Kantian theory claimed that space is a framework already existing intuitively in the human mind, that the axioms and postulates of Euclidean geometry are *a priori* judgements imposed on the mind, and that without these axioms and postulates no consistent reasoning about space can be possible. The invention of a non-Euclidean geometry rendered this viewpoint completely untenable.

The invention of a non-Euclidean geometry, by puncturing a traditional belief and breaking a centuries-long habit of thought, dealt a severe blow to the *absolute truth* viewpoint of mathematics. Indeed, the invention not only liberated geometry, but had a similar effect on mathematics as a whole. Mathematics emerged as an arbitrary creation of the human mind, and not as something essentially dictated to us of necessity by the world in which we live.

The present chapter concerns itself with the first non-Euclidean geometry that was invented—the so-called Lobachevskian geometry. We commence with a sketch of the historical background that led to this momentous discovery; probably nowhere in the whole study of mathematics is a discussion of historical origins more necessary than here for a true appreciation of the subject. The broad effect of the discovery on both geometry in particular and mathematics in general will be the subject matter of the next chapter.

7.1 HISTORICAL BACKGROUND

There is evidence that the theory of parallels gave the early Greeks considerable trouble, and that at one time the theory involved an illogical circularity. Euclid met the difficulty by defining parallel lines as coplanar straight lines which do not meet one another however far they may be produced in either direction, and by adopting as a basic assumption his now famous fifth, or parallel, postulate:

If a straight line falling on two straight lines makes the interior angles on the same side together less than two right angles, the two straight lines, if produced indefinitely, meet on that side on which the angles are together less than two right angles.

If the reader will turn to the Appendix, where Euclid's initial explanations, preliminary definitions, postulates, and axioms are listed, he will instantly note a marked difference between the fifth postulate and the other four; the fifth postulate lacks the terseness and the simple comprehensibility of the other four. Recall (see Section 1.3) that, in a discourse conducted by the Greek method of material axiomatics, a postulate is supposed to be a primitive statement which is felt to be *acceptable as immediately true* on the basis of the properties suggested by the initial explanations. It must be confessed that the fifth postulate hardly satisfies this requirement of a postulate.

Proclus tells us that the fifth postulate was attacked from the very start. A more studied examination reveals that it is actually the converse of Euclid's

Proposition I 17. Moreover, Euclid himself apparently tried to avoid the postulate as long as he could, for he makes no use of it in his proofs until he reaches Proposition I 29. It is not surprising that many Greek geometers felt that the postulate smacked more of the nature of a proposition than of a postulate. From the Greek point of view of material axiomatics, it was certainly very natural to wonder whether the fifth postulate was really needed at all, and to think that perhaps it could be derived as a theorem from the remaining nine "axioms" or "postulates," or, at least, that it could be replaced by a more acceptable equivalent.

Of the many substitutes that have been devised to replace Euclid's fifth postulate, perhaps the most popular is that made well known in modern times by the Scottish physicist and mathematician, John Playfair (1748–1819), although this particular alternative had been advanced by others and had been stated as early as the fifth century by Proclus. It is the substitute usually encountered in present-day American high school texts, namely: *Through a given point not lying on a given line there can be drawn only one line parallel to the given line.** Some other alternatives for the parallel postulate that have been either proposed or tacitly assumed over the years are these: (1) *There exists a pair of coplanar straight lines everywhere equally distant from one another.* (2) *There exists a pair of similar noncongruent triangles.* (3) *If in a quadrilateral a pair of opposite sides are equal and if the angles adjacent to a third side are right angles, then the other two angles are also right angles.* (4) *If in a quadrilateral three angles are right angles, the fourth angle is also a right angle.* (5) *There exists at least one triangle having the sum of its three angles equal to two right angles.* (6) *Through any point within an angle less than $60°$ there can always be drawn a straight line intersecting both sides of the angle.* (7) *A circle can be passed through any three noncollinear points.* (8) *There is no upper limit to the area of a triangle.*

It makes an interesting and challenging collection of exercises for the student to try to show the equivalence of the above alternatives to the original postulate stated by Euclid. To show the equivalence of Euclid's postulate and a particular one of the alternatives, one must show that the alternative follows as a theorem from Euclid's assumptions, and also that Euclid's postulate follows as a theorem from Euclid's system of assumptions with the parallel postulate replaced by the considered alternative.

The attempts to derive the parallel postulate as a theorem from the remaining nine "axioms" and "postulates" occupied geometers for over two thousand years and culminated, as we shall see, in some of the most far-reaching developments of modern mathematics. Many "proofs" of the postulate were offered, but each was sooner or later shown to rest upon a tacit assumption equivalent to the postulate itself.

It was not until 1733 that the first really scientific investigation of the parallel postulate was published. In that year the Italian Jesuit priest Girolamo

* Propositions I 27 and I 28 guarantee, under the assumption of the infinitude of straight lines, the existence of at least *one* parallel.

Saccheri (1667–1733), while Professor of Mathematics at the University of Pavia, published a little book entitled *Euclides ab omni naevo vindicatus* (Euclid Freed of Every Flaw). In an earlier work on logic, Saccheri had become charmed with the powerful method of *reductio ad absurdum* and conceived the idea of applying this method to an investigation of the parallel postulate. Without using the parallel postulate Saccheri easily showed, as can any high school geometry student, that if, in a quadrilateral *ABCD* (see Figure 7.1a), angles *A* and *B* are right angles and sides *AD* and *BC* are

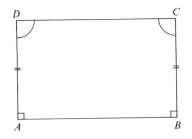

Figure 7.1a

equal, then angles *D* and *C* are equal. There are, then, three possibilities: angles *D* and *C* are equal acute angles, equal right angles, or equal obtuse angles. These three possibilities were referred to by Saccheri as the *hypothesis of the acute angle*, the *hypothesis of the right angle*, and the *hypothesis of the obtuse angle*. The plan of the work was to show that the assumption of either the hypothesis of the acute angle or the hypothesis of the obtuse angle would lead to a contradiction. Then, by *reductio ad absurdum*, the hypothesis of the right angle must hold, and this hypothesis, Saccheri showed, carried with it a proof of the parallel postulate. Tacitly assuming, as in fact did Euclid, the infinitude of the straight line, Saccheri readily eliminated the hypothesis of the obtuse angle, but the case of the hypothesis of the acute angle proved to be much more difficult. After painstakingly obtaining many of the now classical theorems of so-called non-Euclidean geometry, Saccheri lamely forced into his development an unconvincing contradiction involving hazy notions about infinite elements. Had he not seemed so eager to exhibit a contradiction here, but rather had admitted his inability to find one, Saccheri would today unquestionably be credited with the discovery of non-Euclidean geometry. It seems that shortly after the publication of Saccheri's little work, it was suddenly removed from the market, with the result that Saccheri's efforts had little effect on his contemporaries.

Thirty-three years after Saccheri's publication, Johann Heinrich Lambert (1728–1777) of Germany wrote a similar investigation entitled *Die Theorie der Parallellinien*, which, however, was not published until eleven years after his death. Lambert chose a quadrilateral containing three right angles (half of a Saccheri quadrilateral) as his fundamental figure, and considered three hypotheses according as the fourth angle is acute, right, or obtuse. He went considerably beyond Saccheri in deducing propositions under the hypotheses of the acute and obtuse angles. Thus, with Saccheri, he showed that in the three hypotheses the sum of the angles of a triangle is less than, equal to, or greater than two right angles respectively, and then, in addition, that the deficiency below two right angles in the hypothesis of the acute angle, or the excess above two right angles in the hypothesis of the obtuse angle, is proportional to the area of the triangle. He observed the resemblance of the geometry following from the hypothesis of the obtuse angle to spherical geometry, where the area of a triangle is proportional to its spherical excess, and conjectured that the geometry following from the hypothesis of the acute angle could perhaps be verified on a sphere of imaginary radius. The hypothesis of the obtuse angle was eliminated by making the same tacit assumption as had Saccheri, but his conclusions with regard to the hypothesis of the acute angle were indefinite and unsatisfactory, which, indeed, was the reason his work was never published during his lifetime.

A third distinguished effort to establish Euclid's parallel postulate by the *reductio ad absurdum* method was essayed, over a long period of years, by the eminent Italian-French analyst Adrien-Marie Legendre (1752–1833). He began anew and considered three hypotheses according to whether the sum of the angles of a triangle is less than, equal to, or greater than two right angles. Tacitly assuming the infinitude of a straight line, he was able to eliminate the third hypothesis, but, although he made repeated attempts, he could not dispose of the first hypothesis. These various endeavors appeared in the successive editions of his very popular *Éléments de géométrie,** which ran from a first edition in 1794 to a twelfth in 1823. Legendre perhaps holds the record for persistence in attempting to prove the famous postulate. The simple and straightforward style of his proofs, widely circulated because of their appearance in his *Éléments,* and his high eminence in the world of mathematics, created marked popular interest in the problem of the parallel postulate.

It is no wonder that no contradiction was found under the hypothesis of the acute angle, for, as we shall show later, it is now known that the geometry developed under this hypothesis is as consistent as Euclidean geometry; that is, the parallel postulate is independent of the remaining postulates and cannot be deduced from them. Of course there are theorems in the new geometry which contradict theorems in Euclidean geometry, but there apparently are no two theorems in the new geometry which contradict

---

* This work is an attempted pedagogical improvement of Euclid's *Elements* made by considerably rearranging and simplifying the propositions. The work won high regard in continental Europe and was so favorably received in the United States that it became the prototype of the elementary geometry textbooks in this country. The first English translation was made in the United States in 1819 by John Farrar of Harvard University. The next English translation was made in 1824 by the famous Scottish litterateur, Thomas Carlyle, who early in life was a teacher of mathematics. Carlyle's translation ran through 33 American editions.

one another. The first to suspect this fact were Karl Friedrich Gauss (1777–1855) of Germany, János Bolyai (1802–1860) of Hungary, and Nicolai Ivanovitch Lobachevsky (1793–1856) of Russia. These men approached the subject through the Playfair form of the parallel postulate by considering the three possibilities: Through a given point not on a given line can be drawn *more than one*, or *just one*, or *no* line parallel to a given line. These situations are equivalent, respectively, to the hypotheses of the acute, the right, and the obtuse angles. Again assuming the infinitude of a straight line, the third case was easily eliminated. Suspecting, in time, a consistent geometry under the first possibility, each of these three mathematicians independently carried out extensive geometric and trigonometric developments of the hypothesis of the acute angle.

Gauss was, without doubt, the first to reach penetrating conclusions concerning the hypothesis of the acute angle, but since throughout his life he failed to publish anything on the matter, the honor of discovering this particular non-Euclidean geometry must be shared with Bolyai and Lobachevsky. Bolyai published his findings in 1832 in an appendix to a mathematical work of his father. Later it was learned that Lobachevsky, separated from the rest of the scientific world by barriers of distance and language, had published similar findings as early as 1829–30. Because of Lobachevsky's priority in publishing, the geometry of the hypothesis of the acute angle has come to be called *Lobachevskian geometry*.

The actual independence of the parallel postulate from the other postulates of Euclidean geometry was not unquestionably established until consistency proofs of the hypothesis of the acute angle were furnished. These were now not long in coming and were supplied by Eugenio Beltrami, Arthur Cayley, Felix Klein, Henri Poincaré, and others. The method was to set up a model within Euclidean geometry so that the abstract development of the hypothesis of the acute angle could be given a Euclidean interpretation in the model. Then any inconsistency in the non-Euclidean geometry would imply a corresponding inconsistency in Euclidean geometry. We shall consider such a model, due to Poincaré, in the next chapter.

We have seen that the hypothesis of the obtuse angle was discarded by all who did research in this subject because it contradicted the assumption that a straight line is infinite in length. Recognition of a second non-Euclidean geometry, based on the hypothesis of the obtuse angle, was not fully achieved until some years later, when Bernhard Riemann (1826–1866), in his famous probationary lecture of 1854, discussed the concepts of boundlessness and infiniteness. With the difference between these concepts clarified, one can realize an equally consistent geometry satisfying the hypothesis of the obtuse angle if Euclid's Postulates 1, 2, and 5 are modified to read:

1′. Two distinct points determine at least one straight line.
2′. A straight line is boundless.
5′. Any two straight lines in a plane intersect.

Actually, the modified statements (1′) and (2′) are precisely Euclid's first two postulates taken at their face value; though Euclid meant two distinct points to determine one *and only one* straight line, and though he meant straight lines to be *infinite*, he never really asserted this much in his postulates.

Riemann's lecture inaugurated a second period in the development of non-Euclidean geometry, a period characterized by the employment of the methods of differential geometry rather than the previously used methods of elementary synthetic geometry. To this lecture we owe a considerable generalization of the concept of space which has led, in more recent times, to the extensive and important theory of abstract spaces; some of this theory has found application in the physical theory of relativity.

Since the discovery of the two non-Euclidean geometries which result from the hypotheses of the acute and obtuse angles, other nontraditional geometries have been invented. Indeed, Riemann was the originator of a whole class of nontraditional geometries. These have received intensive study in recent times and are referred to as *Riemannian geometries.* Another non-traditional geometry is one devised by Max Dehn in which the Postulate of Archimedes is suppressed; such a geometry is referred to as a *non-Archimedean geometry.* The invention of these new geometries not only liberated geometry from its traditional Euclidean mold, but, as we shall see in the next chapter, considerably modified former conceptions of mathematics in general and led to a profound study of the foundations of the subject and to a further development of the axiomatic method.

The deductive consequences of the Euclidean postulational basis with the parallel postulate extracted constitute what is called *absolute geometry*; it contains those propositions which are common to both Euclidean and Lobachevskian geometry. Among these common propositions are the first twenty-eight propositions of Euclid's Book I. The reader will note that in Appendix I, where all forty-eight propositions of the first book of Euclid's *Elements* are listed, the first twenty-eight are, for convenience, separated from the remaining twenty. When working in Lobachevskian geometry, the student may safely appeal to any of these first twenty-eight propositions.

PROBLEMS

1. Show that Playfair's Postulate and Euclid's fifth postulate are equivalent. (One may use any of Euclid's first 28 propositions.)

2. Prove that each of the following statements is equivalent to Playfair's Postulate:
   (a) If a straight line intersects one of two parallel lines, it will intersect the other also.
   (b) Straight lines which are parallel to the same straight line are parallel to one another.

3. Show that Playfair's Postulate and the statement, "The sum of the angles of a triangle is always equal to two right angles," are equivalent.

4. Find the fallacy in the following "proof," given by B. F. Thibaut (1809), of Euclid's fifth postulate: Let a straightedge be placed with its edge coinciding with side CA of a triangle ABC. Rotate the straightedge successively about the three vertices A, B, C, in the direction ABC, so that it coincides in turn with AB, BC, CA. When the straightedge returns to its original position it must have rotated through four right angles. But the whole rotation is made up of three rotations equal to the exterior angles of the triangle. It now follows that the sum of the angles of the triangle must be equal to two right angles, and from this follows Euclid's parallel postulate.

5. Find the fallacy in the following "proof," given by J. D. Gergonne (1812), of Euclid's fifth postulate: Let PA and QB, lying in the same plane and on the same side of PQ, be perpendicular to PQ. Then PA and QB are parallel. Let PG be the last ray through P, and lying within angle QPA, which intersects QB. Produce QB to a point K beyond the point of intersection of PG with QB, and draw PK. It follows that PG is *not* the last ray through P which meets QB, and therefore all rays through P and lying within angle QPA must meet QB. Thus through P there is only one line parallel to line QB, and Euclid's fifth postulate follows.

6. Find the fallacy in the following "proof," given by J. K. F. Hauff (1819), of Euclid's fifth postulate: Let AD, BE, CF be the altitudes of an equilateral triangle ABC, and let O be the point of concurrency of these altitudes. In right triangle ADC, acute angle CAD equals one half acute angle ACD. Therefore, in right triangle AEO, acute angle OAE equals one half acute angle AOE. A similar treatment holds for each of the six small right triangles of which AEO is typical. It now follows that the sum of the angles of triangle ABC is equal to one half the sum of the angles about O, that is, equal to two right angles. But it is known that the existence of a single triangle having the sum of its angles equal to two right angles is enough to guarantee Euclid's fifth postulate.

7. Show that Propositions I 27 and I 28 guarantee, under the assumption of the infinitude of straight lines, the existence of at least *one* line through a given point parallel to a given line not passing through the point.

8. Prove, by simple congruence theorems (which do not require the parallel postulate), the following theorems about Saccheri quadrilaterals:
(a) The summit angles of a Saccheri quadrilateral are equal to each other.
(b) The line joining the midpoints of the base and summit of a Saccheri quadrilateral is perpendicular to both the base and the summit.
(c) If perpendiculars are drawn from the extremities of the base of a triangle upon the line passing through the midpoints of the two sides, a Saccheri quadrilateral is formed.
(d) The line joining the midpoints of the equal sides of a Saccheri quadrilateral is perpendicular to the line joining the midpoints of the base and summit.

9. Show that a Lambert quadrilateral can be regarded as half a Saccheri quadrilateral.

10. A *spherical degree* for a given sphere is defined to be any spherical area which is equal to (1/720)th of the entire surface of the sphere. The *spherical excess* of a spherical triangle is defined as the excess, measured in degrees of angle, of the sum of the angles of the triangle above 180°.

(a) Show that the area of a lune whose angle is $n°$ is equal to $2n$ spherical degrees.

(b) Show that the area of a spherical triangle, in spherical degrees, is equal to the spherical excess of the triangle.

(c) Show that the area A of a spherical triangle of spherical excess $E°$ is given by

$$A = \pi r^2 E°/180°,$$

where r is the radius of the sphere. This shows that, for a given sphere, the area of a spherical triangle is proportional to its spherical excess.

11. Fill in the details of the following proof of *Legendre's First Theorem*: "The sum of the three angles of a triangle cannot be greater than two right angles." Show that the proof assumes the infinitude of the straight line.

Suppose that the sum of the angles of a triangle ABC is $180° + \theta$, and that angle CAB is not greater than either of the other angles. Join A to D, the midpoint of BC, and produce AD its own length to E. Show that triangles BDA and CDE are congruent; hence that the sum of the angles of triangle AEC is also equal to $180° + \theta$. One of the angles CAE and CEA is not greater than $\frac{1}{2} \measuredangle CAB$. Apply the same process to triangle AEC, obtaining a third triangle whose angle-sum is $180° + \theta$ and one of whose angles is not greater than $(\frac{1}{2})^2 \measuredangle CAB$. By applying the construction n times, a triangle is reached whose angle-sum is $180° + \theta$ and one of whose angles is not greater than $(\frac{1}{2})^n \measuredangle CAB$. But (by the Postulate of Archimedes) there exists an integer k such that $k\theta > \measuredangle CAB$. Choose n so large that $2^n > k$. Then $\theta > (\frac{1}{2})^n \measuredangle CAB$, and the sum of two of the angles of the last triangle must be greater than $180°$. But this conclusion contradicts Proposition I 17.

12. In one effort to eliminate the hypothesis of the acute angle, Legendre tried to obtain, under this hypothesis, a triangle containing a given triangle at least twice. He proceeded as follows. Let ABC be any triangle such that angle A is not greater than either of the other two angles. Construct on side BC a triangle BCD congruent to triangle ABC, with angle DCB equal to angle B, angle DBC equal to angle C, and D on the opposite side of BC from A. Through D draw any line cutting AB and AC produced in E and F, respectively. Then triangle AEF contains triangle ABC at least twice.

(a) Show that this construction assumes that through a point within a given angle less than $60°$ there can always be drawn a straight line intersecting both sides of the angle. (This, we have seen, is equivalent to the fifth postulate.)

(b) If the construction above had been independent of the fifth postulate, how would it have eliminated the hypothesis of the acute angle?

13. Assuming Legendre's First Theorem (see Problem 11, above), prove the following sequence of theorems credited to Legendre.

(a) If the sum of the angles of a triangle is equal to two right angles, then the same is true of any triangle obtained from the given triangle by drawing a cevian line through one of its vertices.

(b) If there exists a triangle with the sum of its angles equal to two right angles, then one can construct an isosceles right triangle having the sum of its angles equal to two right angles and its legs greater in length than any given line segment.

(c) *Legendre's Second Theorem.* If there exists a single triangle having the sum

of its angles equal to two right angles, then the sum of the angles of every triangle
will be equal to two right angles.

(d) If there exists a single triangle having the sum of its angles less than two
right angles, then the sum of the angles of every triangle is less than two right
angles.

14. Show that the assumption of the existence of noncongruent similar triangles is
equivalent to the fifth postulate.

15. Show that the assumption of the existence of a pair of straight lines which are
everywhere equally distant from one another is equivalent to the fifth postulate.

7.2 PARALLELS AND HYPERPARALLELS

We proceed, in this and the following seven sections, to develop some of
the geometry of the Lobachevskian plane. As stated at the end of the previous
section, free use will be made of the first twenty-eight propositions of
Euclid's *Elements*. These twenty-eight propositions are independent of the
parallel postulate and therefore hold in Lobachevskian geometry as well as
in Euclidean geometry. The reader will find these propositions listed in the
Appendix.

In place of Euclid's parallel postulate we make the following assumption:

7.2.1 THE LOBACHEVSKIAN PARALLEL POSTULATE. *If* P *is a point not on
the line* AB *(see Figure 7.2a) and if* Q *is the foot of the perpendicular from*

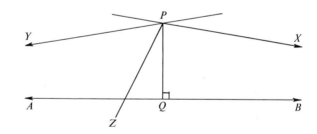

Figure 7.2a

P *on* AB, *there are two rays* PX, PY *from* P, *not in the same line and not
intersecting* AB, *and such that any ray* PZ *from* P *and lying within the* ⋨XPY
containing PQ *intersects* AB.*

We now proceed with the development.

7.2.2 THEOREM. *In Figure 7.2a, any line through* P *and not passing within
the* ⋨XPY *containing* PQ *does not intersect line* AB.

For if such a line did intersect line *AB*, then ray *PX* or ray *PY* would
have to intersect line *AB*.

* For convenience in our brief treatment we have chosen a stronger form of the
Lobachevskian parallel postulate than is really needed.

7.2.3   DEFINITIONS.   We shall call the lines PX and PY of Figure 7.2a *the parallels* through P to line AB. Directed line $\overrightarrow{PX}$ is said to be *parallel* to directed line $\overrightarrow{AB}$, and $\overrightarrow{PY}$ is said to be *parallel* to $\overrightarrow{BA}$. Lines through P and not passing within the $\sphericalangle XPY$ containing PQ will be called *hyperparallels* through P to line AB.

7.2.4   THEOREM.   *If* Q *is the foot of the perpendicular from point* P *on line* AB *and if* PX *and* PY *are the parallels through* P *to line* AB, *then angles* XPQ *and* YPQ *are equal acute angles.*

Suppose $\sphericalangle YPQ > \sphericalangle XPQ$. Lay off (see Figure 7.2b) $\sphericalangle MPQ = \sphericalangle XPQ$.

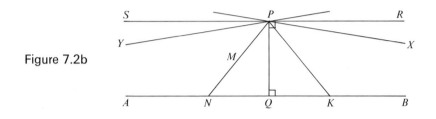

Figure 7.2b

Then PM lies within $\sphericalangle YPQ$, and so must intersect line AB in a point N. Mark off on AB, on the side of Q opposite N, $QK = QN$, and draw PK. Then triangles NPQ and KPQ are congruent (by I 4) and $\sphericalangle KPQ = \sphericalangle MPQ = \sphericalangle XPQ$. Thus PX and PK coincide. But this is impossible, since PX does not intersect AB. Therefore $\sphericalangle YPQ \not> \sphericalangle XPQ$. We may similarly show that $\sphericalangle XPQ \not> \sphericalangle YPQ$. It follows that $\sphericalangle YPQ = \sphericalangle XPQ$.

   Now angles YPQ and XPQ are not right angles, for if they were then PY and PX would lie on the same line, which they do not. Angles YPQ and XPQ are not obtuse angles, for if they were then line SPR through P and perpendicular to PQ would pass within the $\sphericalangle XPY$ containing PQ, and therefore would have to intersect line AB, which it does not (by I 28). It follows that angles YPQ and XPQ are acute angles.

7.2.5   COROLLARY.   *There are infinitely many hyperparallels to a line* AB *through a point* P *not on* AB.

7.2.6   DEFINITION.   Angle XPQ (or angle YPQ) in Figure 7.2b is called the *angle of parallelism* at P for line AB.

7.2.7   THEOREM.   *If* $\overrightarrow{PX}$ *is parallel to* $\overrightarrow{AB}$ *and* R *is any point on* PX *such that* P *and* R *are on the same side of* X, *then* $\overrightarrow{RX}$ *is parallel to* $\overrightarrow{AB}$. (This is called the *transmissibility property* of parallelism.)

   Case 1 (R between P and X): Draw PQ and RS (see Figure 7.2c) perpendicular to AB. Let RT be any ray from R and lying within $\sphericalangle SRX$.

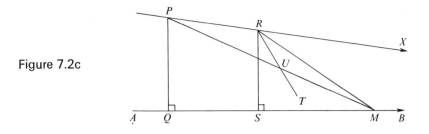

Figure 7.2c

Take any point U on ray RT and on the same side of AB as P. Draw PU. Then PU must cut AB in a point M. It follows that RU must cut SM, and $\overline{RX}$ is parallel to $\overline{AB}$.

Case 2 (P between R and X): The proof is the same as that of Case 1 except we take U as any point on the backward extension of ray RT (see Figure 7.2d).

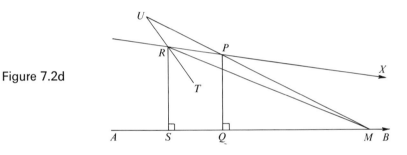

Figure 7.2d

7.2.8 THEOREM. *If* $\overline{CD}$ *is parallel to* $\overline{AB}$, *then* $\overline{AB}$ *is parallel to* $\overline{CD}$. (This is called the *symmetry property* of parallelism.)

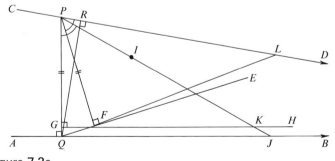

Figure 7.2e

Referring to Figure 7.2e, take P, any point on CD, and draw PQ perpendicular to AB and QR perpendicular to CD. By I 16 it follows that R must be on ray *PD*. Let *QE* be any ray through *Q* and lying within ⨯*RQB*. Draw *PF* perpendicular to line *QE*. Then, again by I 16, *F* lies on ray *QE*. On *PQ* mark off *PG* = *PF*. By I 18, *G* lies between *P* and *Q*. Draw *GH* perpendicular to *PQ*. Lay off ⨯*GPI* = ⨯*FPD* and produce *PI* to cut *AB* in *J*. Since *GH* cuts side *PQ* of triangle *PQJ*, but does not (by I 28) cut side *QJ*, it must cut side *PJ* at some point *K*. On *PD* mark off *PL* = *PK* and draw *FL*. Since triangles *PGK* and *PFL* are congruent (by I 4), it follows that ⨯*PFL* is a right angle. But ⨯*PFE* is a right angle. Hence *QE* falls on *FL*, and thus cuts *PD* in *L*. It now follows that $\overline{AB}$ is parallel to $\overline{CD}$. (This proof is due to Lobachevsky.)

7.2.9 THEOREM. *If* $\overline{AB}$ *and* $\overline{CD}$ *are parallel to* $\overline{EF}$, *then* $\overline{AB}$ *is parallel to* $\overline{CD}$. (This is called the *transitivity property* of parallelism.)

Case 1 (*EF* between *AB* and *CD*): Referring to Figure 7.2f, connect a

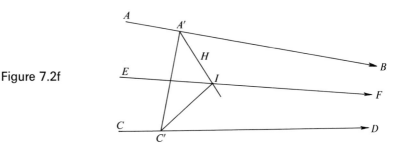

Figure 7.2f

point *A'* on *AB* to a point *C'* on *CD* and let *A'H* be any ray through *A'* and lying within ⨯*C'A'B*. Since $\overline{AB}$ is parallel to $\overline{EF}$, *A'H* cuts *EF* in a point *I*. Draw *C'I*. Since $\overline{EF}$ is parallel to $\overline{CD}$, *A'I* produced must cut *CD*. Since *AB* does not cut *CD* (if it did, it would have to cut *EF*), but every ray *A'H* lying within ⨯*C'A'B* does cut *CD*, it follows that $\overline{AB}$ is parallel to $\overline{CD}$.

Case 2 (*AB* and *CD* on the same side of *EF*): Referring to Figure 7.2g, let $\overline{GH}$ be the parallel to $\overline{CD}$ through a point *G* of *AB*. Then, by Case 1,

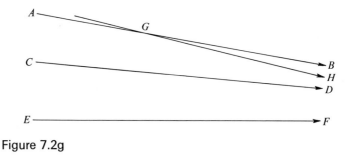

Figure 7.2g

$\overline{GH}$ is parallel to $\overline{EF}$. It follows that AB coincides with GH, and thus $\overline{AB}$ is parallel to $\overline{CD}$. (This proof is due to Gauss.)

7.3 LIMIT TRIANGLES

Very useful for the further development of the geometry of the Lobachevskian plane is the concept of a "limit triangle."

7.3.1 DEFINITIONS. A figure (see Figure 7.3a) consisting of two parallel

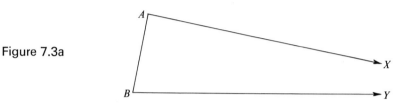

Figure 7.3a

rays and the line segment connecting the origins of the rays is called a *limit triangle*. The segment connecting the origins of the rays is called the *finite side* of the limit triangle, and the angles at the extremities of the finite side are called the *angles* of the limit triangle.

7.3.2 THEOREM. *An exterior angle of a limit triangle is greater than the opposite interior angle.*

Suppose (see Figure 7.3b) $\sphericalangle CAX < \sphericalangle CBY$. Lay off $\sphericalangle CAD = \sphericalangle CBY$.

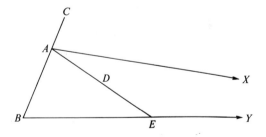

Figure 7.3b

Then AD lies within $\sphericalangle BAX$, and therefore must meet BY in some point E. Then, in triangle ABE, exterior angle CAE is equal to opposite interior angle ABE. But this contradicts I 16. It follows that $\sphericalangle CAX \nless \sphericalangle CBY$.

Suppose (see Figure 7.3c) $\sphericalangle CAX = \sphericalangle CBY$. Let L be the midpoint of AB. Draw NLM perpendicular to AX. Then triangles LAM and LBN are congruent (by I 26). It follows that $\sphericalangle BNL = \sphericalangle AML =$ a right angle. Then $\sphericalangle AML$ is the angle of parallelism at M for line BY. But this is impossible,

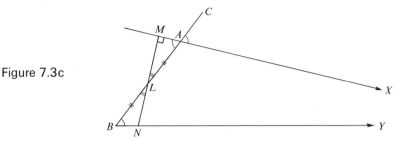

Figure 7.3c

since an angle of parallelism is acute. It follows that $\angle CAX \neq \angle CBY$.
We now conclude that $\angle CAX > \angle CBY$, and the theorem is established.

7.3.3 THEOREM. *If, in two limit triangles, the finite sides are equal, and an angle of one is equal to an angle of the other, then the two remaining angles are also equal.*

Referring to Figure 7.3d, let $AB = A'B'$ and $\angle A = \angle A'$. Suppose $\angle B >$

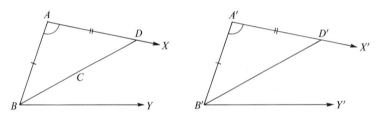

Figure 7.3d

$\angle B'$. Lay off $\angle ABC = \angle B'$. Then BC must intersect AX in some point D.
Mark off on $A'X'$, $A'D' = AD$, and draw $B'D'$. Then triangles BAD and
$B'A'D'$ are congruent (by I 4), and $\angle A'B'D' = \angle ABD = \angle B'$. But this is
impossible. It follows that $\angle B \not> \angle B'$. Similarly, $\angle B' \not> \angle B$. That is,
$\angle B = \angle B'$.

7.3.4 THEOREM. *If, in two limit triangles, the two angles of one are equal to the two angles of the other, then the two finite sides of the triangles are also equal.*

Referring to Figure 7.3e, suppose $AB > A'B'$. Mark off $AC = A'B'$ and
let $\overline{CZ}$ be the parallel through C to $\overline{AX}$ and $\overline{BY}$. By Theorem 7.3.3, $\angle ACZ$
$= \angle B' = \angle B$. But this contradicts Theorem 7.3.2. It follows that
$AB \not> A'B'$. Similarly, $A'B' \not> AB$. Therefore $AB = A'B'$.

7.3.5 THEOREM. *The angle of parallelism at P for a line AB depends only on the distance PQ of P from AB, and it decreases as PQ increases.*

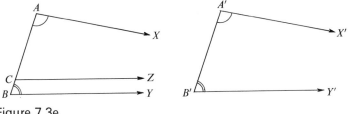

Figure 7.3e

This is an immediate consequence of Theorems 7.3.3 and 7.3.2.

7.3.6 NOTATION. If P (see Figure 7.3f) is at a distance h from line AB,

Figure 7.3f

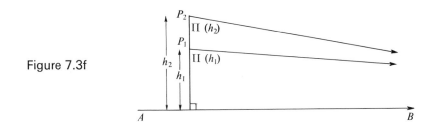

following Lobachevsky we denote the angle of parallelism at P for line AB by the symbol $\Pi(h)$.

We shall later show that $\Pi(h) = 2 \ \mathrm{arc} \ \tan \ e^{-h}$ if the unit of length is chosen as the distance corresponding to the angle of parallelism $\alpha = 2 \ \mathrm{arc} \ \tan \ e^{-1}$.

The relationship between the distance h of a point P from a line AB and the angle $\Pi(h)$ of parallelism at P for line AB reveals an interesting feature of Lobachevskian geometry that is not possessed by Euclidean geometry. In both Euclidean and Lobachevskian geometry, angles possess a natural unit of measure (the right angle, or some given fractional part of a right angle) which is capable of geometrical definition and which, if ever lost, could be geometrically reconstructed. The fact that there exists a unit of angle having a structural connection of this sort with the geometry is expressed by mathematicians by saying that angles are *absolute* in the two geometries. Now in Euclidean geometry, lengths clearly are not absolute; there is no natural unit of length structurally connected with the geometry. Lengths must be measured in terms of some arbitrarily chosen unit of length, and if this unit of length becomes expunged it cannot be geometrically reconstructed. Mathematicians express this fact by saying that in Euclidean geometry lengths are *relative*. The interesting feature of Lobachevskian geometry is that not only angles, but lengths as well, are absolute. For to each angle $\Pi(h)$ of parallelism is associated a definite distance h, and thus one can obtain from a unit of angular measure a corresponding unit of linear measure.

PROBLEMS

1. If PX and PY are both parallel to line AB, show that the bisector of $\measuredangle XPY$ is perpendicular to AB.

2. A limit triangle is said to be *isosceles* if its two angles are equal. Prove that if the finite sides of two isosceles limit triangles are equal, then the angles of one limit triangle are equal to those of the other.

3. Prove that the sum of the angles of a limit triangle is always less than two right angles.

4. Prove that if a transversal cuts two lines, making the sum of the interior angles on the same side equal to two right angles, then the two lines are hyperparallel to one another.

5. Given four segments AC, BD, $A'C'$, $B'D'$. If $\overline{AC}$ is parallel to $\overline{BD}$, $AB = A'B'$, $\measuredangle BAC = \measuredangle B'A'C'$, and $\measuredangle ABD = \measuredangle A'B'D'$, prove that $\overline{A'C'}$ is parallel to $\overline{B'D'}$.

6. Prove that the perpendicular bisector of the finite side of an isosceles limit triangle is parallel to the two parallel sides of the limit triangle, and that any point on the perpendicular bisector is equally distant from the parallel sides of the limit triangle.

7. If the perpendicular bisector of the finite side of a limit triangle is parallel to the parallel sides of the limit triangle, show that the limit triangle is isosceles.

8. If in two limit triangles, an angle of the first is equal to an angle of the second, but the finite side of the first is greater than the finite side of the second, prove that the other angle of the first is smaller than the other angle of the second.

9. (a) Express a distance h in terms of its associated angle of parallelism. (b) Show that as h increases from 0 to ∞, the corresponding angle of parallelism decreases from 90° to 0°. (c) Show that, "in the small," Lobachevskian geometry approximates Euclidean geometry.

10. (a) Why does the Bureau of Standards preserve a standard unit of length but no standard unit of angle? (b) In spherical geometry (that is, geometry on the surface of a given sphere) are lengths absolute or relative?

7.4 SACCHERI QUADRILATERALS AND THE ANGLE-SUM OF A TRIANGLE

We continue our study of Lobachevskian plane geometry with an examination of some properties of Saccheri quadrilaterals and of some important consequences of those properties.

7.4.1 DEFINITIONS. A quadrilateral $ABCD$ in which $\measuredangle A = \measuredangle B = 90°$ and $AD = BC$ is called a *Saccheri quadrilateral*. AB is called the *base*, DC the *summit*, and angles D and C the *summit angles*.

7.4.2 THEOREM. *The summit angles of a Saccheri quadrilateral are equal and acute.*

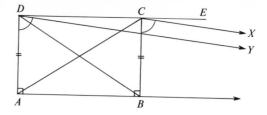

Figure 7.4a

Referring to Figure 7.4a, draw the diagonals AC and BD. Since triangles DAB and CBA are congruent (by I 4), we have $AC = BD$. It then follows that triangles ADC and BCD are congruent (by I 8), whence $\angle ADC = \angle BCD$. To show that these angles are acute, let $\overline{CX}$ and $\overline{DY}$ be the parallels through C and D to $\overline{AB}$. By Theorem 7.3.2, $\angle ECX > \angle EDY$. But, by Theorem 7.3.5, $\angle BCX = \angle ADY$. It follows that $\angle BCE > \angle ADE$. But $\angle ADE = \angle BCD$. Hence $\angle BCD$ is acute.

7.4.3 THEOREM. *The line joining the midpoints of the base and summit of a Saccheri quadrilateral is perpendicular to both of them.*

Let M (see Figure 7.4b) be the midpoint of base AB and N the midpoint

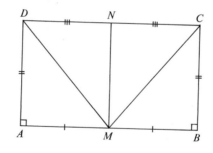

Figure 7.4b

of summit DC. Draw MC, MD. Since triangles DAM and CBM are congruent (by I 4), we have $MD = MC$. We then have (by I 8) triangle DNM congruent to triangle CNM, whence $\angle DNM = \angle CNM = 90°$. Similarly, by drawing NA, NB, we can prove that $\angle AMN = \angle BMN = 90°$.

7.4.4 THEOREM. *Two Saccheri quadrilaterals are congruent if they have equal summits and equal summit angles.*

Suppose, in Figure 7.4c, $AD > A'D'$. On DA and CB mark off $DR = D'A'$ and $CS = C'B'$. Draw RS. Then $RSCD$ is congruent to $A'B'C'D'$. It follows that $\angle ARS = \angle BSR = 90°$, which contradicts Theorem 7.4.2 when applied to Saccheri quadrilateral $ABSR$. This proves the theorem.

7.4.5 THEOREM. *The sum of the angles of any triangle is less than two right angles.*

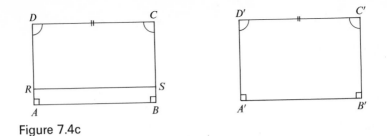

Figure 7.4c

Let L, M (see Figure 7.4d) be the midpoints of the sides AC, BC of a triangle ABC. Let AD, BE, CF be perpendiculars from A, B, C on line LM. By I 26, triangles ADL and CFL and triangles BEM and CFM are congruent. It follows that $AD = CF = BE$, and $DEBA$ is a Saccheri quadrilateral with base DE. Therefore $\angle DAB$ and $\angle ABE$ are acute angles. But $\angle \overline{DAB} = \angle \overline{DAL} + \angle \overline{LAB} = \angle \overline{FCL} + \angle \overline{LAB}$, and $\angle \overline{ABE} = \angle \overline{ABM} + \angle \overline{MBE} = \angle \overline{ABM} + \angle \overline{MCF}$. Consequently $180° > \angle \overline{DAB} + \angle \overline{ABE} = \angle \overline{LAB} + \angle \overline{FCL} + \angle \overline{MCF} + \angle \overline{ABM} = \angle A + \angle C + \angle B$.

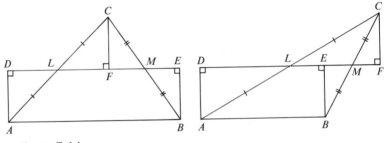

Figure 7.4d

7.4.6 DEFINITION. The deficiency of the sum of the angles of a triangle below two right angles is called the *defect* of the triangle.

7.4.7 THEOREM. *Two triangles are congruent if the three angles of one are equal to the three angles of the other.*

In Figure 7.4e, let $\angle A = \angle A'$, $\angle B = \angle B'$, $\angle C = \angle C'$, and suppose the triangles are not congruent. Then $AB \neq A'B'$. Suppose, without loss of generality, $AB > A'B'$. On AB and AC mark off AD and AE equal to $A'B'$ and $A'C'$ respectively. Now E must fall on C, on AC produced, or between A and C. If E falls on C, then $A'C' = AC$, and the triangles are congruent (by I 26), which contradicts our supposition that they are not. If E falls on AC produced we have a situation contradicting I 16. Thus E must fall between A and C. It then follows that the sum of the angles of the quadrilateral $BCED$ is four right angles. But this is impossible by Theorem 7.4.5, inasmuch as a quadrilateral can be cut into two triangles. The theorem now follows.

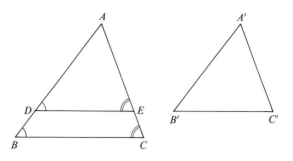

Figure 7.4e

7.4.8 REMARK. Theorem 7.4.7 proves that in Lobachevskian geometry, similar noncongruent figures cannot exist.

PROBLEMS

1. Prove that two Saccheri quadrilaterals with equal bases and equal summits are congruent.

2. Prove that two Saccheri quadrilaterals with equal bases and equal summit angles are congruent.

3. Prove that a quadrilateral with two right angles and the other two angles equal is a Saccheri quadrilateral.

4. Prove that at least two angles of every triangle are acute.

5. Prove that the angle-sum of a convex polygon of n sides is less than $n - 2$ straight angles.

7.5   AREA OF A TRIANGLE

We now consider the concept of the area of a triangle in the Lobachevskian plane; it will be seen to be intimately related to the excess of the triangle. Appeal will be made to some of the early theorems in the previous chapter on dissection theory.

7.5.1 THEOREM. *If a triangle is divided into two subtriangles by a cevian line, then the defect of the triangle is equal to the sum of the defects of the two subtriangles.*

Referring to Figure 7.5a we have

$$\begin{aligned}
\text{defect } ABC &= 180° - (\alpha_1 + \alpha_2 + \beta_2 + \gamma_1) \\
&= 360° - (\alpha_1 + \beta_1 + \gamma_1 + \alpha_2 + \beta_2 + \gamma_2) \\
&= [180° - (\alpha_1 + \beta_1 + \gamma_1)] + [180° - (\alpha_2 + \beta_2 + \gamma_2)] \\
&= \text{defect } ADC + \text{defect } ABD.
\end{aligned}$$

7.5.2 THEOREM. *If a triangle is dissected into subtriangles in any way, the defect of the triangle is equal to the sum of the defects of the subtriangles.*

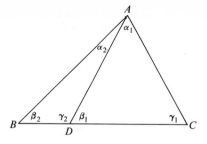

Figure 7.5a

By additional cuts, the given dissection can be converted into a cevian dissection (see Theorem 5.2.9). The theorem now follows from Theorem 7.5.1.

7.5.3 THEOREM. *If two triangles* T_1 *and* T_2 *are congruent by addition, then they have equal defects.*

Since $T_1 \cong T_2$ (+), it follows that T_1 and T_2 can be dissected into the same set of subtriangles. Therefore, by Theorem 7.5.2, T_1 and T_2 have equal defects.

7.5.4 LEMMA. *Any triangle is congruent by addition to a Saccheri quadrilateral whose summit is equal to any given side of the triangle and each of whose summit angles is equal to half the sum of the angles of the triangle.*

Denote the triangle by ABC and let AB be the given side. Let L and M be the midpoints of sides AC and BC respectively, and let D, E, F be the feet of the perpendiculars dropped from A, B, C on line LM. Then, as in the proof of Theorem 7.4.5, $DEBA$ is a Saccheri quadrilateral with summit AB and with each summit angle equal to half the sum of the angles of the triangle. We leave it to the reader to show that triangle ABC is congruent by addition to the Saccheri quadrilateral $DEBA$. There are several cases to consider, two of which are illustrated in Figure 7.4d.

7.5.5 THEOREM. *Two triangles with a side of one equal to a side of the other and having equal defects are congruent by addition.*

Let the triangles be ABC and $A'B'C'$, where $AB = A'B'$. By Lemma 7.5.4, triangle ABC is congruent by addition to a Saccheri quadrilateral of summit AB and summit angles each equal to half the angle-sum of triangle ABC. Similarly, triangle $A'B'C'$ is congruent by addition to a Saccheri quadrilatera of summit $A'B'$ and summit angles each equal to half the angle-sum of triangle $A'B'C'$. By Theorem 7.4.4, the two Saccheri quadrilaterals are congruent. It follows (see Theorem 5.1.3) that triangle ABC is congruent by addition to triangle $A'B'C'$.

7.5.6 THEOREM. *The line through the midpoint of one side of a triangle, perpendicular to the perpendicular bisector of a second side, bisects the third side.*

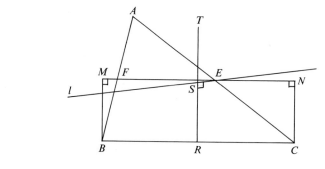

Figure 7.5b

Let E (see Figure 7.5b) be the midpoint of side AC, and RT the perpendicular bisector of side BC, of a triangle ABC. Draw line l through E perpendicular to RT and cutting RT in S. But the line through the midpoints F and E of AB and AC is perpendicular to RT (by Theorem 7.4.3 applied to Saccheri quadrilateral $MNCB$). It follows that l coincides with FE, or that l bisects side AB.

7.5.7 THEOREM. *Any two triangles with the same defect are congruent by addition.*

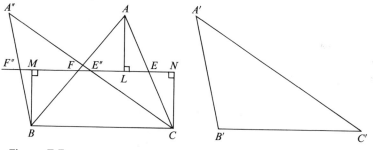

Figure 7.5c

Let ABC, $A'B'C'$ (see Figure 7.5c) be two triangles with the same defect. It has already been proved (Theorem 7.5.5) that if a side of one triangle is equal to a side of the other, the two triangles are congruent by addition. Assume, then, that no side of one is equal to a side of the other; in particular, assume $A'C' > AC$. Join the midpoints F, E of AB, AC and draw perpendiculars AL, BM, CN to line FE. Mark on line FE a point E'' such that $CE'' = (A'C')/2$. This can be done since $(A'C')/2 > CE \geqq CN$. Draw CE'' and extend it to A'' so that $E''A'' = CE''$. Draw $A''B$ to cut line FE in F''.

Since (by Theorem 7.4.3) *FE* is perpendicular to the perpendicular bisector of *BC*, and passes through the midpoint *E″* of *CA″*, it follows (by Theorem 7.5.6) that *F″* is the midpoint of *BA″*. Now triangles *ABC*, *A″BC* have the same angle-sum (namely $\angle MBC + \angle NCB$) and a common side *BC*; they are therefore congruent by addition. But triangles *A″BC* and *A′B′C′* also have the same angle-sum, and side *CA″* of the one is equal to side *C′A′* of the other; they are therefore congruent by addition. It now follows (by Theorem 5.1.3) that triangles *ABC* and *A′B′C′* are congruent by addition.

7.5.8 REMARK. We naturally define two triangles which are congruent by addition to be triangles of *equal area*. We may then take the defect of a triangle as a measure of its area, since the essential properties of a measure of area are that two triangles with the same area have the same measure, that two triangles with the same measure have the same area, and that the measure of a whole is the sum of the measures of its parts. Hence we may say, in Lobachevskian geometry, that the area of a triangle is proportional to its defect.

PROBLEMS

1. Show that Theorems 5.1.3 and 5.2.9 are theorems of absolute geometry.

2. Complete the proof of Lemma 7.5.4.

3. If, in quadrilateral *ABCD*, $\angle A = \angle B = 90°$, prove that $\angle C \gtreqless \angle D$ according as $AD \gtreqless BC$.

4. A *Lambert quadrilateral* is a quadrilateral possessing three right angles.
   (a) Prove that the fourth angle of a Lambert quadrilateral is acute.
   (b) Prove that the sides adjacent to the fourth angle of a Lambert quadrilateral are greater than their respective opposite sides.

5. Which is greater, the base or the summit of a Saccheri quadrilateral?

6. Prove that the line joining the midpoints of the equal sides of a Saccheri quadrilateral is perpendicular to the line joining the midpoints of the base and summit, and that it bisects both diagonals of the quadrilateral.

7. Prove that the segment connecting the midpoints of two sides of a triangle is less than half the third side. (This fact can be used as the characteristic postulate in a development of Lobachevskian geometry.)

7.6 IDEAL AND ULTRA-IDEAL POINTS

We have seen, in plane Euclidean geometry, the convenience of introducing certain ideal points—the so-called *points at infinity*. By this device, many theorems which formerly had exceptions became universally true. A similar, but more complicated, situation exists in plane Lobachevskian geometry. Whereas in plane Euclidean geometry we have only two types of pairs of lines, namely intersecting pairs and parallel pairs, in plane Lobachevskian geometry we have three types of pairs of lines—intersecting pairs, parallel pairs, and hyperparallel pairs. To banish exceptional cases to certain theorems in plane Lobachevskian geometry we must try to introduce a class of fictitious points in which pairs of parallel lines will be said to meet, and another class of fictitious points in which pairs of hyperparallel lines will be said to meet. These two classes of fictitious points will be known as *ideal* and *ultra-ideal* points, respectively. We must first establish an important theorem concerning a pair of hyperparallel lines; the proof we give is due to David Hilbert (1862–1943), one of the world's foremost mathematicians during the first half of the twentieth century.

7.6.1 THEOREM. *Two hyperparallel lines have one and only one common perpendicular.*

Referring to Figure 7.6a, let *l* and *m* be a pair of hyperparallel lines.

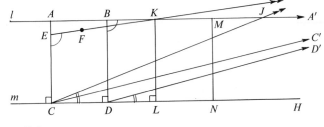

Figure 7.6a

Select two points A and B on l and drop perpendiculars AC and BD on m. If $AC = BD$, then $CDBA$ is a Saccheri quadrilateral, and the line joining the midpoints of AB and CD will be perpendicular to both l and m (by Theorem 7.4.3).

Suppose $AC \neq BD$ and assume, without loss of generality, that $AC > BD$. On CA mark off $CE = DB$. At E draw EF on the side of AC on which BD lies and such that $\angle CEF = \angle DBA'$, where A' is on l on the side of B opposite to A.

We now show that EF, if sufficiently produced, will cut l in a point K. To accomplish this, consider rays $\overline{CC'}$ and $\overline{DD'}$ parallel to $\overline{AA'}$. These rays must lie within angles ACH and BDH respectively, where H is any point on m on the side of D opposite to C. Since $\angle HDD' > \angle HCC'$ (by Theorem 7.3.2), a line CJ drawn such that $\angle JCH = \angle D'DH$ must cut l in a point J. Now figure $FECJ$ is congruent to figure $A'BDD'$, whence $\overline{EF}$ is parallel to $\overline{CJ}$, and EF must intersect l in some point K between A and J.

Draw KL perpendicular to m. On l and m, on the side of BD opposite to AC, mark off $BM = EK$ and $DN = CL$; draw MN. By drawing CK and DM we may easily prove triangle KEC congruent to triangle MBD (by I 4), and then triangle KCL congruent to triangle MDN (by I 4), whence $\angle DNM = \angle CLK = 90°$ and $MN = KL$. Thus $LNMK$ is a Saccheri quadrilateral and the line joining the midpoints of KM and LN is perpendicular to both l and m.

We have now shown that l and m have a common perpendicular. They cannot have two common perpendiculars, for in that case there would result a quadrilateral having four right angles, which is impossible.

7.6.2   REMARK. It is to be noted that not only does the argument above prove the existence of a unique common perpendicular to l and m, but it also supplies a method of constructing this common perpendicular. There are many interesting construction problems in Lobachevskian geometry which are by no means evident. The reader might like to try, for example, to construct: (1) the parallels through a given point to a given line, (2) the common parallel of two intersecting lines. Solutions to these problems would permit us to construct $\Pi(h)$ given h, and to construct h given $\Pi(h)$.

7.6.3   DEFINITIONS AND NOTATION. We assign to each family of parallel rays a common *ideal point*. We assign to each family of hyperparallel lines all perpendicular to a common line a common *ultra-ideal point*. In general we shall denote ideal and ultra-ideal points by upper case Greek letters, in the latter case frequently attaching a subscript to denote the associated common perpendicular line (see Figure 7.6b).

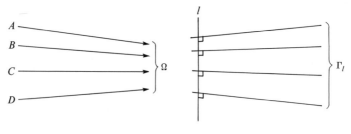

Figure 7.6b

Much of the information concerning ordinary, ideal, and ultra-ideal points can be neatly summarized graphically as follows. Let us represent the ordinary points of the Lobachevskian plane by points interior to a given circle K of an extended Euclidean plane (see Figure 7.6c), ideal points of the Lobachevskian plane by points on circle K, and ultra-ideal points of the Lobachevskian plane by the finite and infinite points outside circle K. Let straight lines of the Lobachevskian plane be represented by straight lines in the extended Euclidean plane which cut into the interior of circle K. Let the line associated with an ultra-ideal point be the polar line of that point with respect to circle K.

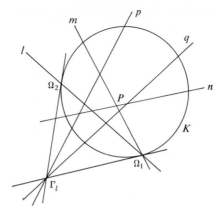

Figure 7.6c

In Figure 7.6c, lines m and n intersect in an ordinary point; lines l and m intersect in an ideal point; lines p and q intersect in the ultra-ideal point having line l as its associated line.

We observe that any two lines intersect in a point—ordinary, ideal, or ultra-ideal. On the other hand, not every pair of points determines a line. Two ordinary points determine a line; an ordinary and an ideal point determine a line; two ideal points determine a line; an ordinary point and an ultra-ideal point determine a line. Two ultra-ideal points sometimes determine a line and sometimes do not; an ideal point and an ultra-ideal point sometimes determine a line and sometimes do not. The first situation fails to yield a line when the associated lines of the two ultra-ideal points "intersect" or are "parallel"; the second situation fails to yield a line when the ideal point lies on the associated line of the ultra-ideal point.

It was Arthur Cayley (1821–1895) who called the locus of the ideal points of a Lobachevskian plane the *absolute* of that plane. The above representation of the absolute of a Lobachevskian plane by a circle in an extended Euclidean plane is, for the time being, to be regarded as simply a useful device for easily recalling and summarizing much of the information about the ordinary, ideal, and ultra-ideal points of the Lobachevskian plane. With the development of a proper background, the model can be used to show an interesting tie-up between Lobachevskian geometry and projective geometry.

PROBLEMS

Verify, in the graphic model of the absolute, the following facts about the Lobachevskian plane.

1. Two lines are perpendicular if and only if their representations in the model are conjugate lines with respect to the absolute.

2. Through an ordinary point not on a given line there exist two lines parallel to the given line.

3. Two lines have a common perpendicular if and only if they are hyperparallel.

4. The altitudes of a triangle are concurrent in an ordinary, ideal, or ultra-ideal point.

5. The orthogonal projection of a line m on a nonparallel line n is an open segment of line n.

6. It is possible to have an angle and an ordinary point within it such that no line through the point will intersect both sides of the angle in ordinary points.

7. Given two hyperparallel lines a and b, there exist two intersecting lines c and d each parallel to both a and b, and two hyperparallel lines e and f each parallel to both a and b.

8. In a limit triangle there is a line parallel to the parallel sides and perpendicular to the finite side.

9. There exist four lines parallel to two given intersecting lines.

10. If two lines a and b intersect at an acute angle, there are two lines parallel to a and perpendicular to b. If a and b intersect at a right angle, there is no line parallel to a and perpendicular to b. If a and b are parallel, there is one line parallel to a and perpendicular to b. If a and b are hyperparallel, there are two lines parallel to a and perpendicular to b.

11. Given two parallel lines, there exists a unique line parallel to each but in opposite senses.

12. Given six lines 1, 2, 3, 4, 5, 6 such that 1 is parallel to 2, 2 to 3, 3 to 4, 4 to 5, 5 to 6, 6 to 1. If no three of the lines are parallel to one another, and if the pairs of lines 1 and 4, 2 and 5, 3 and 6 intersect, then the three points of intersection are collinear.

13. Given six lines 1, 2, 3, 4, 5, 6 such that 1 is parallel to 2, 2 to 3, 3 to 4, 4 to 5, 5 to 6, 6 to 1. If no three of the lines are parallel to one another, and if the pairs of lines 1 and 4, 2 and 5, 3 and 6 are hyperparallel, then their three common perpendiculars are concurrent.

7.7 AN APPLICATION OF IDEAL AND ULTRA-IDEAL POINTS

If we draw the three perpendicular bisectors of the sides of an ordinary triangle, pairs of these perpendicular bisectors may be intersecting lines, parallel lines, or hyperparallel lines. Without the convention of ideal and ultra-ideal points, we cannot, then, say that the three perpendicular bisectors of the sides of an ordinary triangle are concurrent. It is interesting, however, that with the fictitious ideal and ultra-ideal points, the general theorem can be stated. We now proceed to establish this general theorem, thereby illustrating the utility of the fictitious points.

7.7.1 THEOREM. *The three perpendicular bisectors of the sides of an ordinary triangle are concurrent.*

Case 1 (two of the perpendicular bisectors intersect in an ordinary point): Suppose the perpendicular bisectors of AB and BC (see Figure 7.7a) intersect in an ordinary point O. Connect O with A, B, C. Then triangles $AC'O$

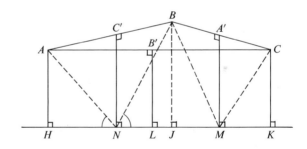

Figure 7.7a

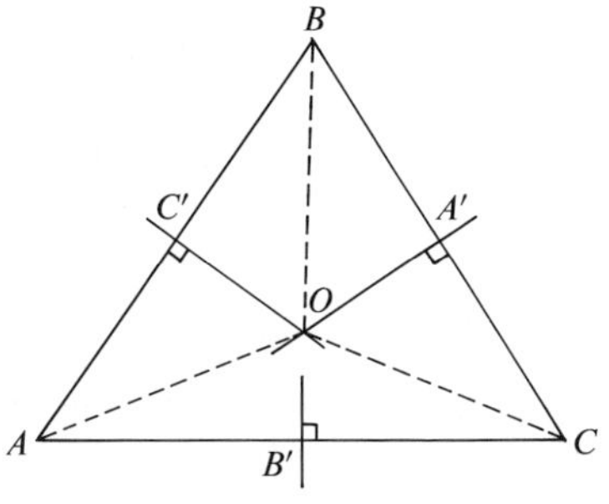

and $BC'O$ are congruent, and triangles $BA'O$ and $CA'O$ are congruent. It follows that $AO = BO = CO$. Connect O with B', the midpoint of AC. Then triangles $AB'O$ and $CB'O$ are congruent, whence $\angle AB'O = \angle CB'O = 90°$, and $B'O$ is the perpendicular bisector of AC.

Case 2 (two of the perpendicular bisectors intersect in an ultra-ideal point): Let the perpendiculars to AB and BC at the midpoints C' and A' be hyperparallel (see Figure 7.7b). They then have a common perpendicular

Figure 7.7b

MN. Draw AH, BJ, CK perpendicular to MN. Draw AN, NB, BM, MC. Then triangles $AC'N$ and $BC'N$ are congruent (by I 4), whence $AN = BN$ and $\angle ANH = \angle BNJ$. It now follows that triangles AHN and BJN are congruent (by I 26), whence $AH = BJ$. Similarly, $CK = BJ$. Therefore $AH = CK$ and $HKCA$ is a Saccheri quadrilateral, and $B'L$, joining the midpoint B' of AC to the midpoint L of HK, is perpendicular to both AC and MN. It follows that $A'M$, $B'L$, $C'N$ intersect in an ultra-ideal point Γ_{MN}.

Case 3 (two of the perpendicular bisectors intersect in an ideal point): Let the perpendicular bisectors $A'M$ and $C'N$ of BC and BA be parallel (see Figure 7.7c). Then the perpendicular bisector $B'L$ of AC cannot

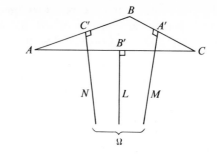

Figure 7.7c

intersect either $A'M$ or $C'N$ (for if it did, then, by Case 1, $A'M$ and $C'N$ would have to intersect). Also, $B'L$ cannot be hyperparallel to either $A'M$ or $C'N$ (for if it were, then, by Case 2, $A'M$ and $C'N$ would have to be hyperparallel). It follows that $A'M$, $B'L$, $C'N$ all pass through a common ideal point Ω, or they form a triangle having three ideal vertices Ω_1, Ω_2, Ω_3. We show that the second situation cannot occur.

Consider a triangle having three ideal vertices Ω_1, Ω_2, Ω_3 (see Figure 7.7d). We shall prove that no straight line can cut all three sides of

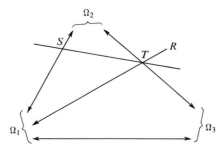

Figure 7.7d

such a triangle. Let ST, for example, cut $\Omega_1\Omega_2$ in S and $\Omega_2\Omega_3$ in T. Draw $T\Omega_1$ and produce it in the opposite sense to a point R. Then ST lies within the vertical angles $\Omega_2 T\Omega_1$ and $RT\Omega_3$, and consequently cannot intersect $\Omega_1\Omega_3$.

We now show that there is always at least one line which intersects all three of the perpendicular bisectors of the sides of an ordinary triangle. Let $\angle A$ (see Figure 7.7e) be not less than any other angle of triangle ABC. Mark off $\angle BAK = \angle B$, $\angle CAL = \angle C$. Then AK, AL cut side BC in points K, L. But K lies on the perpendicular bisector of side AB, and L lies on the perpendicular bisector of side AC. Thus side BC cuts all three of the perpendicular bisectors.

It now follows that the perpendicular bisectors of the sides of an ordinary triangle cannot form a triangle having three ideal vertices, and Case 3 is completed.

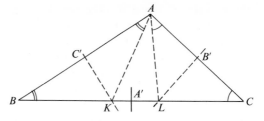

Figure 7.7e

7.8 MAPPING THE PLANE ONTO THE INTERIOR OF A CIRCLE

In this short section we consider an interesting mapping which, when carried out in the Euclidean plane, maps the plane onto itself, but when carried out in the Lobachevskian plane, maps the plane onto the interior of a circle.

We first establish an important theorem that has come to be called Hjelmslev's Theorem (see Problem 5, Section 3.5). This theorem has been known for a long time, but it was J. T. Hjelmslev (1873–1950) who, in 1907, made the remarkable discovery that it belongs to absolute geometry and that it has numerous applications in Lobachevskian geometry. Following is a proof, by absolute geometry, of Hjelmslev's Theorem.

7.8.1 HJELMSLEV'S THEOREM. *The midpoints of the segments joining pairs of corresponding points of two congruent point rows of either a Euclidean or a Lobachevskian plane are collinear or coincident.*

Let $ABC \cdots$ and $A'B'C' \cdots$ (see Figure 7.8a) be two congruent point

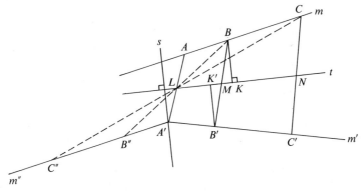

Figure 7.8a

rows on lines m and m' such that $AB = A'B'$, $BC = B'C'$, etc., and let L, M, N be the midpoints of AA', BB', CC', respectively. It is easy to see that if two of the midpoints coincide, then they all do. Consequently we assume that the midpoints are distinct. Let m'' be the image of m under the reflection $R(L)$, B'' and C'' being the images of B and C under the reflection. Let s be the bisector of $\angle B'A'B''$. Then the transformation $R(s)R(L)$ is an isometry which maps the point row on m onto the point row on m'. Let t be the line through L perpendicular to s. Since both $R(L)$ and $R(s)$ map t onto itself, it follows that $R(s)R(L)$ maps t onto itself. Let K be the foot of the perpendicular from B on t. Then $R(s)R(L)$ carries K into some point K' on t. Since BK is carried into $B'K'$ and $\angle BKK'$ into $\angle B'K'K$, it follows that K' is the foot of the perpendicular from B' on t, whence (since B and B' lie on opposite sides of t) the midpoint M of BB' is on t. Similarly, the midpoint of CC' is on t, etc., and the theorem is proved.

7.8.2 Notation. Let O be a fixed point of either a Euclidean or a Lobachevskian plane, and let θ be a fixed sensed acute angle. The transformation which maps any point P of the plane (see Figure 7.8b) into the point P' such that $\angle POP' = \theta$ and $\angle OP'P = 90°$ will be denoted by S. The transformation $R(O, -\theta)S$ will be denoted by T.

Figure 7.8b

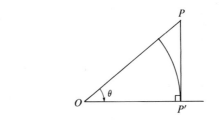

7.8.3 Theorem. *Under transformation* T: (1) *point* O *is invariant*, (2) *straight line segments map into straight line segments*, (3) *angles with vertex at* O *map into equal angles with vertex at* O, (4) *right angles with one side passing through* O *map into right angles with one side passing through* O, (5) *circles with center* O *map into circles with center* O.

If the five properties hold for transformation S, then they also hold for transformation T, since $T = R(O, -\theta)S$ and $R(O, -\theta)$ is an isometry. Consider, then, the plane subjected to transformation S.

Property (1) is obvious. It is also clear that any straight line passing through O maps into a straight line passing through O. Consider (see Figure 7.8c) a straight line m not passing through O and let A, B, C be any three points on m. The rotation $R(O, 2\theta)$ carries line m and its point row A, B, C into a line m'' and a congruent point row A'', B'', C''. By Hjelmslev's Theorem, the midpoints L, M, N of AA'', BB'', CC'' are collinear (and, in this case, noncoincident). But L, M, N are the images A', B', C' of A, B, C under the transformation S. It is now clear that transformation S maps straight line segments into straight line segments, and property (2) is established. We leave it to the reader to establish the easy properties (3), (4), (5).

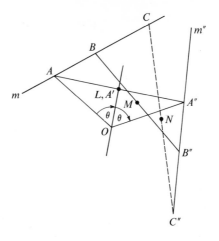

Figure 7.8c

7.8.4 THEOREM. *In the Euclidean plane, transformation* T *is the homothety* H(O, *cos* θ), *which therefore maps the plane onto itself.*

Because, under *T*, points *P*, *P'*, *O* are collinear and, in the Euclidean plane, $OP' = OP \cos \theta$.

7.8.5 THEOREM. *In the Lobachevskian plane, transformation* T *maps the plane onto the interior of the circle of center* O *and radius* h, *where* Π(h) = θ.

Let *r* be any ray from *O* (see Figure 7.8d). Then, under *S*, ray *r* maps

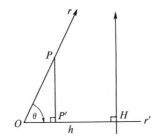

Figure 7.8d

into ray *r'* making an angle θ with ray *r*. On *r'* mark off *OH* equal to *h*, the distance corresponding to θ as angle of parallelism. Then *S* maps ray *r* onto the segment *OH*, point *H* excluded. It follows that *S*, and therefore also *T*, maps the Lobachevskian plane onto the interior of the circle having *O* as center and *h* as radius.

Many properties of the Lobachevskian plane can be easily obtained from its map under transformation *T*. The reader has undoubtedly observed that the map is similar to the graphical model of the Lobachevskian plane described at the end of Section 7.6. Space forbids us from developing this connection further.

PROBLEMS

1. Show that if two of the midpoints of the segments joining pairs of corresponding points of two congruent point rows of either a Euclidean or a Lobachevskian plane are coincident, then they are all coincident.

2. Show, in the proof of Theorem 7.8.3, that L, M, N cannot be coincident.

3. (a) Give a proof, by absolute geometry, of the theorem: If a quadrilateral $OABC$ has right angles at A and C, and if D is the foot of the perpendicular from O on AC, then $\measuredangle AOD = \measuredangle BOC$.
   (b) Give a simple Euclidean proof of the theorem of part (a).

4. (a) Give a proof, by absolute geometry, of the theorem: Any two plane angles of a dihedral angle are equal.
   (b) Look up, in an elementary solid geometry text, the customary proof of the theorem of part (a) and note that it utilizes the Euclidean parallel postulate.

5. With the aid of the mapping induced by transformation T in the Lobachevskian plane, prove the following theorems of Lobachevskian geometry:
   (a) If $\overline{PX}$ is parallel to $\overline{AB}$ and R is any point on PX such that P and R are on the same side of X, then $\overline{RX}$ is parallel to $\overline{AB}$.
   (b) If $\overline{CD}$ is parallel to $\overline{AB}$, then $\overline{AB}$ is parallel to $\overline{CD}$.
   (c) If $\overline{AB}$ and $\overline{CD}$ are parallel to $\overline{EF}$, then $\overline{AB}$ is parallel to $\overline{CD}$.
   (d) As h increases, $\Pi(h)$ decreases, passing through all values between 90° and 0°.
   (e) The theorem of Problem 3 (a) also holds if B is an ideal point.

7.9 GEOMETRY AND PHYSICAL SPACE

Euclidean geometry is based upon a set of postulates, one of which, the parallel postulate, can be taken in the form: "In the plane of a given non-incident point and line, *only one line* can be drawn through the given point and not intersecting the given line." Lobachevskian geometry is based upon the same set of postulates, except the parallel postulate is rejected and replaced by: "In the plane of a given nonincident point and line, *more than one line* can be drawn through the given point and not intersecting the given line."* It can be shown (and we shall do so in the next chapter) that this alternative postulate is entirely compatible with all the other Euclidean postulates.

One might think that a third geometry can similarly be based upon the Euclidean postulate set with the parallel postulate rejected and replaced by: "In the plane of a given nonincident point and line, *no line* can be drawn through the given point and not intersecting the given line." It can easily be shown, however, that this replacement of the parallel postulate is not compatible with the remaining Euclidean postulates,† and, to get a consistent

---

* In Section 7.2 we actually adopted a somewhat stronger postulate than this, but, as was pointed out at the time, we did so merely for convenience. In fact, it can be shown that a postulate even weaker than the one stated above will suffice.
† At least as Euclid interpreted them; the replacement contradicts, for example, Proposition I 27.

postulate set for a third geometry based on this second alternative, some of the other Euclidean postulates must also be altered. But these adjustments of the other postulates can be made, and geometers have in this way developed a second non-Euclidean geometry. This second non-Euclidean geometry is the Riemannian non-Euclidean geometry mentioned toward the end of Section 7.1. In some respects the Riemannian non-Euclidean geometry is more complicated than is the Lobachevskian non-Euclidean geometry, and accordingly we do not take the space to develop it here.

Since we have a number of geometries of space—the Euclidean and the two classical non-Euclidean geometries—the question is often asked, "Which is the true geometry?" By this question is meant, of course, "Which geometry properly describes physical space?" Now when it comes to the application of several mathematical theories to a given physical situation, we are interested in that mathematical theory which best explains, or more closely agrees with, the observed facts of the physical situation, and which will stand the kinds of tests customarily placed on hypotheses in any field of scientific enquiry. In the present case, then, we are interested in which of the Euclidean and two classical non-Euclidean systems of geometry most closely agrees with the observed facts of physical space. Now it is not difficult to show that all three geometries under consideration fit our very limited portion of physical space equally well, and so it would seem we must be content with an indeterminate answer until some crucial experimental test on a great scale can be devised to settle the matter. Such a crucial test would appear to be the measurement of the sum of the three angles of a large physical triangle. To date no deviation, exceeding expected errors in measurement, from 180° has been found in the sum of the angles of any physical triangle. But, we recall, the discrepancy of the sum of the angles of a triangle from 180° in the two non-Euclidean geometries is proportional to the area of the triangle, and the area of any triangle so far measured may be so small that any existing discrepancy is swallowed by the allowed errors in measurement.

Because of the apparently inextricable entanglement of space and matter, there are reasons for believing it may be impossible to determine by physical experiments whether our space is Euclidean or non-Euclidean. Since all measurements involve both physical and geometrical assumptions, an observed result can be explained in many different ways by merely making suitable compensatory changes in our assumed qualities of space and matter. For example, it is quite possible that a discrepancy observed in the angle sum of a triangle could be explained by preserving the assumptions of Euclidean geometry but at the same time modifying some physical law, such as some law of optics. And again, the absence of any such discrepancy might be compatible with the assumptions of a non-Euclidean geometry, together with some suitable adjustments in our assumptions about matter. On these grounds Henri Poincaré (1854–1912) maintained the impropriety of asking which geometry is the true one. To clarify this viewpoint, Poincaré devised an imaginary universe Σ occupying the interior of a sphere of radius R immersed in Euclidean space, and in which he assumed the following physical laws to hold:

(1) At any point P of Σ the absolute temperature T is given by $T = k(R^2 - r^2)$, where r is the distance of P from the center of Σ and k is a constant.

(2) The linear dimensions of a material body vary directly with the absolute temperature of the body's locality.

(3) All material bodies in Σ immediately assume the temperatures of their localities.

Now it is quite possible for the inhabitants of Σ to be unaware of the above three physical laws holding in their universe. For example, one way to detect a change in temperature as we go from a warm room, say, into a cool one, is that we *feel* the change by carrying our warmed bodies into the cooler room and then awaiting our skins to adjust to the new temperature. In Σ, Law 3 would neutralize this test. A thermometer would not serve either; by Law 2 all material bodies thermally expand and contract in the same way. Law 2 also prohibits an inhabitant of Σ from detecting his change of size with a measuring stick that he carries about with him.

An inhabitant of Σ would feel that his universe is infinite in extent on the simple grounds that he never reaches a boundary after taking any finite number N of steps, no matter how large N may be chosen. Of course he does not know that this is because he himself, and the length of his steps along with him, are becoming smaller and smaller as he advances from the center of his universe.

It is easy to see that geodesics (paths of shortest length joining pairs of points) in Σ, as measured by an inhabitant of Σ, are curves bending toward the center of Σ. As a matter of fact, it can be shown that the geodesic through two points A and B of Σ is an arc of the circle through A and B which cuts the bounding sphere of Σ orthogonally. Let us now impose one further physical law on the universe Σ, by supposing:

(4) Light travels along the geodesics of Σ.

This condition might be physically realized by filling Σ with a gas having a proper index of refraction at each point of Σ. Now, because of Law 4, geodesics of Σ actually "look straight" to an inhabitant of Σ. But the reader can easily show (see Figure 7.9a, where three "lines" are drawn through point P and not intersecting "line" l) that in the geometry of geodesics in Σ the Lobachevskian parallel postulate holds, so that an inhabitant of Σ would believe that he lives in a non-Euclidean world. Here we have a piece of ordinary, and supposedly Euclidean, space, which, because of different (and actually undetectable) physical laws, appears to be non-Euclidean.

We would do better, then, to ask, not which is the *true* geometry, but which is the *most convenient* geometry, and this convenience might depend

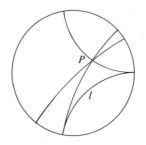

Figure 7.9a

upon the application at hand. Certainly, for drafting, for terrestrial survey-ing, and for the construction of ordinary buildings and bridges, the Euclidean geometry is probably the most convenient simply because it is the easiest with which to work.

There are physical studies where geometries other than the Euclidean have been found to be more acceptable. For example, Einstein found in his study of the general theory of relativity that none of the three geometries that we have been considering is adequate, and he adopted another kind of non-Euclidean geometry that had been first broached by Bernhard Riemann in his probationary lecture of 1854—a non-Euclidean geometry whose discrepancies from Euclidean geometry are not uniform, but vary from place to place in space depending on the concentration of matter pre-sent in space at the place. Again, a recent study* of *visual space* (the space psychologically observed by persons of normal vision) came to the conclusion that such space can most conveniently be described by the Lobachevskian non-Euclidean geometry. Other examples can be given.

We here conclude the present chapter on non-Euclidean geometry. We already see, in this last section, that non-Euclidean geometry has vitally effected earlier concepts of geometry, and we are now naturally led to con-sider the foundations of geometry. This we do in the next chapter, where, among other things, we shall look into the matter of the consistency of non-Euclidean geometry.

BIBLIOGRAPHY

ADLER, C. F., *Modern Geometry*. New York: McGraw-Hill Book Company, 1967.

BOLYAI, JOHN, "The Science of Absolute Space," tr. by G. B. Halsted (1895). See BONOLA.

BONOLA, ROBERTO, *Non-Euclidean Geometry*, tr. by H. S. Carslaw. New York: Dover Publications, Inc., 1955. Containing BOLYAI and LOBACHEVSKI. Original Italian volume published in 1906.

CARSLAW, H. S., *Non-Euclidean Plane Geometry and Trigonometry*. New York: Longmans, Green and Company, 1916. Reprinted in *String Figures, and Other Monographs*, Chelsea Publishing Company, 1960.

* R. K. Luneberg, *Mathematical Analysis of Binocular Vision*, Princeton, N.J.: Princeton University Press (published by the Dartmouth Eye Institute), 1947.

COOLIDGE, J. L., *The Elements of Non-Euclidean Geometry*. New York: Oxford University Press, 1909.

COXETER, H. S. M., *Non-Euclidean Geometry*. Toronto: University of Toronto Press, 1947.

EVES, HOWARD, and C. V. NEWSOM, *An Introduction to the Foundations and Fundamental Concepts of Mathematics*. New York: Holt, Rinehart and Winston, revised ed., 1965.

FORDER, H. G., *Geometry*. New York: Hutchinson's University Library, 1950.

JAMES, GLENN, ed., *The Tree of Mathematics*. Pacoima, Calif.: The Digest Press, 1957.

KULCZYCKI, STEFAN, *Non-Euclidean Geometry*, tr. by Stanislaw Knapowski. New York: Pergamon Press, 1961.

LEVI, HOWARD, *Topics in Geometry*. Boston, Mass.: Prindle, Weber & Schmidt, Inc., 1968.

LIEBER, L. R., and H. G. LIEBER, *Non-Euclidean Geometry, or Three Moons in Mathesis*. Brooklyn, N.Y.: The Galois Institute of Mathematics and Art, 1931.

LOBACHEVSKI, NICHOLAS, "Geometrical Researches on the Theory of Parallels," tr. by G. B. Halsted (1891). See BONOLA.

MANNING, H. P., *Non-Euclidean Geometry*. Boston: Ginn and Company, 1901.

MESCHKOWSKI, HERBERT, *Noneuclidean Geometry*. New York: Academic Press, 1964.

MOISE, E. E., *Elementary Geometry from an Advanced Standpoint*. Reading, Mass.: Addison-Wesley Publishing Company, Inc., 1963.

POINCARÉ, HENRI, *The Foundations of Science*, tr. by G. B. Halsted. Lancaster, Pa.: The Science Press, 1913.

RAINICH, G. Y., and S. M. DOWDY, *Geometry for Teachers*. New York: John Wiley and Sons, Inc., 1968.

RANSOM, W. R., *3 Famous Geometries*. Medford, Mass.: priv. ptd., 1959.

SHIROKOV, P. A., *A Sketch of the Fundamentals of Lobachevskian Geometry*, tr. by L. F. Boron and W. D. Baiwsma. Gronigen, The Netherlands: P. Noordhoff, Ltd., 1964.

SOMMERVILLE, D. M. Y., *The Elements of Non-Euclidean Geometry*. London: G. Bell and Sons, Ltd., 1914.

TULLER, ANNITA, *A Modern Introduction to Geometries*. Princeton, N.J.: D. Van Nostrand Company, Inc., 1967.

WOLFE, H. E., *Introduction to Non-Euclidean Geometry*. New York: The Dryden Press, Inc., 1945.

YOUNG, J. W. A., ed., *Monographs on Topics of Modern Mathematics Relevant to the Elementary Field*. New York: Longmans, Green and Company, 1911. Reprinted by Dover Publications, Inc., 1955.

8

The Foundations of
Geometry

The discovery of a non-Euclidean geometry had a marked effect on the subsequent development of geometry, and, indeed, on the subsequent development of much of mathematics in general. Prior to the discovery it was believed that mathematics was concerned with finding unique and necessary truths about the real world, and that the postulates and theorems of mathematics were essentially observed and derived laws of nature. In particular, the postulates and theorems of geometry were thought to constitute an unequivocal description of physical space. The discovery of a non-Euclidean geometry compelled mathematicians to adopt a new viewpoint of their subject. Clearly, Euclidean and Lobachevskian geometry could not each be such a description of physical space, for many theorems of these two geometries contradict each other. If Lobachevskian geometry was to be considered as a part of mathematics, then mathematics could not be concerned only with apodictic truths about the physical world, and a whole new viewpoint of the nature of mathematics needed to be developed. Not only this, but the foundations of the subject had to be carefully reconstructed. Many gifted individuals were attracted to this latter task, and an enormous amount of foundational research was carried out. It has been reported that in the thirty-year period from 1880 to 1910, some 1,385 articles appeared devoted to the foundations of geometry alone, and activity in this field is still very great.

319

The present chapter is devoted to the above matters. It will attempt to describe a present-day view of the nature of mathematics; it will introduce the reader to the important and highly interesting subject of the foundations of geometry; and it will point up the purely hypothetico-deductive nature of much of current geometrical study.

8.1 SOME LOGICAL SHORTCOMINGS OF EUCLID'S *ELEMENTS*

It would be very surprising indeed if Euclid's *Elements*, being such an early and colossal attempt at the postulational method of presentation, should be free of logical blemishes. It is therefore no great discredit to the work that critical investigations have revealed a number of defects in its logical structure. Probably the gravest of these defects are certain tacit assumptions that are employed in the deductions but which are not granted by the postulates and axioms of the work. This danger exists in any deductive study when the subject matter is overly familiar to the author. Usually a thorough grasp of the subject matter in a field of study is regarded as an indispensable prerequisite to serious work, but in developing a deductive system such knowledge can be a definite disadvantage unless proper precautions are taken.

A deductive system differs from a mere collection of statements in that it is organized in a very special way. The key to the organization lies in the fact that all statements of the system other than the original assumptions must be deducible from these initial hypotheses, and that if any additional assumption should creep into the work the desired organization is not realized. Now anyone formulating a deductive system knows more about his subject matter than just the initial assumptions he wishes to employ. He has before him a set of statements belonging to his subject matter, some of which he selects for postulates and the rest he presumably deduces from his postulates as theorems. But, with a large body of information before one, it is very easy to employ in the proofs some piece of this information which is not embodied in the postulates. Any piece of information used in this way may be so apparently obvious or so seemingly elementary that it is assumed unconsciously. Such a tacit assumption, of course, spoils the rigidity of the organization of the deductive system. Moreover, should that piece of information involve some misconception, its introduction may lead to results which not only do not strictly follow from the postulates but which may actually contradict some previously established theorem. Herein, then, lies the pitfall of too great a familiarity with the subject matter of the discourse; at all times in building up a deductive system one must proceed *as if* one were completely ignorant of the developing material.

The tacit assumption by Euclid of something that is not contained in his basic assumptions is exemplified in the very first deduced proposition of

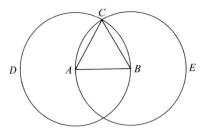

Figure 8.1a

the *Elements*. In order to examine the difficulty we shall quote Proposition
I 1 verbatim from Heath's translation.*

> *On a given finite straight line to construct an equilateral triangle.*
> Let *AB* [see Figure 8.1a] be the given finite straight line.
> Thus it is required to construct an equilateral triangle on the straight line *AB*.
> With center *A* and distance *AB*, let the circle *BCD* be described. [Postulate 3]
> Again, with center *B* and distance *BA*, let the circle *ACE* be described.
> [Postulate 3]
> And from the point *C*, in which the circles cut one another, to the points *A*, *B*,
> let the straight lines *CA*, *CB* be joined. [Postulate 1]
> Now, since the point *A* is the center of the circle *CDB*, *AC* is equal to *AB*.
> [Definition 15]
> Again, since the point *B* is the center of the circle *CAE*, *BC* is equal to *BA*.
> [Definition 15]
> But *CA* was also proved equal to *AB*; therefore each of the straight lines *CA*,
> *CB* is equal to *AB*. And things which are equal to the same thing are also equal
> to one another; therefore *CA* is also equal to *CB*. [Axiom 1]
> Therefore the three straight lines *CA*, *AB*, *BC* are equal to one another.
> Therefore the triangle *ABC* is equilateral; and it has been constructed on the
> given finite straight line *AB*.
>
> (Being) what it was required to do.

Now the construction of the two circles in this demonstration is certainly
justified by Postulate 3, but there is nothing in Euclid's first principles which
explicitly guarantees that the two circles shall intersect in a point *C*, and
that they will not, somehow or other, slip through each other with no
common point. The existence of this point, then, must be either postulated
or proved, and it can be shown that Euclid's postulates are insufficient to
permit the latter. Only by the introduction of some additional assumption
can the existence of the point *C* be established. Therefore the proposition
does not follow from Euclid's first principles, and the proof of the proposition
is invalid.

The fallacy here lies, not in assuming something contrary to our concept
of circles, but in assuming something which is not implied by our accepted

* T. L. Heath, *The Thirteen Books of Euclid's Elements*, 3 vols., 2nd ed. New York:
Dover Publications, Inc., 1956.

basic assumptions. This is an example where the tacit assumption is so evident and elementary that there does not appear to be any assumption. The fallacy is a subtle one, but had Euclid known nothing more about circles than what his first principles say of them he certainly could not have fallen into this error.

What is needed here is some additional postulate which will guarantee that the two circles concerned will intersect. Postulate 5 gives a condition under which two straight lines will intersect. We need similar postulates telling when two circles will intersect and when a circle and a straight line will intersect. What is essentially involved here is the continuity of circles and straight lines, and in modern treatments of geometry the existence of the desired points of intersection is taken care of by some sort of continuity postulate.

Another tacit assumption made by Euclid is that the straight line is of infinite extent. Although Postulate 2 asserts that a straight line may be produced indefinitely, it does not necessarily imply that a straight line is infinite in extent, but merely that it is endless, or boundless. The arc of a great circle joining two points on a sphere may be produced indefinitely along the great circle, making the prolonged arc endless, but certainly it is not infinite in extent. Now it is conceivable that a straight line may behave similarly, and that after a finite prolongation it, too, may return on itself. It was Bernhard Riemann who, in his famous lecture, *Über die Hypothesen welche der Geometrie zu Grunde liegen*, of 1854, distinguished between the boundlessness and the infinitude of straight lines. There are numerous occasions where Euclid unconsciously assumed the infinitude of a straight line. Let us briefly consider, for example, Proposition I 16:

In any triangle, if one of the sides be produced, the exterior angle is greater than either of the interior and opposite angles.

A précis of Euclid's proof runs as follows. Let *ABC* (see Figure 8.1b) be the given triangle, with *BC* produced to *D*. Let *E* be the midpoint of *AC*. Draw *BE* and extend it its own length to *F*. Draw *CF*. Then triangles *BEA* and *FEC* can easily be shown to be congruent, whence $\angle FCE = \angle BAC$.

Figure 8.1b

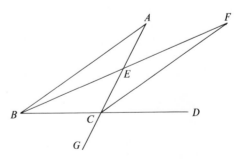

But $\angle ACD > \angle FCE$, whence $\angle ACD > \angle BAC$. By producing AC to G we may similarly show that $\angle BCG$, which is equal to $\angle ACD$, is also greater than $\angle ABC$.

Now if a straight line should return on itself, like the great circle arc considered above, BF may be so long that F will coincide with B or lie on the segment BE. Should this be the case the proof would certainly fail. Euclid was misled by his visual reference to the figure rather than to the principles that should be the basis of his argument. Clearly, then, to make the proof universally valid we must either prove or postulate the infinitude of straight lines.

One can point out other tacit assumptions which, like the above, were unconsciously made by Euclid and which vitiate the true deductive character of his work. For example, in the proof of Proposition I 21, Euclid unconsciously assumed that if a straight line enters a triangle at a vertex it must, if sufficiently produced, intersect the opposite side. It was Moritz Pasch (1843–1930) who recognized the necessity of a postulate to take care of this situation. Again, Euclid made no provision for *linear order*, and his concept of " betweenness " is without any postulational foundation, with the result that paradoxes are possible. Also, his Postulate 1, which guarantees the existence of at least one straight line joining two points A and B, probably was meant to imply uniqueness of this line, but the postulate fails to assert so much. And the objections that can be raised against the principle of superposition, still employed in some textbooks, are only partially met by Euclid's Axiom 4.

In short, the truth of the matter is that Euclid's initial assumptions are simply not sufficient for the derivation of all of the 465 propositions of the *Elements*; his set of postulates and axioms needs to be considerably amplified. The work of perfecting Euclid's initial assumptions, so that all of his geometry can rigorously follow, occupied mathematicians off and on for more than two thousand years. Not until the end of the nineteenth century and the early part of the twentieth century, after the foundations of geometry had been subjected to an intensive study, were satisfactory sets of postulates supplied for Euclidean plane and solid geometry.

Not only is Euclid's work marred by numerous tacit assumptions, but some of the preliminary definitions also are open to criticism. Euclid made some sort of attempt to give an explicit definition, or at least an explanation, of all the terms of his discourse. Now some terms of a logical discourse must deliberately be chosen as primitive or undefined terms, for it is as impossible to define all the terms of the discourse *explicitly* as it is to prove all of the statements of the discourse. The postulates of the discourse are, in final analysis, assumed statements about the primitive terms. From this point of view, the primitive terms may be regarded as being defined *implicitly*, in the sense that they are any things which satisfy the postulates.

In Euclid's development of geometry the terms *point* and *straight line*, for example, could well have been included in a set of primitive terms for the discourse. At any rate, Euclid's definition of a point as "that which has no part" and a straight line as "a line which lies evenly with the points on itself" are, from a logical viewpoint, woefully inadequate. One distinction, as we shall see, between the Greek and the modern conception of the postulational method lies in this matter of primitive terms; in the Greek conception there is no simple listing of the primitive terms. Other differences between the Greek and the modern conception of the postulational method will become apparent when the modern revisions of Euclid's work are discussed.

PROBLEMS

1. If an assumption tacitly made in a deductive development should involve a misconception, its introduction may lead not only to a result which does not follow from the postulates of the deductive system, but to one which may actually contradict some previously established theorem of the system. From this point of view, criticize the following three geometrical paradoxes:

(a) *To prove that any triangle is isosceles.*

Let ABC be any triangle (see Figure 8.1c). Draw the bisector of $\angle C$ and the

Figure 8.1c

perpendicular bisector of side AB. From their point of intersection E, drop perpendiculars EF and EG on AC and BC, respectively, and draw EA and EB. Now right triangles CFE and CGE are congruent, since each has CE as hypotenuse and since $\angle FCE = \angle GCE$. Therefore $CF = CG$. Again, right triangles EFA and EGB are congruent, since leg EF of one equals leg EG of the other (any point E on the bisector of an angle C is equidistant from the sides of the angle) and since hypotenuse EA of one equals hypotenuse EB of the other (any point E on the perpendicular bisector of a line segment AB is equidistant from the extremities of that line segment). Therefore $FA = GB$. It now follows that $CF + FA = CG + GB$, or $CA = CB$, and the triangle is isosceles.

(b) *To prove that a right angle is equal to an obtuse angle.*

Let $ABCD$ be any rectangle (see Figure 8.1d). Draw BE outside the rectangle and equal in length to BC, and hence to AD. Draw the perpendicular bisectors of DE and AB; since they are perpendicular to nonparallel lines, they must intersect in a point P. Draw AP, BP, DP, EP. Then $PA = PB$ and $PD = PE$ (any point on the perpendicular bisector of a line segment is equidistant from

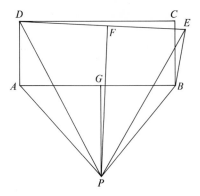

Figure 8.1d

the extremities of the line segment). Also, by construction, $AD = BE$. Therefore triangles APD and BPE are congruent, since the three sides of one are equal to the three sides of the other. Hence $\angle DAP = \angle EBP$. But $\angle BAP = \angle ABP$, since these angles are base angles of the isosceles triangle APB. By subtraction it now follows that right angle $DAB =$ obtuse angle EBA.

(c) *To prove that there are two perpendiculars from a point to a line.*

Let any two circles intersect in A and B (see Figure 8.1e). Draw the diameters

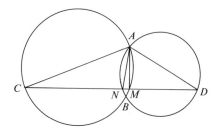

Figure 8.1e

AC and AD, and let the join of C and D cut the respective circles in M and N. Then angles AMC and AND are right angles, since each is inscribed in a semi-circle. Hence AM and AN are two perpendiculars to CD.

2. To guarantee the existence of certain points of intersection (of line with circle and circle with circle) Richard Dedekind (1831–1916) introduced into geometry the following continuity postulate: *If all the points of a straight line fall into two classes, such that every point of the first class lies to the left of every point of the second class, then there exists one and only one point of the line which produces this division of all points into two classes, that is, this severing of the straight line into two portions.*

(a) Complete the details of the following indicated proof of the theorem: *The straight line segment joining a point* A *inside a circle to a point* B *outside the circle has a point in common with the circle.*

Let O be the center and r the radius of the given circle (see Figure 8.1f), and let C be the foot of the perpendicular from O on the line determined by A and B. The points of the segment AB can be divided into two classes: those points P for which $OP < r$ and those points Q for which $OQ \geq r$. It can be shown that,

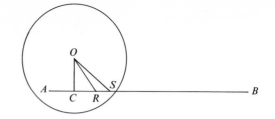

Figure 8.1f

in every case, $CP < CQ$. Hence, by Dedekind's Postulate, there exists a point R of AB such that all points which precede it belong to one class and all which follow it belong to the other class. Now $OR \not< r$, for otherwise we could choose S on AB, between R and B, such that $RS < r - OR$. But, since $OS < OR + RS$, this would imply the absurdity that $OS < r$. Similarly, it can be shown that $OR \not> r$. Hence we must have $OR = r$, and the theorem is established.

(b) How might Dedekind's Postulate be extended to cover angles?

(c) How might Dedekind's Postulate be extended to cover circular arcs?

3. Let us, for convenience, restate Euclid's first three postulates in the following equivalent forms:

(1) *Any two distinct points determine a straight line.*

(2) *A straight line is boundless.*

(3) *There exists a circle having any given point as center and passing through any second given point.*

Show that Euclid's postulates, partially restated above, hold if the points of the plane are restricted to those whose rectangular Cartesian coordinates for some fixed frame of reference are rational numbers. Show, however, that under this restriction a circle and a line through its center need not intersect each other.

4. Show that Euclid's postulates (as partially restated in Problem 3 above) hold if we interpret the plane as the surface of a sphere, straight lines as great circles on the sphere, and points as points on the sphere. Show, however, that in this interpretation the following are true:

(a) Parallel lines do not exist.

(b) All perpendiculars to a given line erected on one side of the line intersect in a point.

(c) It is possible to have two distinct lines joining the same two points.

(d) The sum of the angles of a triangle exceeds two right angles.

(e) There exist triangles having all three angles right angles.

(f) An exterior angle of a triangle is not always greater than each of the two remote interior angles.

(g) The sum of two sides of a triangle can be less than the third side.

(h) A triangle with a pair of equal angles may have the sides opposite them unequal.

(i) The greatest side of a triangle does not necessarily lie opposite the greatest angle of the triangle.

5. In 1882 Moritz Pasch formulated the following postulate: *Let A, B, C be three points not lying in the same straight line, and let m be a straight line lying in the plane of ABC and not passing through any of the points A, B, C. Then, if the line*

m *passes through a point of the segment* AB, *it will also either pass through a point of the segment* BC *or a point of the segment* AC. This postulate is one of those assumptions classified by modern geometers as a *postulate of order*, and it assists in bringing out the idea of "betweenness."

(a) Prove, as a consequence of Pasch's Postulate, that *if a line enters a triangle at a vertex, it must cut the opposite side.*

(b) Show that Pasch's Postulate does not hold for an arbitrary spherical triangle cut by a great circle.

8.2 MODERN POSTULATIONAL FOUNDATIONS FOR EUCLIDEAN GEOMETRY

After the discovery of non-Euclidean geometry, a need was felt for a truly satisfactory postulational treatment of Euclidean geometry. All hidden, or tacit, assumptions had to be ferreted out, and a logically acceptable set of underlying postulates for the subject had to be clearly and unequivocally put forth. Such an organization of Euclidean geometry was first accomplished in 1882 by the German mathematician Moritz Pasch.

In his treatment of Euclidean geometry, Pasch recognized the important distinction between *explicit* and *implicit* definition. Most people are familiar with the concept of explicit definition, inasmuch as this is the type of definition most frequently employed. In an explicit definition a new term is expressed by means of terms which are already accepted in the vocabulary. In a technical sense, then, a new term introduced in such a manner serves merely as an abbreviation for a complex combination of terms already present. Thus a new term introduced by explicit definition is really arbitrary, though convenient, and may be entirely dispensed with, though then the discourse in which the vocabulary is to be employed would immediately increase in complexity.

Implicit definition, on the other hand, is relatively unfamiliar to most people, though such a notion is indispensable in logical theory. The necessity for implicit definition is due to the fact that it is impossible to define all terms explicitly if we wish to avoid circularity. It is impossible, for example, to define word A in terms of word B, then word B in terms of word C, and so on indefinitely, for such a procedure would imply an infinite number of words in the vocabulary. Of course the dictionary makes an attempt to define all words explicitly through the admitted use of circularity, but it is hoped that the person using the dictionary has developed an adequate vocabulary so that the words in terms of which some unknown word is defined are already familiar to him.

Every person is aware of his own attempt to introduce new words into his vocabulary by observing how these words are used by others. This idea of defining a word through the medium of the context in which it occurs is the basic idea of implicit definition. In a logical discourse, since not all technical terms can be explicitly defined, there must be some whose meanings can be realized only by observing the context in which they are employed.

A logical discourse, in other words, must accept a relatively small number of primitive technical terms which can be used for explicitly defining all the other technical terms which occur in the discourse, these primitive terms of the discourse receiving no definitions except those given to them implicitly by their presence in the adopted postulates of the discourse.

Whereas Euclid attempted a kind of explicit definition of the terms *point*, *line*, and *plane*, for example, Pasch accepted these as primitive, or irreducible, terms in his development of Euclidean geometry; he considered them only implicitly defined by the basic propositions that he assumed as postulates in his treatment. These assumed basic propositions were described by Pasch as *nuclear*. Although the origin of the nuclear propositions might be found in empirical considerations, Pasch emphasized that they are to be enunciated without regard to any empirical significance. He declared that the creation of a truly deductive science demands that all logical deductions must be independent of any meanings which might be attached to the various concepts. In fact, if it becomes necessary at any point in the construction of a proof to refer to certain interpretations of the basic terms, then that is sufficient evidence that the proof is logically inadequate. On the other hand, by keeping all of the work purely formal, various applications of the discourse may be obtained by assigning different suitable meanings to the primitive terms employed. From this point of view, Euclidean geometry is essentially a symbolic system whose validity and possibility for further development do not depend upon any specific meanings given to the basic terms employed in the postulates of the geometry; Euclidean geometry is reduced to a pure exercise in logical syntax. Where Euclid appears to have been guided by visual imagery, and thus subjected to the making of tacit assumptions, Pasch attempted to avoid this pitfall by deliberately considering geometry as a purely hypothetico-deductive system. Pasch profoundly influenced postulational thinking in geometry, and later workers in the field attempted to maintain the standards of rigor which he had introduced.

Following Pasch, the Italian mathematician Giuseppi Peano (1858–1932) gave, in 1889, a new postulational development of Euclidean geometry. Like Pasch, Peano based his treatment upon certain primitive terms, among which are an entity called a "point" and a relation among points designated by "betweenness." From many points of view Peano's work is largely a translation of Pasch's treatise into the notation of a symbolic logic which Peano introduced to the mathematical world. In Peano's version no empiricism is found; his geometry is purely formalistic by virtue of the fact that it is constructed as a calculus of relations between variables. Here we have the mathematician's ultimate cloak of protection from the pitfall of over-familiarity with his subject matter. We have seen that Euclid, working with visual diagrams in a field of study with which he was very familiar, unconsciously made numerous hidden assumptions which were not guaranteed to him by his axioms and postulates. To protect himself from similar prejudice, Peano conceived the idea of symbolizing his primitive terms and his logical processes of thought. Clearly, if one is to say, "Two x's determine a y," instead of, "Two points determine a straight line," one is not so likely to be biased by preconceived notions about "points" and "straight lines," and if a symbolic logic is employed in the reasoning, one is not so likely to fall into fallacies stemming from slippery intuition and other modes of loose reasoning. The derivation of theorems becomes an algebraic process in which only symbols and formulas are employed, and geometry is reduced to a strictly formal process which is entirely independent of any interpretations of the symbols involved.

Another Italian mathematician, Mario Pieri (1860–1904), employed, in 1899, in a study of Euclidean geometry, a quite different approach from that of his predecessors. He considered the subject of his study to be an aggregate of undefined elements called "points" and an undefined concept of "motion." Pieri's first five postulates will indicate the important role assigned to the concept of motion. They are:

1. *There is given an aggregate* S *of points containing at least two distinct members.*
2. *A motion establishes a pairing of the points of* S *such that to each point* P *of* S *there corresponds some point* P′ *of* S. *For any motion which establishes a correspondence between the points* P *of* S *and the points* P′ *of* S, *there is an* inverse *motion which establishes a correspondence between the points* P′ *of* S *and the points* P *of* S.
3. *The resultant of two motions performed successively is equivalent to a single motion.*

As a consequence of Postulates 2 and 3, the motion equivalent to some motion followed by its inverse is a motion which makes each point of S correspond to itself, or, in other words, leaves each point of S fixed. This motion is called the *identical motion*. A motion which is not the identical motion is called an *effective motion*.

4. *For any two distinct points* A *and* B, *there exists an effective motion which leaves* A *and* B *fixed. Such a motion may be referred to as a* rotation motion *about the two points* A *and* B.
5. *If there is an effective motion which leaves three points* A, B, C *fixed, then every motion which leaves* A *and* B *fixed also leaves* C *fixed.*

As a result of Postulate 5, it is now possible to define the straight line determined by two points A and B: "The *straight line* AB is the aggregate of points which remain fixed under any effective motion which leaves A and B fixed."

Although Pieri's treatment of Euclidean geometry received no wide acceptance, the development of certain modern notions is apparent in his work. We have, for instance, the idea of *transformation*, or *mapping*. Pieri's motions are the direct isometries, or rigid displacements, which map the set S of all points of space onto itself, and Pieri was considering Euclidean geometry as the study of the properties and relations of configurations of points which remain invariant under the group of direct isometries. This idea had earlier been generalized, in the famed Erlanger Programm, to form the basis of Felix Klein's remarkable codification of geometries. It should incidentally be noted that Pieri's idea of motion can be nicely adapted to the Euclidean superposition proofs.

The modern postulational treatment of Euclidean geometry that has received the widest acceptance is that due to the eminent German mathematician David Hilbert (1862–1943). Professor Hilbert gave a course of lectures on the foundations of Euclidean geometry at the University of Göttingen during the 1898–1899 winter term. These lectures were rearranged and published in a slender volume in June, 1899, under the title *Grundlagen der Geometrie* (Foundations of Geometry). This work, in its various improved revisions, is today a classic in its field; it has done far more than any other single work since the discovery of non-Euclidean geometry to promote the modern postulational method and to shape the character of a good deal of present-day mathematics. The influence of the book was immediate. A French edition appeared soon after the German publication, and an English version, translated by E. J. Townsend, appeared in 1902. The work went through seven German editions during the author's lifetime, the seventh edition appearing in 1930.* By developing a postulate set for Euclidean geometry that does not depart too greatly in spirit from Euclid's own, and by employing a minimum of symbolism, Hilbert succeeded in convincing mathematicians to a far greater extent than had Pasch and Peano, of the purely hypothetico-deductive nature of geometry. But the influence of Hilbert's work went far beyond this, for, backed by the author's great mathematical authority, it firmly implanted the postulational method, not only in the field of geometry, but also in essentially every other branch of mathematics. The stimulus to the development of the foundations of mathematics provided by Hilbert's little book is difficult to overestimate. Lacking the strange symbolism of the works of Pasch and Peano, Hilbert's work can be read, in great part, by any intelligent student of high school geometry.

Whereas Euclid made a distinction between "axioms" and "postulates," modern mathematicians consider these two terms as synonymous, and designate all the assumed propositions of a logical discourse by either term. From this point of view, Hilbert's treatment of plane and solid Euclidean geometry rests on 21 axioms or postulates, and these involve six primitive, or undefined, terms. For simplicity we shall consider only those postulates of Hilbert's set which apply to *plane* geometry. Under this limitation there are 15 postulates and five primitive terms.

The primitive terms in Hilbert's treatment of plane Euclidean geometry are *point*, *line* (meaning *straight line*), *on* (a relation between a point and a

---

* An eighth German edition, a revision and enlargement by Paul Bernays, appeared in 1956, a ninth German edition, by Bernays in 1962, and a tenth German edition by Bernays in 1968. The last has been translated into English by Leo Unger (La Salle, Illinois: Open Court Publishing Company, 1971).

line), *between* (a relation between a point and a pair of points), and *congruent* (a relation between pairs of points and between configurations called *angles*, which are explicitly defined in the treatment). For convenience of language, the phrase "point A is on line m" is frequently stated alternatively by the equivalent phrases, "line m passes through point A" or "line m contains point A."

In Appendix 2 are listed, for purposes of discussion and for future reference, Hilbert's 15 postulates for plane geometry. The reader will note that the postulates are interspersed with occasional definitions when needed. The statements of the postulates are taken, with some slight modifications for the sake of clarity, from the seventh (1930) edition of Hilbert's *Grundlagen der Geometrie*. Following Hilbert, the postulates are presented in certain related groups.

Upon these fifteen postulates rests the extensive subject of plane Euclidean geometry! To develop the geometry appreciably from these postulates is too long a task for us to undertake here, but we shall add a few words concerning the significance of some of the postulates.

The postulates of the first group define implicitly the idea expressed by the primitive term "on," and they establish a connection between the two primitive entities, "points" and "lines."

The postulates of the second group were first studied by Pasch, and they define implicitly the idea expressed by the primitive term "between." In particular, they assure us of the existence of an infinite number of points on a line and that a line is not terminated at any point, and they guarantee that the order of points on a line is serial rather than cyclical. Postulate II-4 (Pasch's Postulate) differs from the other postulates of the group, for, since it involves points not all on the same line, it gives information about the plane as a whole. The postulates of order are of historical interest inasmuch as Euclid completely failed to recognize any of them. It is this serious omission on Euclid's part that permits one, using only Euclid's list of assumptions, to derive paradoxes that arise from applying sound reasoning to misconceived figures.

The postulates of the third group define implicitly the idea expressed by the primitive term "congruent" as applied to pairs of points and to angles. These postulates are included in order to circumvent the necessity of dealing with the concept of motion. For example, it is interesting to note how, in Postulate III-6, Hilbert introduces the congruency of triangles without employing Euclid's method of superposition, still found in some high school textbooks.

The Playfair parallel postulate appears as the only postulate of Group IV; it is, of course, equivalent to Euclid's parallel postulate. Using the postulates of the first three groups one can prove that there is at least one line through the given point A and not intersecting the given line m.

The first postulate of the last group (the Postulates of Archimedes) corresponds to the familiar process of estimating the distance from one point of a line to another by the use of a measuring stick; it guarantees that if we start at the one point and lay off toward the second point a succession of equal distances (equal to the length of the measuring stick) we will ultimately pass the second point. Upon this postulate can be made to depend the entire theory of measurement and, in particular, Euclid's theory of proportion. The final postulate (the Postulate of Completeness) is not required for the derivation of the theorems of Euclidean geometry, but it makes possible the establishment of a one-to-one correspondence between the points of any line and the set of all real numbers, and is necessary for the free use of the real number system in analytic, or coordinate, geometry. It can be shown that, in the presence of the other thirteen postulates, these last two postulates are equivalent to the Postulate of Dedekind (see Problem 2, Section 8.1), and therefore, if we should wish, they can be replaced by this postulate.

Considerably more was accomplished by Hilbert in his *Grundlagen der Geometrie* than just the establishment of a satisfactory set of postulates for Euclidean geometry. In showing the logical consistency and partial independence of his postulates, Hilbert had to devise many interesting models, or interpretations, for various subsets of the postulates. This amounted to introducing various new systems of geometry and to creating a number of unusual algebras of segments. The significance of several important postulates and theorems in the development of Euclidean geometry is clearly shown in the work, and examples of various kinds of nontraditional geometries are illustrated. For example, to show the independence of the Postulate of Archimedes from the other postulates of the treatment, an example of a non-Archimedean system is offered in which all the postulates except the Postulate of Archimedes are shown to hold. There is also developed in this work a theory of proportion and a theory of areas which are independent of the postulate of Archimedes. These accompanying investigations by Hilbert virtually inaugurated the twentieth-century study of abstract geometry and successfully convinced many mathematicians of the hypothetico-deductive nature of mathematics. By implanting the postulational method in nearly all of mathematics since 1900, Hilbert's *Grundlagen der Geometrie* represents a definite landmark in the history of mathematical thought.

Other postulational treatments of Euclidean geometry followed Hilbert's effort. In 1904, the American mathematician Oswald Veblen (1880–1960) furnished a new postulate set in which he replaced the primitive notion of "betweenness," as used by Peano and Hilbert, by a more pervasive primitive relation of "order."* With this new primitive relation, the terms "line," "plane," "on," and "congruent," can receive explicit definition, and thus the list of primitive terms be reduced to just two, namely, "point" and "order." There is a feeling among mathematicians that the smaller one can make the number of primitive terms in a postulational development,

---

* Oswald Veblen, "A system of axioms for geometry," *Transactions of the American Mathematical Society*, 5 (1904), 343–384.

the more aesthetically pleasing is that postulational development—a principle that was emphasized by Peano. A second study of the foundations of Euclidean geometry was made by Veblen in 1911, in which his original treatment was slightly revised to accord with some ideas put forth by R. L. Moore.*

A very satisfying combination of the postulates of Hilbert and Veblen has been employed by Gilbert de B. Robinson.† Robinson's postulates are essentially Veblen's postulates of order combined with Hilbert's postulates of congruence and continuity. In 1913, E. V. Huntington (1874–1952) offered a treatment of three-dimensional Euclidean geometry based upon "sphere" and "inclusion" (one sphere lying within another) as primitive terms.‡ This unusual approach exemplifies the fact that it is possible to characterize Euclidean geometry by systems of postulates which are superficially very different from one another.

An excellent and detailed abstract postulational examination of Euclidean geometry appeared in 1927 in a work of Henry George Forder.§ Here we find many alternative postulate sets compared with one another. For example, Forder considers nine different parallel postulates, which vary in the strengths of their assumptions. By adopting a strong parallel postulate and by using Dedekind's Postulate as a postulate of continuity, Forder gives a postulate set for Euclidean geometry based on only the two primitive terms "point" and "order." He also gives an abstract treatment of a 1909 postulate set of Pieri's and based on the two primitive terms "point" and "congruence."

Whether or not it is wise to attempt a rigorous postulational treatment of Euclidean geometry at the high school level is a matter of pedagogical opinion. George Bruce Halsted made an unsuccessful effort in 1904, when he published an elementary geometry textbook based upon Hilbert's postulate set.‖ More successful, and certainly worthy of examination, is an attempt made in 1940 by Professors George David Birkhoff and Ralph Beatley of Harvard University.¶ Here a teachable high school course

---

* Oswald Veblen, "The foundations of geometry," in *Monographs on Topics of Modern Mathematics Relevant to the Elementary Field*, edited by J. W. A. Young. New York: Dover Publications, Inc., 1955.
† G. de B. Robinson, *The Foundations of Geometry*, 2nd ed. Mathematical Expositions No. 1. Toronto: University of Toronto Press, 1946.
‡ E. V. Huntington, "A set of postulates for abstract geometry, expressed in terms of the simple relation of inclusion," *Mathematische Annalen*, 73 (1913), 522–559. Also see G. de B. Robinson, *loc. cit.*, Appendix, pp. 157–160.
§ H. G. Forder, *The Foundations of Euclidean Geometry*. New York: Cambridge University Press, 1927. Reprinted by Dover Publications, Inc.
‖ G. B. Halsted, *Rational Geometry*. New York: John Wiley and Sons, Inc., 1904. Certain logical criticisms of the text were met by Halsted in 1907 in a thoroughly revised second edition; this edition has been translated into French.
¶ G. D. Birkhoff and R. Beatley, *Basic Geometry*. Chicago: Scott, Foresman and Company, 1940. Reprinted by Chelsea Publishing Company Inc., 1959.

in plane Euclidean geometry is evolved from five postulates based upon an ability to measure line segments and angles. Although for pedagogical reasons certain subtler mathematical and logical details are either ignored or slurred over, the work does stem from a rigorous mathematical presentation that had been made earlier by Birkhoff.*

Since about 1960 a number of authors and writing groups have engaged in the task of attempting to produce textual materials for the high school geometry class wherein geometry is developed rigorously from a postulational basis. In these attempts, usually either the Hilbert postulate set or the Birkhoff postulate set (sometimes slightly altered or augmented) is adopted.

PROBLEMS

1. (a) Consider the following definitions taken from an elementary geometry text:
    (1) The *diagonals* of a quadrilateral are the two straight line segments joining the two pairs of opposite vertices of the quadrilateral.
    (2) *Parallel lines* are straight lines that lie in the same plane and that never meet, however far they are extended in either direction.
    (3) A *parallelogram* is a quadrilateral having its opposite sides parallel.
    Now, without using any of the italicized words above, restate the proposition, "The diagonals of a parallelogram bisect each other."
   (b) By means of appropriate explicit definitions reduce the following sentence to one containing only five words: "The movable seats with four legs and a back were restored to a sound state by the person who takes care of the building."
    These exercises illustrate the convenience of explicit definitions.

2. Trace the following words through a standard dictionary until a circular chain has been established: (a) dead, (b) noisy, (c) line (in the mathematical sense).

3. In each of the following, is the given conclusion a valid deduction from the given pair of premises?
   (a) If today is Saturday, then tomorrow will be Sunday.
      But tomorrow will be Sunday.
      Therefore, today is Saturday.
   (b) Germans are heavy drinkers.
      Germans are Europeans.
      Therefore, Europeans are heavy drinkers.
   (c) If *a* is *b*, then *c* is *d*.
      But *c* is *d*.
      Therefore, *a* is *b*.
   (d) All *a*'s are *b*'s.
      All *a*'s are *c*'s.
      Therefore, all *c*'s are *b*'s.
    These exercises illustrate how a person may allow the meanings which he associates with words or expressions to dominate his logical analysis. There is a greater tendency to go wrong in (a) and (b) than in (c) and (d), which are symbolic counterparts of (a) and (b).

* G. D. Birkhoff, "A set of postulates for plane geometry, based on scale and protractor," *Annals of Mathematics*, 33 (1932), 329–345.

4. Consider the following set of postulates about certain objects called "dabbas" and certain collections of dabbas called "abbas":

P1: Every abba is a collection of dabbas.

P2: There exist at least two dabbas.

P3: If p and q are two dabbas, then there exists one and only one abba containing both p and q.

P4: If L is an abba, then there exists a dabba not in L.

P5: If L is an abba, and p is a dabba not in L, then there exists one and only one abba containing p and not containing any dabba that is in L.

(a) What are the primitive terms of this postulate set?

(b) Deduce the following theorems from the postulate set:

(1) Every dabba is contained in at least two abbas.

(2) Every abba contains at least two dabbas.

(3) There exist at least four distinct dabbas.

(4) There exist at least six distinct abbas.

(c) Restate the postulates by interpreting "abba" as "straight line" and "dabba" as "point." Note that P5 is now Playfair's Postulate.

(d) Define a *kurple* as any three dabbas not contained in the same abba. What is a kurple in the interpretation of part (c)?

5. (a) Establish the following consequences of the first five postulates of Pieri's postulate set for Euclidean geometry (as given in Section 8.2).

(1) If C and D are two distinct points of the straight line AB, then A and B are points of the straight line CD.

(2) If three points are on a straight line, then the three points corresponding to them in any motion are also on a straight line.

(b) Pieri defines a sphere as follows: "If A and B are two distinct points, then the aggregate of all points P such that for each P there exists a motion which leaves A fixed but makes P correspond to B is called *the sphere of center* A *passing through* B." Establish the following consequences of this definition and Pieri's first five postulates:

(1) A sphere transforms into a sphere in every motion.

(2) A motion which leaves the center of a sphere fixed transforms the sphere into itself.

(3) If two spheres with centers A and B have only one point C in common, then the three points A, B, C lie on a line.

(c) Try to formulate, in terms of motion, a suitable definition of *perpendicularity*.

6. Prove the theorem that follows Postulate III-3 in Hilbert's postulate set for plane Euclidean geometry.

7. Assuming that if point B is between points A and D, and point C is between B and D, then C is between A and D, deduce the following theorems from Hilbert's postulate set for plane Euclidean geometry:

(a) There is no limit to the number of distinct points between two given distinct points.

(b) If neither of two distinct lines, a and b, intersects a third line c, then a and b do not intersect.

(c) If two sides and the included angle of one triangle are congruent, respectively, to two sides and the included angle of another triangle, then the third side of the first triangle is congruent to the third side of the second triangle.

(d) If two angles and the included side of one triangle are congruent, respectively, to two angles and the included side of another triangle, then all the parts (angles and sides) of the first triangle are congruent to the corresponding parts of the second triangle.

8. Try to deduce the following proposition from Hilbert's postulate set for plane Euclidean geometry: Given any four points on a line, it is always possible to denote them by letters A, B, C, D in such a way that B is between A and C and also between A and D, and that C is between A and D and also between B and D.

This proposition was included as a postulate in the first edition of Hilbert's work, but was later proved by E. H. Moore to be a consequence of Hilbert's other postulates.*

9. (a) Consider the configuration formed by the positive x axis and a directed circular arc C of radius r radiating from the origin O and such that the arc C is convex when viewed from its right-hand side. We shall call such a configuration a (special kind of) *horn angle*, and denote it by h. Let T be the directed tangent to C at O, and designate the positive (counterclockwise) angle from the positive x axis around to T by θ. We shall compare two such horn angles h and h' in the following way. If $\theta = \theta'$ and $r = r'$, then we say $h = h'$; if $\theta > \theta'$, we say $h > h'$; if $\theta = \theta'$ but $r < r'$, then again we say $h > h'$. We further say $h' = nh$, where n is a positive integer, if and only if $\theta' = n\theta$ and $r' = r/n$. Show that our horn angles now form a non-Archimedean system of entities; that is, show that there exist horn angles h and h' such that $nh < h'$ for every positive integer n.

(b) Show that a horn angle for which $\theta = 0$ can be trisected with Euclidean tools.

(c) Consider pairs of power series of the form $y = a_1x + a_2x^2 + a_3x^3 + \cdots$ and $y' = a_1'x + a_2'x^2 + a_3'x^3 + \cdots$, where the coefficients are real numbers. We shall compare two such power series as follows. We say $y = y'$ if and only if $a_i = a_i'$ for all i; we say $y > y'$ if and only if there exists some positive integer k such that $a_1 = a_1', a_2 = a_2', \ldots, a_k = a_k', a_{k+1} > a_{k+1}'$. We further say that $y' = ny$, where n is a positive integer, if and only if $a_i' = na_i$ for all i. Show that the set of all such power series form, under the above definitions, a non-Archimedean system of entities. [This can be interpreted as a generalization of part (a), where the circular arcs C have been replaced by analytic curves passing through O.]

(d) Consider a set of entities of which a typical member M is composed of the segment $-a \leq x \leq 0$ $(a > 0)$, and the isolated points $x = 1, 2, \ldots, k$. Devise a method of comparing such entities, and define nM, where n is a positive integer, such that the entities form a non-Archimedean system.

(e) Let $z = a + ib$ and $z' = a' + ib'$ (a, b, a', b' real and $i = \sqrt{-1}$) be two complex numbers. Set $z = z'$, if and only if $a = a'$ and $b = b'$. If $a > a'$, set $z > z'$; if $a = a'$, but $b > b'$, again set $z > z'$. Define $nz = na + i(nb)$. Show that the complex numbers, under the above definitions, form a non-Archimedean system of entities.

(f) Show that the set of all coplanar vectors radiating from a point O can be formed into a non-Archimedean system.

---

* E. H. Moore, "On the projective axioms of geometry," *Transactions of the American Mathematical Society*, 3 (1902), 142–158.

10. It is pointed out in Section 8.2 that Huntington has given a postulate set for Euclidean geometry of space in which *sphere* is taken as a primitive element and *inclusion* as a primitive relation. Try to formulate, in terms of these primitive terms, suitable definitions for *point*, *segment*, *ray*, and *line*.

8.3 FORMAL AXIOMATICS

The discovery of a non-Euclidean geometry (and not long after, of a non-commutative algebra) led to a deeper study and refinement of axiomatic procedure, and from the material axiomatics of the ancient Greeks there evolved the formal axiomatics of the twentieth century. To help clarify the difference between the two forms of axiomatics we first introduce the modern concept of *propositional function*, the fundamental importance of which was first brought to notice by the English mathematician and philosopher Bertrand Russell (1872–1970).

Consider the three statements:

(1) Spring is a season.
(2) 8 is a prime number.
(3) x is a y.

Each of these statements has form—the same form; statements (1) and (2) have content as well as form; statement (3) has form only. Clearly, statements (1) and (2) are propositions, one true and the other false. Equally clearly, statement (3) is *not* a proposition, for, since it asserts nothing definite, it is neither true nor false, and a proposition, by definition, is a statement which is true or false. Statement (3), however, though not a proposition, does have the form of a proposition. It has been called a *propositional function*, for if in the form

$$x \text{ is a } y,$$

we substitute terms of definite meaning for the variables x and y, we may obtain propositions, true propositions if the substituted terms should verify the propositional function, false propositions if the substituted terms should falsify the propositional function. It is apparent that some substitutions for x and y convert the propositional function into so much nonsense; such meanings for the variables are considered inadmissible. The form considered above is a propositional function in two variables, and it has infinitely many verifiers.

A propositional function may contain any number of variables. An example having but one is: x is a volume in the Library of Congress. Here x evidently has as many verifying values as there are volumes in the Library of Congress. Evidently, too, the variable x has many falsifying values.

There is no need for the variables in a propositional function to be denoted by symbols, such as x, y, ...; they may be ordinary words. Thus, should a statement whose terms are ordinary words appear in a discourse with no indication as to the senses in which the words are to be understood, then in that discourse the statement is really a propositional function, rather than a proposition, and in the interests of clarity the ambiguous or undefined terms might better be replaced by such symbols as $x, y, \ldots$

With the idea of a propositional function firmly in mind, let us return to a discussion of axiomatic procedure. We recall that any logical discourse, in an endeavor to be clear, tries to define explicitly the elements of the discourse, the relations among these elements, and the operations to be performed upon them. Such definitions, however, must employ other elements, relations, and operations, and these, too, are subject to explicit definition. If these are defined, it must again be by reference to further elements, relations, and operations. There are two roads open to us; either the chain of definitions must be cut short at some point, or else it must be circular. Since circularity is not to be tolerated in a logical discourse, the definitions must be brought to a close at some point; thus it is necessary that one or more elements, relations, and operations receive no explicit definition. These are known as the *primitive terms* of the discourse. There is likewise an effort logically to deduce the statements of the discourse, and, again, in order to get started and also to avoid the vicious circle, one or more of the statements must remain entirely *unproved*. These are known as the *postulates* (or axioms, or primary statements) of the discourse. Clearly, then, any logical discourse such as we are considering must conform to the following pattern.

Pattern of Formal Axiomatics

(A) The discourse contains a set of technical terms (elements, relations among elements, operations to be performed on elements) which are deliberately chosen as undefined terms. These are the *primitive terms* of the discourse.

(B) The discourse contains a set of statements about the primitive terms which are deliberately chosen as unproved statements. These are called the *postulates* (or *axioms*), *P*, of the discourse.

(C) All other technical terms of the discourse are defined by means of previously introduced terms.

(D) All other statements of the discourse are logically deduced from previously accepted or established statements. These derived statements are called the *theorems*, *T*, of the discourse.

(E) For each theorem T_i of the discourse there exists a corresponding statement (which may or may not be formally expressed) asserting that theorem T_i is logically implied by the postulates *P*. (Often a corresponding statement appears at the end of the proof of the theorem in some such words as, "Hence the theorem," or "This completes the proof of the theorem"; in some elementary geometry textbooks the statement appears, at the end of the proof of the theorem, as "Q.E.D." (Quod erat demonstrandum). The modern symbol ▮, or some variant of it, suggested by Paul R. Halmos, is frequently used to signalize the end of a proof.)

The first thing to notice in the above pattern is that the primitive terms, being undefined terms, might just as well (if such is not already the case) be replaced by symbols like x, y, ... Let us suppose this substitution is made. Then the primitive terms are clearly variables. The second thing to notice is that the postulates, P, since they are statements about the primitive terms, are nothing less than propositional functions. And the third thing to notice is that the theorems, T, since they are but logical implications of the postulates P, also are propositional functions. We are thus brought to a fact of cardinal importance, namely, that once the primitive terms are realized to be variables, both the postulates and the theorems of a logical discourse are not propositions but propositional functions.

Since the postulates and the theorems of a logical discourse are propositional functions, that is, are statements of form only and without content, it would seem that the whole discourse is somewhat vacuous and entirely devoid of truth or falseness. Such, however, is not the case, for by (E) of the postulational pattern we have the all-important statement,

(F) The postulates P imply the theorems T.

Now (F) asserts something definite; it is true or false, and so is a proposition —a true one if the theorems T are in fact implied by the postulates P, and a false one if they are not. The statement (F) is precisely what the discourse is designed for; it is the discourse's sole aim and excuse for being.

A discourse conducted according to the above pattern has been called, by some mathematicians, a *branch of pure mathematics*, and the grand total of all such existing branches of pure mathematics, the *pure mathematics of to-date*.

If, for the variables (the primitive terms) in a branch of pure mathematics we should substitute terms of definite meaning which convert all the postulates of the branch into true propositions, then the set of substituted terms is called an *interpretation* of the branch of pure mathematics. The interpretation will also, provided all deductions have been correctly performed, convert the theorems of the discourse into true propositions. The result of such an interpretation is called a *model* of the branch of pure mathematics.

A model of a branch of pure mathematics has been called a *branch of applied mathematics*, and the grand total of all existing branches of applied mathematics, the *applied mathematics of to-date*. Thus the difference between applied and pure mathematics is not one of applicability and inapplicability, but rather of concreteness and abstractness. Behind every branch of applied mathematics lies a branch of pure mathematics, the latter being an abstract development of what formerly was a concrete development. It is conceivable (and indeed such is often the case) that a single branch of pure mathematics may have several models, or associated branches of applied mathematics. This is the "economy" feature of pure mathematics, for the establishment of a branch of pure mathematics automatically assures the simultaneous establishment of all of its branches of applied mathematics.

The abstract development of some branch of pure mathematics is an instance of *formal axiomatics*, whereas the concrete development of a given branch of applied mathematics is an instance of *material axiomatics*. In the former case we think of the postulates as prior to any specification of the primitive terms, and in the latter we think of the objects that interpret the primitive terms as being prior to the postulates. In the former case a postulate is simply a basic assumption about some undefined primitive terms; in the latter case a postulate expresses some property of the basic objects which is taken as initially evident. This latter is the older view of a postulate, and was the view held by the ancient Greeks. Thus, to the Greeks, geometry was thought of as a study dealing with a unique structure of physical space, in which the elements *points* and *lines* are regarded as idealizations of certain actual physical entities, and in which the postulates are readily accepted statements about these idealizations. From the modern point of view, geometry is a purely abstract study devoid of any physical meaning or imagery.

The notion of pure mathematics gives considerable sense to Bertrand Russell's facetious remark that "mathematics may be defined as the subject in which we never know what we are talking about, nor whether what we are saying is true." It also accords with Henri Poincaré's saying that mathematics is "the giving of the same name to different things," and with Benjamin Peirce's (1809–1880) remark that "mathematics is the science which draws necessary conclusions."

PROBLEMS

1. (a) Construct an example of a propositional function containing three variables.
   (b) Obtain, by appropriate substitutions, three true propositions from the propositional function.
   (c) Obtain, by appropriate substitutions, three false propositions from the propositional function.

2. Consider the following postulate set, in which *bee* and *hive* are primitive terms:

   P1: Every hive is a collection of bees.
   P2: Any two distinct hives have one and only one bee in common.
   P3: Every bee belongs to two and only two hives.
   P4: There are exactly four hives.

   Deduce the following theorems:

   T1: There are exactly six bees.
   T2: There are exactly three bees in each hive.
   T3: For each bee there is exactly one other bee not in the same hive with it.

3. Consider a set *K* of undefined elements, which we shall denote by lower case letters, and let *R* denote an undefined dyadic relation (that is, a relation connecting two elements) which may or may not hold between a given pair of elements of *K*. If element *a* of *K* is related to element *b* of *K* by the *R* relation we shall

write *a R b*. We now assume the following four postulates concerning the elements of *K* and the dyadic relation *R*.

P1: If *a* and *b* are any two distinct elements of *K*, then either *a R b* or *b R a*.

P2: If *a* and *b* are any two elements of *K* such that *a R b*, then *a* and *b* are distinct.

P3: If *a, b, c* are any three elements of *K* such that *a R b* and *b R c*, then *a R c*.

P4: *K* consists of exactly four distinct elements.

Deduce the following seven theorems from the above four postulates.

T1: If *a R b*, then we do not have *b R a*.

T2: If *a R b* and if *c* is distinct from *a*, then either *a R c* or *c R b*.

T3: There is at least one element of *K* not *R*-related to any element of *K*. (This is an existence theorem.)

T4: There is only one element of *K* not *R*-related to any element of *K*. (This is a uniqueness theorem.)

Definition 1. If *b R a*, we say *a D b*.

T5: If *a D b* and *b D c*, then *a D c*.

Definition 2. If *a R b* and there is no element *c* such that also *a R c* and *c R b*, then we say *a F b*.

T6: If *a F c* and *b F c*, then *a* and *b* are identical.

T7: If *a F b* and *b F c*, then we do not have *a F c*.

Definition 3. If *a F b* and *b F c*, then we say *a G c*.

4. (a) Establish the following theorem of the branch of pure mathematics of Problem 3 above: *If* a G c *and* b G c, *then* a = b.

(b) Define the triadic relation *B(abc)* to mean either (*a R b* and *b R c*) or (*c R b* and *b R a*). Now prove: *If* B(abc) *holds, then* B(acb) *does not hold.*

5. Consider the following set of postulates about a certain collection *S* of primitive elements and certain primitive subcollections of *S* called m-*classes*:

P1: If *a* and *b* are distinct elements of *S*, then there is one and only one *m*-class containing *a* and *b*.

D1: Two *m*-classes having no elements in common are called *conjugate* *m*-classes.

P2: For every *m*-class there is one and only one conjugate *m*-class.

P3: There exists at least one *m*-class.

P4: Every *m*-class contains at least one element of *S*.

P5: Every *m*-class contains only a finite number of elements of *S*.

Establish the following theorems:

T1: Every *m*-class contains at least two elements.

T2: *S* contains at least four elements.

T3: *S* contains at least six *m*-classes.

T4: No *m*-class contains more than two elements.

6. Construct all geometries satisfying the following postulates:

P1: Space *S* is a set of *n* points, *n* a positive integer.

P2: A line is a non-null subset of *S*.

P3: Any two distinct lines have exactly one common point.

P4: Every point lies in exactly two distinct lines.

8.4 METAMATHEMATICS

It must not be thought, in building up a branch of pure mathematics, that we may set down a collection of symbols for undefined terms and then list for postulates an arbitrary system of assumed statements about these terms. There are certain required and certain desired properties which our system of assumed statements—our postulates—should possess. This section will accordingly be devoted to a brief examination of some of the properties of postulate sets. Such a study is technically known as *metamathematics*, and was first brought into prominence by Hilbert's *Grundlagen der Geometrie*. Of the properties of postulate sets, we shall consider the four known as *equivalence, consistency, independence*, and *categoricalness*. The first property applies to pairs of postulate sets, and the remaining three apply to individual postulate sets.

Two postulate systems $P^{(1)}$ and $P^{(2)}$ are said to be *equivalent* if each system implies the other, that is, if the primitive terms in each are definable by means of the primitive terms of the other, and if the postulates of each are deducible from the postulates of the other. If two postulate systems are equivalent, then the two abstract studies implied by them are, of course, the same, and it is merely a matter of " saying the same thing in different ways." The idea of equivalent postulate systems arose in ancient times when geometers, dissatisfied with Euclid's parallel postulate, tried to substitute for it a more acceptable equivalent. The modern studies of Euclidean geometry, with their various and quite different postulational bases, clearly illustrate that a postulate system is by no means uniquely determined by the study in question, but depends upon which technical terms of the study are chosen as primitive and which statements of the study are taken as unproved.

A postulate set is said to be *consistent* if contradictory statements are not implied by the set. This is the most important and most fundamental property of a postulate set; without this property the postulate set is worthless.

The most successful method so far invented for establishing consistency of a postulate set is the method of models. A model of a postulate set, recall, is obtained if we assign meanings to the primitive terms of the set which convert the postulates into true statements about some concept. There are two types of models—concrete models and ideal models. A model is said to be *concrete* if the meanings assigned to the primitive terms are objects and relations adapted from the real world, whereas a model is said to be *ideal* if the meanings assigned to the primitive terms are objects and relations from some other postulate system.

Where a concrete model has been exhibited we feel that we have established the *absolute* consistency of our postulate system, for if contradictory theorems are implied by our postulates, then corresponding contradictory statements would hold in our concrete model. But contradictions in the real world we accept as being impossible.

It is not always feasible to try to set up a concrete model of a given postulate set. Thus, if the postulate set contains an infinite number of primitive elements, a concrete model would certainly be impossible, for the real world does not contain an infinite number of objects. In such instances we attempt to set up an ideal model, by assigning to the primitive terms of postulate system A, say, concepts of some other postulate system B, in such a way that the interpretations of the postulates of system A are logical consequences of the postulates of system B. But now our test of consistency of the postulate set A can no longer claim to be an absolute test, but only a *relative* test. All we can say is that postulate set A is consistent if postulate set B is consistent, and we have reduced the consistency of system A to that of another system B.

Relative consistency is the best we can hope for when we apply the method of models to many branches of mathematics, for many of the branches of mathematics contain an infinite number of primitive elements. This is true, for example, of plane Lobachevskian geometry. In the next section we shall, however, by setting up a model of plane Lobachevskian geometry within plane Euclidean geometry, show that the former geometry is consistent if the latter geometry is.

A postulate of a postulate set is said to be *independent* if it is not a logical consequence of the other postulates of the set, and the entire postulate set is said to be *independent* if each of its postulates is independent. The most famous consideration in the history of mathematics of the independence of a postulate is that associated with the study of Euclid's parallel postulate. For centuries mathematicians had difficulty in regarding the parallel postulate as independent of Euclid's other postulates (and axioms), and accordingly made repeated attempts to show that it was a consequence of these other assumptions. It was the discovery of, and the ultimate proof of the relative consistency of, Lobachevskian non-Euclidean geometry that finally established the independence of Euclid's parallel postulate. In fact, it is no exaggeration to say that the historical consideration of the independence of Euclid's parallel postulate is responsible for initiating the entire study of properties of postulate sets and hence for shaping much of the modern axiomatic method.

A test for the independence of a postulate consists in finding an interpretation of the primitive terms which fails to verify the concerned postulate but which does verify each of the remaining postulates. If we are successful in finding such an interpretation, then the concerned postulate cannot be a logical consequence of the remaining postulates, for if it were a logical consequence of the remaining postulates, then the interpretation which converts all the other postulates into true propositions would have to convert it also into a true proposition. A test, along these lines, of the independence of an entire set of postulates can apparently be a lengthy business, for if there are n postulates in the set, n separate tests (one for each postulate) will have to be formulated.

Independence of a postulate set is by no means necessary, and a postulate set clearly is not invalidated just because it lacks independence. Generally speaking, a mathematician prefers a postulate set to be independent, for he wants to build his theory on a minimum amount of assumption. A postulate set which is not independent is merely redundant in that it contains one or more statements which can appear as theorems instead of as postulates. Sometimes, for pedagogical reasons, it may be wise to develop a subject from a postulate set which is not independent—for example, in developing plane geometry in high school from a postulational foundation.

There are some well-known postulate sets which, when first published, unknowingly contained postulates that were not independent. Such was the situation with Hilbert's original set of postulates for Euclidean geometry. This set was later shown to possess two postulates which are implied by the others. The finding of these two dependent postulates in no way invalidated Hilbert's system; in a subsequent amendment these postulates were merely changed to theorems, and their proofs supplied.

Similarly, R. L. Wilder was able to show that R. L. Moore's famous set of eight postulates, which virtually inaugurated modern set-theoretic topology, could be reduced to seven by the elimination of Moore's sixth postulate. The suspicion that the sixth postulate was not independent arose from the fact that the independence proof for this postulate was found to be at fault, and a subsequent search for a satisfactory proof turned out to be fruitless. Of course, Moore's mathematical theory remained intact in spite of Wilder's discovery, but the reduction of an eight-postulate system to an equally effective seven-postulate system has an aesthetic appeal to the mathematician.

The property of *categoricalness* is more recondite than the three properties already described, and we must precede its definition by first introducing the notion of *isomorphic interpretations* of a postulate system.

Among the primitive terms of a postulate set P we have a collection of E's, say, which denote elements, perhaps some relations $R_1, R_2, \ldots$ among the elements, and perhaps some operations $O_1, O_2, \ldots$ upon the elements. Accordingly, an interpretation of the postulate set is composed at least in part by element constants (the meanings assigned to the E's), perhaps in part of relation constants (the meanings assigned to the R's), and perhaps in part of operation constants (the meanings assigned to the O's). Now in any given interpretation I of P, let a collection of e's be the element constants (representing the E's), $r_1, r_2, \ldots$ the relation constants (representing the R's), and $o_1, o_2, \ldots$ the operation constants (representing the O's); and in any other interpretation I' of P let the element constants be a collection of e''s, the relation constants be $r'_1, r'_2, \ldots,$ and the operation constants be $o'_1, o'_2, \ldots.$ If it is possible to set up a one-to-one correspondence between the elements e of I and the elements e' of I' in such a way that, if two or more of the e's are related by some r, the corresponding e''s are related by the corresponding r', and if an o operating on one or more of the e's yields an e, the corresponding o' operating on the corresponding e''s yields the corresponding e', then we say that the two interpretations I and I' of P are *isomorphic.* This definition is often more briefly stated by saying that two interpretations I and I' of a postulate set P are isomorphic if one can set up a one-to-one correspondence between the elements of I and those of I' in such a way as "to be preserved by the relations and the operations of P." It follows that if two interpretations I and I' of a postulate set P are isomorphic, then any true (false) proposition p in interpretation I becomes a true (false) proposition p' in interpretation I' when we replace the e's, r's, and o's in p by their corresponding e''s, r''s, and o''s. Two isomorphic interpretations of a postulate set P are, except for superficial differences in terminology and notation, identical; they differ from each other no more than does the multiplication table up to 10×10 when correctly written first in English and then in French.

With the notion of isomorphic interpretations of a postulate set established, we are prepared to define categoricalness of a postulate set. A postulate set P, as well as the resulting branch of mathematics, is said to be *categorical* if every two interpretations of P are isomorphic.

Categoricalness of a postulate set is usually established by showing that any interpretation of the postulate set is isomorphic to some given interpretation. This procedure has been applied to Hilbert's postulate set for plane Euclidean geometry; it can be shown that any interpretation of Hilbert's postulates is isomorphic to the algebraic interpretation provided by Descartes' analytic geometry. Plane Lobachevskian geometry has also been shown to be a categorical system.

There are advantages and disadvantages in having a system categorical. Perhaps the most desirable feature of a noncategorical postulate set is its wide range of applicability—there is not essentially only one model for the system. For example, the theorems of absolute plane geometry—those theorems common to Euclidean and Lobachevskian plane geometry—may be obtained from Hilbert's postulate set for Euclidean plane geometry with the parallel postulate deleted. This truncated postulate set is, of course, noncategorical; it is satisfied by nonisomorphic interpretations. One advantage of categoricalness, on the other hand, is that often theorems of a categorical system may be more easily established by establishing their counterparts in some model. Thus, in Section 8.5 we shall obtain a model of Lobachevskian plane geometry within Euclidean plane geometry; since Lobachevskian plane geometry is a categorical system, and since we are much more conversant with Euclidean plane geometry than with Lobachevskian plane geometry, there arises the very real possibility of establishing theorems in the Lobachevskian plane geometry by establishing their counterparts in the Euclidean model. We shall illustrate this procedure in Section 8.6.

PROBLEMS

1. If p, q, r represent propositions, show that the following set of four statements is inconsistent:
   (1) If q is true, then r is false.
   (2) If q is false, then p is true.

(3) r is true.

(4) p is false.

2. Compare the concept of consistency and inconsistency of a set of simultaneous equations with the concept of consistency and inconsistency of a postulate set.

3. (a) Show that the postulate set of Problem 2, Section 8.3, is absolutely consistent.

   (b) Show that Postulates P2, P3, P4 of Problem 2, Section 8.3, are independent.

   (c) Show that the postulate set of Problem 2, Section 8.3, is categorical.

4. (a) Establish the absolute consistency of the postulate set of Problem 3, Section 8.3 by means of each of the following interpretations:

   1. Let K consist of a man, his father, his father's father, and his father's father's father, and let $a\ R\ b$ mean "a is an ancestor of b."

   2. Let K consist of four distinct points on a horizontal line, and let $a\ R\ b$ mean "a is to the left of b."

   3. Let K consist of the four integers 1, 2, 3, 4, and let $a\ R\ b$ mean "$a < b$."

   (b) Write out statements of the theorems and definitions in Problem 3, Section 8.3, for each of the interpretations of part (a), thus obtaining three branches of applied mathematics from the one branch of pure mathematics.

5. Establish the independence of the postulate set of Problem 3, Section 8.3, by means of the following four partial interpretations:

   1. Let K consist of two brothers, their father, and their father's father, and let $a\ R\ b$ mean "a is an ancestor of b."

   2. Let K consist of the four integers 1, 2, 3, 4, and let $a\ R\ b$ mean "$a \leqq b$."

   3. Let K consist of the four integers 1, 2, 3, 4, and let $a\ R\ b$ mean "$a \neq b$."

   4. Let K consist of the five integers 1, 2, 3, 4, 5, and let $a\ R\ b$ mean "$a < b$."

6. Show that, in Problem 3, Section 8.3, P1, T1, P3, P4 constitute a postulate set equivalent to P1, P2, P3, P4.

7. Show that the postulate set of Problem 3, Section 8.3, is categorical.

8. Clearly, if a postulate p of a consistent postulate set P contains a primitive term which is not among the primitive terms occurring in any other postulates of P, then p is independent.

   (a) Using the above principle, show that the additional postulate: "All hives lie in the same apiary," where *apiary* is a primitive term, is independent of the four postulates of Problem 2, Section 8.3.

   (b) Also show the independence of the additional postulate by the method of models.

9. Show that the postulate set of Problem 5, Section 8.3, is categorical.

10. Let S be a set of elements and F a dyadic relation satisfying the following postulates:

   P1: If a and b are elements of S and if $b\ F\ a$, then we do not have $a\ F\ b$.

   P2: If a is an element of S, then there is at least one element b of S such that $b\ F\ a$.

   P3: If a is an element of S, then there is at least one element b of S such that $a\ F\ b$.

   P4: If a, b, c are elements of S such that $b\ F\ a$ and $c\ F\ b$, then $c\ F\ a$.

   P5: If a and b are elements of S such that $b\ F\ a$, then there exists at least one element of c of S such that $c\ F\ a$ and $b\ F\ c$.

Show that the statement, "If a is an element of S, then there is at least one element b of S, distinct from a, such that we do not have $b\,F\,a$ and we do not have $a\,F\,b$," is both consistent with and independent of the above postulates.

(This set of postulates, augmented by the above statement, has been used in relativity theory, where the elements of S are interpreted as *instants* of time and F as meaning "follows."*)

8.5 THE POINCARÉ MODEL AND THE CONSISTENCY OF LOBACHEVSKIAN PLANE GEOMETRY

A satisfactory postulate set for Lobachevskian plane geometry may be obtained from the Hilbert postulate set for plane Euclidean geometry by simply replacing the postulate of parallels (Postulate IV–1) by

IV'–1 *Through a given point* A *not on a given line* m *there pass at least two lines which do not intersect line* m.

Developments of Lobachevskian plane geometry have been made using such a postulate set as a foundation.†

It is our purpose, in the present section, to show that Lobachevskian plane geometry is consistent if Euclidean plane geometry is consistent. To accomplish this we shall represent the primitive terms of Lobachevskian plane geometry by certain entities of Euclidean plane geometry which, when substituted for the primitive terms in the postulates of Lobachevskian plane geometry, will convert these postulates into theorems of Euclidean plane geometry. In other words, we shall give an interpretation of Lobachevskian plane geometry within Euclidean plane geometry. The interpretation that we shall employ was devised by Henri Poincaré, and leads to the so-called *Poincaré model* of Lobachevskian plane geometry.

We choose a fixed circle Σ in the Euclidean plane and call it the *fundamental circle*, and then set up in the Euclidean plane the following representations of the primitive terms of Lobachevskian plane geometry (see Figure 8.5a). For clarity we designate the representations by means of boldface type.

point: point in the interior of Σ
line: the part interior to Σ of any "circle" (straight line or circle) orthogonal to Σ
point on a line (**line through a point**, **line containing a point**): the obvious interpretation
point between two points: the obvious interpretation

---

* See A. A. Robb, *A Theory of Time and Space*. New York: Cambridge University Press, 1914.
† See, for example, G. Verriest, *Introduction à la géométrie non-Euclidienne par la méthode élémentaire*. Paris: Gauthier-Villars, 1951.

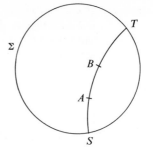

Figure 8.5a

DEFINITION. **length of segment** $AB = \log (AB,TS) = \log [(AT/BT)(BS/AS)]$, where S and T are the points in which the "circle" containing the **segment** AB cuts Σ, S and T being labeled so that A is between S and B. It should be noted that $(AB,TS) > 1$, whence $\log (AB,TS) > 0$.

DEFINITION. **measure of an angle between two intersecting lines** = radian measure of the angle between the two "circles" containing the two **lines.**

congruent segments: segments of equal **length**
congruent angles: angles of equal **measure**

We now show that, with the above representations, the postulates for Lobachevskian plane geometry become theorems in Euclidean plane geometry; it will be noted that many of the resulting theorems are obvious.

Group I: Postulates of Connection

I–1. *There is one and only one* **line passing through** *any two given distinct* **points.** This is Theorem 2.9.5.

I–2. *Every* **line contains** *at least two distinct* **points,** *and for any given* **line** *there is at least one* **point** *not* **on** *the* **line.** Obvious.

Group II: Postulates of Order

II–1. *If* **point** C *is* **between points** A *and* B, *then* A, B, C *are all* **on** *the same* **line,** *and* C *is* **between** B *and* A, *and* B *is not* **between** C *and* A, *and* A *is not* **between** C *and* B. Obvious.

II–2. *For any two distinct* **points** A *and* B *there is always a* **point** C *which is* **between** A *and* B, *and a* **point** D *which is such that* B *is* **between** A *and* D. Obvious.

II–3. *If* A, B, C *are any three distinct* **points on** *the same* **line,** *then one of the* **points** *is* **between** *the other two.* Obvious.

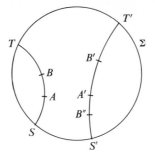

Figure 8.5b

II–4. (Pasch's Postulate) *A* **line** *which* **intersects** *one* **side** *of a* **triangle** *but does not* **pass through** *any of the* **vertices** *of the* **triangle** *must also* **intersect** *another* **side** *of the* **triangle.**

Obvious, because of I–1.

Group III: Postulates of Congruency

III–1. *If* A *and* B *are distinct* **points** *and if* A′ *is a* **point on a line** m, *then there are two and only two* **points** B′ *and* B″ **on** m *such that the pair of* **points** A′, B′ *is* **congruent** *to the pair* A, B *and the pair of* **points** A′, B″ *is* **congruent** *to the pair* A, B; *moreover,* A′ *is* **between** B′ *and* B″.

We note (see Figure 8.5b) that $(A'T'/B'T')(B'S'/A'S')$ increases continuously from 1 to ∞ as B' moves along m from A' to T'. Similarly, $(A'S'/B''S')(B''T'/A'T')$ increases continuously from 1 to ∞ as B'' moves along m from A' to S'. It follows that **length** $A'B'$ and **length** $A'B''$ increase continuously from 0 to ∞ in the two cases. There are therefore unique positions of B' and B'' such that **length** $A'B' = $ **length** $A'B'' = $ **length** AB.

III–2. *If two pairs of* **points** *are* **congruent** *to the same pair of* **points**, *then they are* **congruent** *to each other.*

Obvious.

III–3. *If* **point** C *is* **between points** A *and* B *and* **point** C′ *is* **between points** A′ *and* B′, *and if the pair of* **points** A, C *is* **congruent** *to the pair* A′, C′, *and the pair of* **points** C, B *is* **congruent** *to the pair* C′, B′, *then the pair of* **points** A, B *is* **congruent** *to the pair* A′, B′.

For we have (see Figure 8.5c)

$$
\begin{aligned}
\textbf{length } AB &= \log[(AT/BT)(BS/AS)] \\
&= \log[(AT/CT)(CS/AS)(CT/BT)(BS/CS)] \\
&= \log[(AT/CT)(CS/AS)] + \log[(CT/BT)(BS/CS)] \\
&= \textbf{length } AC + \textbf{length } CB.
\end{aligned}
$$

Similarly, **length** $A'B' = $ **length** $A'C' + $ **length** $C'B'$. But **length** $AC = $ **length** $A'C'$ and **length** $CB = $ **length** $C'B'$. Therefore **length** $AB = $ **length** $A'B'$, and AB is **congruent** to $A'B'$.

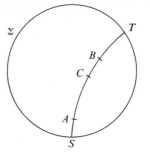

Figure 8.5c

III–4. *If* BAC *is an* **angle** *whose* **sides** *do not* **lie** *in the same* **line**, *and if* A′ *and* B′ *are two distinct* **points**, *then there are two and only two distinct* **rays** A′C′ *and* A′C″, *such that* **angle** B′A′C′ *is* **congruent** *to* **angle** BAC *and* **angle** B′A′C″ *is* **congruent** *to* **angle** BAC; *moreover, if* D′ *is any* **point on** *the* **ray** A′C′ *and* D″ *is any* **point on** *the* **ray** A′C″, *then the* **segment** D′D″ **intersects** *the* **line** *determined by* A′ *and* B′.

This is a consequence of Theorem 2.9.6.

III–5. *Every* **angle** *is* **congruent** *to itself.*
Obvious.

LEMMA. **Length** AB *is invariant under inversion in any circle orthogonal to* Σ.

Let A', B' be the inverses of A, B for a circle orthogonal to Σ and note that A', B' lie inside Σ. Then (see Figure 8.5d)

$$\begin{aligned}
\textbf{length } A'B' &= \log (A'B', T'S') \\
&= \log (AB, TS) \quad \text{(by Theorem 3.7.6)} \\
&= \textbf{length } AB.
\end{aligned}$$

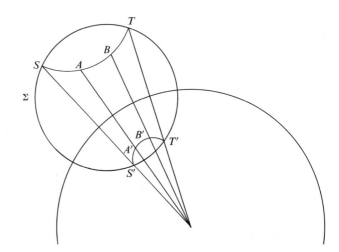

Figure 8.5d

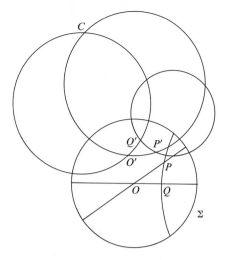

Figure 8.5e

III–6. *If two* **sides** *and the* **included angle** *of one* **triangle** *are* **congruent**, *respectively, to two* **sides** *and the* **included angle** *of another* **triangle**, *then each of the remaining* **angles** *of the first* **triangle** *is* **congruent** *to the corresponding* **angle** *of the second* **triangle**.

Consider any **triangle** $O'P'Q'$ where, without loss of generality, we assume O' is not the center O of Σ (see Figure 8.5e), and let the "circles" along **sides** $O'P'$ and $O'Q'$ intersect again in C. Invert the figure for C as center and with a power that carries Σ into itself. Since the size of an angle is preserved under inversion, the "circles" $CP'O'$ and $CQ'O'$ (being orthogonal to Σ and passing through the center C of inversion) map into two diametral lines of Σ, and **triangle** $O'P'Q'$ maps into a **triangle** OPQ, where OP, OQ are radial lines of Σ. Since both **measures** of **angles** and **lengths** of **segments** are preserved under inversion, it follows that **triangles** $O'P'Q'$ and OPQ are **congruent**.

Now let $O_1'P_1'Q_1'$ be any **triangle** in which **length** $O_1'P_1' =$ **length** $O'P'$, **length** $O_1'Q_1' =$ **length** $O'Q'$, and **angle** $P_1'O_1'Q_1' =$ **angle** $P'O'Q'$. As above, **triangle** $O_1'P_1'Q_1'$ is **congruent** to a **triangle** OP_1Q_1, where OP_1, OQ_1 are radial lines of Σ. But **triangle** OP_1Q_1 is congruent (in the Euclidean sense) to **triangle** OPQ. Therefore **triangle** $O_1'P_1'Q_1'$ is **congruent** to **triangle** $O'P'Q'$.

Group IV: Postulate of parallels

IV'–1. *Through a given* **point** A *not* **on** *a given* **line** m *there* **pass** *at least two* **lines** *which do not* **intersect line** m.

Let the "circle" containing **line** m cut Σ in X and Y (see Figure 8.5f). Through A draw the unique "circles" orthogonal to Σ and passing respectively through X and Y. These "circles" are distinct, and the desired result follows.

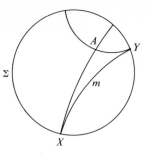

Figure 8.5f

Alternative Group V

V'–1. *If the* **points** *of an* **ordered segment** *of origin* A *and extremity* B *are separated into two classes in such a way that*

(1) *each* **point** *of* AB *belongs to one and only one of the classes,*

(2) *the* **points** A *and* B *belong to different classes (which we shall respectively call the* first *class and the* second *class),*

(3) *each* **point** *of the first class* **precedes** *each* **point** *of the second class,*

then there exists a **point** C *on* AB *such that every* **point** *which* **precedes** C *belongs to the first class and every* **point** *which* **follows** C *belongs to the second class.*

Invert with respect to any point *D* not on the **segment** *AB* but on the "circle" containing **segment** *AB*. Then the **ordered segment** *AB* becomes a similarly ordered straight line segment *A'B'*. The reader can easily complete the proof.

We may now consider that the purpose of this section—to show that Lobachevskian plane geometry is consistent if Euclidean plane geometry is consistent—has been accomplished. For, should there be any inconsistency in the Lobachevskian plane geometry, there would have to be a corresponding inconsistency in the Euclidean plane geometry of the Poincaré model.

Assuming Euclidean plane geometry is consistent, we have not only established the consistency of Lobachevskian plane geometry, but we have shown that the Euclidean parallel postulate is independent of the other postulates of Euclidean geometry. For in the Poincaré model we have an interpretation of part of plane Euclidean geometry that falsifies the Euclidean parallel postulate but verifies all the other Euclidean postulates. If the Euclidean parallel postulate were implied by the other Euclidean postulates, then it too would have to hold in the model. Since we have seen that it does not, and since we are assuming Euclidean geometry to be consistent, it follows that the Euclidean parallel postulate cannot be implied by the other Euclidean postulates, and is thus independent of them.

Of course, a satisfactory proof of the consistency of Euclidean plane geometry has never been given except (as we shall see in the second volume of our work) by referring back to the concepts of analytic geometry and hence ultimately to the real number system, whose consistency, in turn, is an open question.

PROBLEMS

1. Establish the interpretation of I-1 by inverting the figure with respect to a point on Σ.

2. Show how the interpretation of I-1 proves the interpretation of II-4.

3. Show how the interpretation of III-4 is a consequence of Theorem 2.9.6.

4. Complete the proof of the interpretation of V'-1.

5. (a) Show that an oblique circular cone possesses circular sections whose planes are not parallel to the plane of the generating circle of the cone.
   (b) Show that two circles in different planes but sharing a common diameter are projectively related by a parallel perspectivity.
   (c) Show that there exist infinitely many projectivities which map a circle and its interior onto the same circle and its interior.

6. We describe a model of Lobachevskian plane geometry that was devised by Felix Klein (1849–1929).

   Choose a fixed circle Σ in the Euclidean plane and set up the following representations of the primitive terms of Lobachevskian plane geometry:

   point: point in the interior of Σ
   line: chord of Σ
   point on a line: the obvious interpretation
   point between two points: the obvious interpretation
   congruent segments: segments which can be mapped onto one another by a projectivity of Σ and its interior onto itself
   congruent angles: angles which can be mapped onto one another by a projectivity of Σ and its interior onto itself

   Verify, in the Klein model, the Hilbert postulates for Lobachevskian plane geometry.

7. Show that, in the Klein model, we may define the **length of segment** AB as log (AB,TS), where S and T are the points in which the chord containing segment AB cuts Σ, S and T being labeled so that A is between S and B.

8.6 DEDUCTIONS FROM THE POINCARÉ MODEL

If a model of a categorical system A has been formed within another system B, it is conceivable that some theorems of system A might be more readily established by demonstrating their counterparts in system B—especially if one is more familiar with system B than with system A. Now the Poincaré model is a model of Lobachevskian plane geometry within Euclidean plane geometry. Since we are more familiar with Euclidean plane geometry than with Lobachevskian plane geometry, the idea of trying to establish some

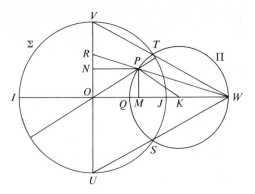

Figure 8.6a

theorems of Lobachevskian plane geometry by establishing their counterparts in the model suggests itself. It is the purpose of the present section to illustrate this procedure. In addition to establishing quite simply from the Poincaré model a few difficult theorems of Lobachevskian plane geometry, we shall show how Lobachevskian trigonometry can be derived from the model.*

As in the proof of the interpretation of III–6 in the previous section, any **triangle** $O'P'Q'$, **right-angled** at Q' is (see Figure 8.6a) **congruent** to a **triangle** OPQ, where OP, OQ are along radii of the fundamental circle Σ and **angle** $OQP = \pi/2$ radians. Let the circle Π determined by **line** PQ cut Σ in S and T and let $IOQJ$ be a diameter of Σ, cutting Π again in W. We now establish a short chain of theorems connected with Figure 8.6a. For brevity we shall denote the **length of a segment** AB by **AB.** Since the **measure of an angle** is the same as the Euclidean measure of the corresponding angle, boldface type is not needed here.

8.6.1 THEOREM. *If* WS *and* WT *cut* Σ *again in* U *and* V, *then* UV *is the diameter of* Σ *perpendicular to diameter* IJ.

Select W as center of inversion and choose a power such that Σ inverts into itself. Then S inverts into U, and T into V. Since Π is orthogonal to both Σ and IJ, it follows that UV is the diameter of Σ perpendicular to diameter IJ.

8.6.2 THEOREM. *Let* WP *cut* UV *in* R, *and designate the lengths of* OW *and* OR *by* m *and* n, *and the radius of* Σ *by* r. *Let* K *be the center of* Π *and let* M *and* N *be the feet of the perpendiculars dropped from* P *on* OW *and* OR *respectively. Then*

$$\text{(a) } KP = (m^2 - r^2)/2m,$$
$$\text{(b) } OM = m(n^2 + r^2)/(m^2 + n^2),$$
$$\text{(c) } OP = (m^2n^2 + r^4)^{1/2}(m^2 + n^2)^{1/2},$$
$$\text{(d) } OQ = r^2/m.$$

* See Howard Eves and V. E. Hoggatt, Jr., "Hyperbolic trigonometry derived from the Poincaré model," *The American Mathematical Monthly*, vol. LVIII, no. 7, Aug.-Sept., 1951.

Since $KP = OW - OK = m - (r^2 + KP^2)^{1/2}$, it follows that
$$KP = (m^2 - r^2)/2m.$$
Also, since $\tan PWO = n/m$, and since $\sphericalangle PKO = 2 \sphericalangle PWO$, it follows that
$$\tan PKO = 2mn/(m^2 - n^2), \quad \sin PKO = 2mn/(m^2 + n^2).$$
Therefore
$$OM = NP = (OW)(NR)/OR = m(n - MP)/n = m(n^2 + r^2)/(m^2 + n^2).$$
Then
$$OP^2 = MP^2 + OM^2 = (m^2n^2 + r^4)/(m^2 + n^2).$$
Finally,
$$OQ = OW - 2KP = m - 2(m^2 - r^2)/2m = r^2/m.$$

8.6.3 Theorem. *The segments* OP, OQ, OR *are connected by the relation*
$$\frac{r^2 + OP^2}{r^2 - OP^2} = \frac{r^2 + OR^2}{r^2 - OR^2} \cdot \frac{r^2 + OQ^2}{r^2 - OQ^2}.$$

For, by Theorem 8.6.2 (c),
$$\frac{r^2 + OP^2}{r^2 - OP^2} = \frac{r^2(m^2 + n^2) + m^2n^2 + r^4}{r^2(m^2 + n^2) - m^2n^2 - r^4} = \frac{(r^2 + n^2)(m^2 + r^2)}{(r^2 - n^2)(m^2 - r^2)}$$
$$= \frac{r^2 + n^2}{r^2 - n^2} \cdot \frac{r^2 + r^4/m^2}{r^2 - r^4/m^2} = \frac{r^2 + OR^2}{r^2 - OR^2} \cdot \frac{r^2 + OQ^2}{r^2 - OQ^2},$$
since $OR = n$, and, by Theorem 8.6.2 (d), $OQ = r^2/m$.

8.6.4 Definitions and notation. For any real x we define
$$\sinh x = (e^x - e^{-x})/2, \quad \cosh x = (e^x + e^{-x})/2,$$
$$\tanh x = \sinh x/\cosh x,$$
where sinh, cosh, and tanh are abbreviations for *hyperbolic sine, hyperbolic cosine*, and *hyperbolic tangent*.

8.6.5 Theorem. *If* OQ *is any radial segment of* Σ, *then*
(a) *cosh* $\mathbf{OQ} = (r^2 + OQ^2)/(r^2 - OQ^2)$,
(b) *sinh* $\mathbf{OQ} = 2rOQ/(r^2 - OQ^2)$,
(c) *tanh* $\mathbf{OQ} = 2rOQ/(r^2 + OQ^2)$.

For, since $\mathbf{OQ} = \log (OQ,IJ)$, we have
$$\cosh \mathbf{OQ} = (e^{OQ} + e^{-OQ})/2$$
$$= [(OQ,IJ) + (OQ,JI)]/2$$
$$= [(OI/IQ)/(OJ/JQ) + (OJ/JQ)/(OI/IQ)]/2$$
$$= (QJ/IQ + IQ/QJ)/2$$
$$= [(r - OQ)/(r + OQ) + (r + OQ)/(r - OQ)]/2$$
$$= (r^2 + OQ^2)/(r^2 - OQ^2),$$

which is relation (a). Relations (b) and (c) follow in a similar manner, or from the identities $\sinh^2 x = \cosh^2 x - 1$ and $\tanh x = \sinh x/\cosh x$.

8.6.6 THEOREM. *In Figure 8.6a, cosh* **OP** $=$ *cosh* **OQ** *cosh* **QP.**

As an immediate consequence of Theorems 8.6.3 and 8.6.5 (a) we have

$$\cosh \mathbf{OP} = \cosh \mathbf{OQ} \cosh \mathbf{OR}.$$

But, by Theorem 8.6.1, $(QP,ST) = W(QP,ST) = (OR,UV)$, whence $\mathbf{OR} = \mathbf{QP}$, and the theorem is established.

8.6.7 THEOREM. *In Figure 8.6a, cos* QOP $=$ *tanh* **OQ**/*tanh* **OP.**

For, by Theorem 8.6.5 (c),

$$\tanh \mathbf{OQ}/\tanh \mathbf{OP} = OQ(r^2 + OP^2)/OP(r^2 + OQ^2).$$

Substituting the expressions for OP and OQ as given by Theorem 8.6.2 (c) and (d), and simplifying, we find

$$\begin{aligned}
\tanh \mathbf{OQ}/\tanh \mathbf{OP} &= m(r^2 + n^2)/(m^2n^2 + r^4)^{1/2}(m^2 + n^2)^{1/2} \\
&= [m(r^2 + n^2)/(m^2 + n^2)]/[(m^2n^2 + r^4)^{1/2}/(m^2 + n^2)^{1/2}] \\
&= OM/OP \text{ (by Theorem 8.6.2 (b) and (c))} \\
&= \cos QOP,
\end{aligned}$$

and the theorem is established.

8.6.8 REMARK. Theorems 8.6.6 and 8.6.7 say that in any Lobachevskian right triangle ABC, right-angled at C, we have

$$\cosh c = \cosh a \cosh b, \quad \cos A = \tanh b/\tanh c,$$

where a, b, c are the sides opposite angles A, B, C respectively. From these two formulas, all the other formulas of Lobachevskian trigonometry can be obtained by purely analytical procedures.

We now use the Poincaré model to re-establish in a simple way two of our former theorems (Theorem 7.4.5 and Theorem 7.6.1) and a new theorem giving the relation between the angle of parallelism of a point P for a line AB and the distance of point P from line AB.

8.6.9 THEOREM. *The sum of the angles of a Lobachevskian triangle is always less than two right angles.*

As in the proof of the interpretation of III–6 in the previous section, any

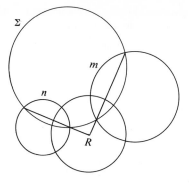

Figure 8.6b

triangle is **congruent** to a **triangle** OPQ, where O is the center of the fundamental circle Σ and OP, OQ are radial lines of Σ. But for this **triangle** it is
obvious that the **angle-sum** is less than two right angles.

8.6.10   THEOREM.   *Two hyperparallel lines have a unique common perpendicular.*

Let m and n (see Figure 8.6b) be two hyperparallel **lines.** The radical
center R of the three "circles" Σ, m, n lies outside Σ; hence the three "circles"
have a radical circle, which is the unique circle orthogonal to each of Σ,
m, and n. This proves the theorem.

8.6.11   THEOREM.   *If α is the angle of parallelism for distance* p, *then*
$e^{-p} = \tan \alpha/2$.

Let the "circle" containing the **perpendicular** from the given **point** P to
the given **line** m and the "circle" containing the given **line** m intersect again
in R (see Figure 8.6c). Then R is outside Σ and Q and R are collinear with
the center of Σ. Invert the figure with respect to R as center of inversion
and such that Σ inverts into itself. Then arcs PQ and m invert into straight

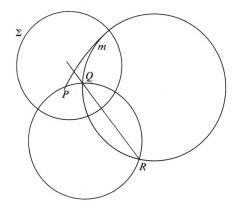

Figure 8.6c

line segments perpendicular to each other, and Q inverts into the center of Σ, as shown in Figure 8.6d. In this figure we have

$$\alpha + 2\beta = \pi/2, \quad \text{or} \quad \beta = \pi/4 - \alpha/2.$$

Now, taking Σ of unit radius, we have

$$p = \log(P'Q', T'S') = \log(P'T'/P'S') = \log \frac{1 + \tan \beta}{1 - \tan \beta},$$

whence

$$e^{-p} = \frac{1 - \tan(\pi/4 - \alpha/2)}{1 + \tan(\pi/4 - \alpha/2)} = \frac{1 - \dfrac{1 - \tan \alpha/2}{1 + \tan \alpha/2}}{1 + \dfrac{1 - \tan \alpha/2}{1 + \tan \alpha/2}}$$

$$= \frac{1 + \tan \alpha/2 - 1 + \tan \alpha/2}{1 + \tan \alpha/2 + 1 - \tan \alpha/2} = \tan \alpha/2.$$

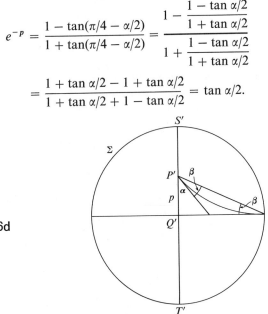

Figure 8.6d

8.6.12 REMARK. Note that $p = \log \cot \alpha/2$ and $\alpha = 2 \tan^{-1} e^{-p}$. It follows that we have chosen our unit of length to be the distance corresponding to $2 \tan^{-1}(1/e)$ as the angle of parallelism. We see that, in Lobachevskian geometry, lengths and areas are absolute. With every linear segment there can be associated a definite angle (namely the angle of parallelism for this segment), and with every area there can be associated a definite angle (namely the acute angle of an equivalent Saccheri quadrilateral with unit base).

PROBLEMS

1. Establish the identity $\sinh^2 x = \cosh^2 x - 1$.

2. Derive relations (b) and (c) of Theorem 8.6.5.

3. Establish, through the Poincaré model, Theorems 7.2.7, 7.2.8, and 7.2.9.

4. Show, by means of the Poincaré model, that as distance p increases from 0 to ∞, the corresponding angle of parallelism decreases from $\pi/2$ to 0.

5. Show, by means of the Poincaré model, that the fourth angle of a Lambert quadrilateral is acute.

6. Show that: (a) $\sinh x + \cosh x = e^x$,
   (b) $\tanh x = (e^{2x} - 1)/(e^{2x} + 1)$.

7. (a) If $x = \sinh y$, show that $y = \log[x + (x^2 + 1)^{1/2}]$.
   (b) If $x = \tanh y$, show that $y = (1/2)\log[(1 + x)/(1 - x)]$.

8. Establish the following identities:
   (a) $\sinh(x \pm y) = \sinh c \cosh y \pm \cosh x \sinh y$
   (b) $\cosh(x \pm y) = \cosh x \cosh y \pm \sinh x \sinh y$
   (c) $\tanh(x \pm y) = (\tanh x \pm \tanh y)/(1 \pm \tanh x \tanh y)$

9. Establish the following identities:
   (a) $\sinh 2x = 2 \sinh x \cosh x$
   (b) $\cosh 2x = \sinh^2 x + \cosh^2 x$
   (c) $\tanh 2x = 2 \tanh x/(1 + \tanh^2 x)$

10. Establish the following identities (where $u = (x + y)/2$ and $v = (x - y)/2$):
    (a) $\sinh x + \sinh y = 2 \sinh u \cosh v$
    (b) $\sinh x - \sinh y = 2 \cosh u \sinh v$
    (c) $\cosh x + \cosh y = 2 \cosh u \cosh v$
    (d) $\cosh x - \cosh y = 2 \sinh u \sinh v$

11. Show that:
    (a) $\sinh(-x) = -\sinh x$
    (b) $\cosh(-x) = \cosh x$
    (c) $\tanh(-x) = -\tanh x$

12. Express $\tanh x$ in terms of: (a) $\sinh x$, (b) $\cosh x$.

13. Derive the following formulas for the Lobachevskian right triangle ABC (right-angled at C) from the two formulas (1) $\cosh c = \cosh a \cosh b$, (2) $\cos A = \tanh b/\tanh a$.
    (a) $\sin A = \sinh a/\sinh c$, $\sin B = \sinh b/\sinh c$
    (b) $\cot A \cot B = \cosh c$
    (c) $\tan A = \tanh a/\sinh b$, $\tan B = \tanh b/\sinh a$
    (d) $\cosh a = \cos A/\sin B$, $\cosh b = \cos B/\sin A$

14. For the general Lobachevskian triangle ABC show that:
    (a) $\sinh a : \sinh b : \sinh c = \sin A : \sin B : \sin C$
    (b) $\cosh a = \cosh b \cosh c - \sinh b \sinh c \cos A$

15. Show that if we had chosen length $AB = k \log (AB,TS)$, where k is a positive constant (thus changing the unit of length), then:
    (a) the formulas of Problems 13 and 14 above would have to be modified by everywhere replacing a, b, c by a/k, b/k, c/k.
    (b) $\tan \alpha/2 = e^{-p/k}$, where α is the angle of parallelism corresponding to distance p.
    (c) $\lim_{k \to \infty} \tan \alpha/2 = 1$, and Lobachevskian geometry becomes sensibly Euclidean if k is chosen very large in comparison with the lengths of the segments involved.

8.7 A POSTULATIONAL FOUNDATION FOR PLANE PROJECTIVE GEOMETRY

In Chapter 6 we considered projective geometry essentially as it historically developed. That is, we noted a certain core of theorems lying within the expanding body of theorems of Euclidean geometry and possessing a common character that we called *descriptive*. There are certain unsatisfactory features of such an approach to projective geometry. In the first place, if the postulational basis of Euclidean geometry is somewhat vague or incomplete (as is usually the case with a high school treatment of the subject), then this vagueness or incompleteness is also reflected in the projective geometry. And even if the postulational basis for Euclidean geometry should be made precise, the theorems of projective geometry are of such a special type that this basis, at least in its entirety, fails to distinguish the projective theorems from the nonprojective ones. Perhaps the most satisfying way out of the difficulty is to reconstruct projective geometry as an independent discipline, with its own primitive terms and postulational basis.

There are many postulate sets for plane projective geometry that we could give, but the following* will most neatly serve the purposes of the present section. Here "point," "line," and "on" are primitive terms.

8.7.1 POSTULATES FOR PLANE PROJECTIVE GEOMETRY

I. *There is one and only one line on every two distinct points, and one and only one point on every two distinct lines.*

II. *There exist two points and two lines such that each of the points is on just one of the lines and each of the lines is on just one of the points.*

III. *There exist two points and two lines, the points not on the lines, such that the point on the two lines is on the line on the two points.*

Now it will be recalled (from Section 6.7) that from a given proposition concerning only the primitive terms *point*, *line*, and *on*, we can form a second proposition, called the *dual* of the first, by simply everywhere interchanging the words *point* and *line*. The principle of duality of plane projective geometry asserts that: *The dual of a true proposition of plane projective geometry is another true proposition of plane projective geometry.* The proof of this principle that was given in Section 6.7 was based on Poncelet's theory of poles and polars, and a feeling of dissatisfaction with this proof was expressed at the time. We are now in a position to give a simple and completely satisfying proof of the principle of duality of plane projective geometry. All we have to do is to show that the postulates for plane projective geometry imply the dual of each postulate, for then the dual of any theorem that has been derived from the postulate set can be established by simply dualizing, one by one, the statements of the proof of the original theorem. The fact that each of the postulates we have selected is actually self-dual saves us the task of deriving as theorems the system of dual statements, and therefore immediately vouchsafes us the principle of duality for plane projective geometry. We might now restate the principle of duality as follows:

* K. Menger, "Self-dual postulates in projective geometry," *The American Mathematical Monthly*, 55 (1948), 195.

8.7.2 THE PRINCIPLE OF DUALITY FOR PLANE PROJECTIVE GEOMETRY. *If a theorem is deducible from the postulates of plane projective geometry, then its dual is also deducible from these postulates.*

It is an interesting and significant fact that the postulates for plane projective geometry given above do not require that the number of points in our geometry be infinite. In fact, consider the situation where there are seven "points," namely the seven letters *A, B, C, D, E, F, G,* and seven "lines," namely the seven letter trios (*AFB*), (*BDC*), (*CEA*), (*AGD*), (*BGE*), (*CGF*), (*DEF*). Postulate I is easily verified by considering separately each of the 21 pairs of "points" and each of the 21 pairs of "lines." Postulate II is verified by the two "points" *B, C* and the two "lines" (*AFB*), (*AEC*). Postulate III is verified by the two "points" *B, C* and the two "lines" (*AGD*), (*DEF*).

A projective geometry containing only a finite number of distinct points is called a *finite projective geometry.* The above model establishes the existence of a finite projective geometry. It also establishes the absolute consistency of our postulate set for plane projective geometry. And, finally, it shows that if we wish our plane projective geometry to resemble more closely the earlier historical concept of the subject we shall have to introduce further postulates. In fact, by gradually adding appropriate further postulates and introducing certain modifications, it is possible to convert the postulate set for plane projective geometry into a postulate set for plane Euclidean geometry, passing through postulate sets for various intermediate geometries on the way. Such a passage from projective geometry to Euclidean geometry is an interesting but lengthy procedure which we must forgo. We shall, however, say a few words in the next section about the enlargement of our postulate set for plane projective geometry so as to make the resulting geometry resemble the classical plane projective geometry considered in Chapter 6.

PROBLEMS

1. Show that the postulate set 8.7.1 and the following postulate set are equivalent (that is, each implies the other).

 P1. *There exists at least one line.*
 P2. *There are at least three points on every line.*
 P3. *Not all points are on the same line.*
 P4. *There is exactly one line on any two distinct points.*
 P5. *There is at least one point on any two distinct lines.*

2. (a) Derive the following theorems from the postulate set of Problem 1 above.

 T1. *There exists at least one point.*
 T2. *There is exactly one point on any two distinct lines.*
 T3. *There are at least three lines on every point.*
 T4. *Not all lines are on the same point.*

(b) Now show that the principle of duality holds in the geometry following from the postulates of Problem 1 above.

3. Deduce the following additional theorems from the postulate set of Problem 1 above.

     T5. *If there are at least* n *points on one line, then there are at least* n *points on any other line.*

     T6. *If there are at most* n *points on one line, then there are at most* n *points on any other line.*

     T7. *If there are exactly* n *points on one line, then there are exactly* n *points on every line.*

     T8. *If there are exactly* m *lines on one point, then there are exactly* m *lines on every point.*

     T9. *If there are exactly* n *points on each line, then there are exactly* n *lines on each point.*

     T10. *If there are exactly* n *points on each line, then there are* $n^2 - n + 1$ *points altogether and* $n^2 - n + 1$ *lines altogether.*

4. Prove that a plane projective geometry must contain at least 7 distinct points and 7 distinct lines.

5. Construct a plane projective geometry with exactly four distinct points on each line.

6. Show that the following is a Euclidean interpretation of the postulates for plane projective geometry. Let O be a fixed point in space; by a "point" we shall mean a line through O; by a "line" we shall mean a plane through O; by "on" we shall mean the obvious interpretation.

7. Establish the independence of the postulates of the postulate set of Problem 1 above.

8. Show that postulates, P1, P2, P3 of the postulate set of Problem 1 above can be replaced by the single postulate:

     P1'. *There exist four points, no three of which are collinear.*

8.8 NON-DESARGUESIAN GEOMETRY

Desargues' two-triangle theorem for the plane (that *copolar triangles in a plane are coaxial, and conversely*) is a statement about the incidence of some points and lines of the plane. It might seem, then, that Desargues' two-triangle theorem for the plane should be deducible from the above postulates for plane projective geometry, for these postulates describe the incident relation of point and line in the plane. We shall now show, however, that Desargues' two-triangle theorem for the plane is not implied by the postulates of plane projective geometry. To accomplish this we construct a model in which the postulates for plane projective geometry are verified but in which Desargues' two-triangle theorem does not hold.

Choose an ordinary line h of an extended Euclidean plane as an x axis of a rectangular Cartesian coordinate system (see Figure 8.8a). By *points* in the model we shall mean the ordinary and ideal points of the extended

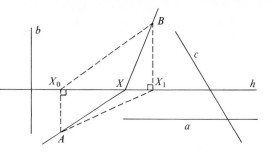

Figure 8.8a

Euclidean plane. By *lines* in the model we shall mean the ideal line at infinity, all ordinary lines of zero, infinite, or negative slope (like lines *a*, *b*, and *c*), plus all broken lines consisting of two half-lines of positive slope meeting on *h* and with the slope of the upper half-line twice that of the lower half-line (like broken line *AXB*). By *on* in the model we shall mean the obvious interpretation.

One may now verify the three postulates of plane projective geometry in our model. Postulates II and III are easily verified. The only part of the verification of Postulate I that requires comment is the verification that two points *A* and *B* of the model, where *A* is in the lower half of the plane and *B* is in the upper half with *B* to the right of *A*, determine a unique line of the model. In Figure 8.8a, let X_0 and X_1 be the feet of the perpendiculars from *A* and *B* on line *h*. As point *X* moves along line *h* from X_0 to X_1, the value of (slope *XB*)/(slope *AX*) increases continuously from 0 to ∞. It follows that there is a unique point *X* between X_0 and X_1 such that (slope *XB*)/(slope *AX*) = 2. If *X* is not between X_0 and X_1, then clearly the broken line *AXB* is not a line in the model. Thus there is one and only one line in the model joining the points *A* and *B*.

Now consider the two copolar triangles *ABC*, *A'B'C'* of Figure 8.8b. Let *AA'*, *BB'*, *CC'* concur in *O*, and let *BC* and *B'C'* intersect in *L*, *CA* and

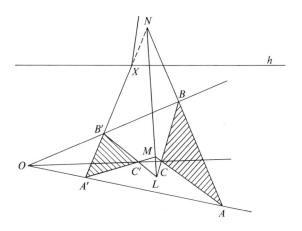

Figure 8.8b

$C'A'$ intersect in M, AB and $A'B'$ intersect in N. Then, since Desargues' two-triangle theorem holds in the Euclidean plane, L, M, N are collinear. But we have drawn the figures so that all the lettered points except N lie below h, and so lines LM and AB have negative slope while line $A'B'$ has positive slope. It follows that, in the model, the two triangles are still copolar at O, but the line determined by A' and B' no longer passes through N, and the two triangles are not coaxial in the model.

8.8.3 THEOREM. *Desargues' two-triangle theorem is independent of the postulates for plane projective geometry.*

The model we have just described is a projective geometry in which Desargues' Theorem does not hold. Such a geometry is called a non-Desarguesian geometry, and this particular example of a non-Desarguesian geometry was given by F. R. Moulton in 1902, and is perhaps the simplest example of a non-Desarguesian geometry yet devised. A more complex example of a non-Desarguesian geometry had been given by Hilbert in his *Grundlagen der Geometrie*, where it is shown that the plane case of Desargues' Theorem cannot be deduced without using all the congruence postulates or the assumption that the plane is embedded in a three-dimensional space— a curious situation inasmuch as the plane case of Desargues' Theorem would seem to be independent of both metrical notions and any higher dimensional space. If the reader will recall the various proofs that we have given of the plane case of Desargues' Theorem, he will be able to note that these proofs use either metrical notions or the fact that the plane lies within three-dimensional space.

Let us, for convenience, denote the following three (planar) propositions by D, P, and F; it will be noted that Proposition D is part of the plane case of Desargues' Theorem and that Proposition P is Pappus' Theorem— Proposition F is usually called the *Fundamental Theorem of Classical Projective Geometry.*

8.8.4 PROPOSITION D. *If ABC, A'B'C' are two triangles in which the six sides and six vertices are distinct, and if AA', BB', CC' concur, then the three points of intersection of BC and B'C', CA and C'A', AB and A'B' are collinear.*

8.8.5 PROPOSITION P. *If A, B, C and A', B', C' are two sets of collinear points on distinct lines, then the three points of intersection of BC' and B'C, CA' and C'A, AB' and A'B are collinear.*

8.8.6 PROPOSITION F. *If a projectivity carries each of three collinear points into itself, then it carries every point of their line into itself.*

It can be shown that in the presence of our three postulates I, II, III for plane projective geometry: (1) Proposition F implies Proposition P, (2)

Propositions P and D imply Proposition F, (3) Proposition D implies the converse of Proposition D, (4) Proposition P implies Proposition D. It follows that Propositions P and F are equivalent and that Proposition P implies Propositions D and F and the converse of Proposition D. In fact, if in addition to our three postulates I, II, III we assume only Proposition P, then the whole of classical plane projective geometry can be deduced.

To state this otherwise, we formulate the following definitions.

8.8.7 DEFINITIONS. A *projective plane* is the composite notion of two nonempty sets of objects called *points* and *lines*, along with a relation *on* between points and lines, for which our three postulates I, II, III hold. According as Proposition D, P, or F also holds, the projective plane is said to be *Desarguesian, Pappian,* or *classical.*

We then have the following theorem.

8.8.8 THEOREM. *A Pappian projective plane is Desarguesian and classical.*

We do not take the space to consider here the important parts played by Propositions D and P in the setting up of a coordinate system in a projective plane.

PROBLEMS

1. Show that the principle of duality holds in a Desarguesian projective plane.
2. Actually verify Postulates I, II, III of Section 8.7 in the model set up in Section 8.8 of a non-Desarguesian plane projective geometry.
3. (a) Show that the nonvertical modified lines through the point $(a,0)$ in the model of the non-Desarguesian plane projective geometry of Section 8.7 may be defined analytically by the equation

$$y = m(x - a)f(y,m),$$

where (i) $f(y,m) = 1$ if $m \leq 0$, (ii) $f(y,m) = 1$ if $m > 0$ and $y \leq 0$, (iii) $f(y,m) = 2$ if $m > 0$ and $y > 0$.
   (b) Let $P(r,s)$, $Q(t,u)$ be two points of the Cartesian plane such that $r < t$, $s < 0$, $u > 0$. Show that the unique modified line determined by P and Q cuts the x axis in the point where $x = (ur - 2st)/(u - 2s)$.
4. If in a trigonometric equation each trigonometric function which appears is replaced by its cofunction, the new equation obtained is called the *dual* of the original equation. Establish the following *principle of duality* of trigonometry: *If a trigonometric equation involving a single angle is an identity, then its dual is also an identity.*

8.9 FINITE GEOMETRIES

A *finite geometry* is a geometry containing only a finite number of points, lines, and planes. The existence of such a geometry containing exactly 7 points, 7 lines, and 1 plane was established in Section 8.7. It was G. Fano who, in 1892, first considered a finite geometry—a three-dimensional geometry containing 15 points, 35 lines, and 15 planes, each plane containing 7 points and 7 lines. But finite geometries were not brought into prominence until about 1906, when Oswald Veblen and W. H. Bussey made a study of finite projective geometries. Since then the study of finite geometries has grown considerably and has given rise to many unsolved problems that are engaging the attention of current researchers. For example, it has been shown that there exist finite projective geometries having n points on a line if $n - 1$ is a power of a prime, but it is not known if n can have any other value. It has been shown that there is essentially only one finite projective geometry with n points on a line if $n = 3, 4, 5,$ or 6, no such geometry for $n = 7$, exactly one for $n = 8$, at least one for $n = 9$, and at least four for $n = 10$ (one of which is Desarguesian and the other three non-Desarguesian). A systematic study of finite geometries requires a long excursion into abstract algebra, and is thus beyond the limits imposed on this book. The existence of finite geometries furnishes additional support of the hypothetico-deductive nature of much of present-day geometric study. In this section we shall very briefly consider a few special plane finite geometries.

8.9.1 *Fano's seven-point projective geometry.* Postulates for the 7 point finite geometry of Section 8.7 may be stated as follows:

P1. *There exists at least one line.*

P2. *There are exactly three points on every line.*

P3. *Not all points are on the same line.*

P4. *There is exactly one line on any two distinct points.*

P5. *There is at least one point on any two distinct lines.*

The finite geometry is a projective geometry (see Problem 1, Section 8.7) and can be represented by the following symbolic diagram, where the numbers $1, \ldots, 7$ represent points and the vertical columns represent lines:

$$
\begin{array}{ccccccc}
1 & 2 & 3 & 4 & 5 & 6 & 7 \\
2 & 3 & 4 & 5 & 6 & 7 & 1 \\
4 & 5 & 6 & 7 & 1 & 2 & 3
\end{array}
$$

8.9.2 *A finite projective geometry of thirteen points and thirteen lines.* As a postulational basis for this geometry we may take the postulates of 8.9.1 with Postulate P2 replaced by:

P2'. *There are exactly four points on every line.*

This geometry can be represented by the following symbolic diagram:

1	2	3	4	5	6	7	8	9	10	11	12	13
2	3	4	5	6	7	8	9	10	11	12	13	1
4	5	6	7	8	9	10	11	12	13	1	2	3
10	11	12	13	1	2	3	4	5	6	7	8	9

8.9.3 *Young's finite geometry of nine points and twelve lines.* If in the finite projective geometry of 8.9.2 we omit one of the lines of four points, we obtain a geometry of 9 points and 12 lines. This is equivalent to projecting one of the lines to infinity and thereby converting the projective geometry of 13 points and 13 lines without parallel lines into a Euclidean geometry of 9 points and 12 lines with parallel lines.* This geometry can be defined by the following postulates:

P1. *There exists at least one line.*

P2. *There are exactly three points on every line.*

P3. *Not all points are on the same line.*

P4. *There is exactly one line on any two distinct points.*

P5. *Given a line* l *and a point* P *not on* l, *there is exactly one line on* P *and not on any point on* l.

Two lines not containing a point in common are said to be *parallel,* and this finite geometry is Euclidean in the sense that through any point not on a line there is one and only one line parallel to that line. The principle of duality does not hold in this geometry inasmuch as any two distinct points determine a line but two distinct lines do not always determine a point. The reader might care to try to deduce the following theorems of this geometry from the above postulational basis.

T1. *There exist exactly nine points in the geometry.*

T2. *There exist exactly twelve lines in the geometry.*

T3. *Every line has precisely two lines parallel to it.*

T4. *Two lines parallel to a third line are parallel to each other.*

T5. *Pappus' Theorem holds for the six points on any pair of parallel lines.*

The geometry can be represented by the following symbolic diagram:

1	1	1	1	2	2	2	3	3	3	4	7
2	4	5	6	5	4	6	4	5	6	5	8
3	7	9	8	8	9	7	8	7	9	6	9

8.9.4 *The Pappus finite geometry.* This finite geometry of 9 points and

* This geometry is used to illustrate a complete logical system in M. R. Cohen and E. Nagel, *Introduction to Logic and Scientific Method.* New York: Harcourt, Brace and Company, Inc., 1934.

9 lines with 3 points on each line and 3 lines on each point can be defined by the following postulates:

P1. *There exists at least one line.*

P2. *There are exactly three points on every line.*

P3. *Not all points are on the same line.*

P4. *There is at most one line on any two distinct points.*

P5. *Given a line l and a point P not on l, there is exactly one line on P and not on any point on l.*

P6. *Given a point P and a line l not on P, there is exactly one point on l and not on any line on P.*

Two lines without a point in common are said to be *parallel lines*, and two points without a line in common are said to be *parallel points*. The principle of duality holds in this geometry, which has the Euclidean property of parallelism of lines and also the dual property of parallelism of points. The following theorems can be established.

T1. *There exists at least one point.*

T2. *Not all lines pass through the same point.*

T3. *There is at most one point on any two distinct lines.*

T4. *There are exactly three lines on every point.*

T5. *The principle of duality holds.*

T6. *Pappus' Theorem holds for the six points on any pair of parallel lines.*

T7. *There are exactly nine points in the geometry.*

T8. *There are exactly nine lines in the geometry.*

The geometry can be represented by the following symbolic diagram or by the graphical diagram of Figure 8.9a.

$$
\begin{array}{ccccccccc}
1 & 2 & 3 & 4 & 5 & 6 & 1 & 2 & 7 \\
2 & 3 & 4 & 5 & 6 & 1 & 3 & 4 & 8 \\
7 & 8 & 9 & 7 & 8 & 9 & 5 & 6 & 9
\end{array}
$$

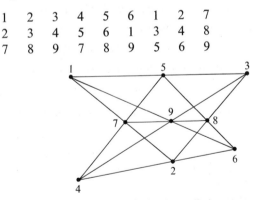

Figure 8.9a

8.9.5 *The Desargues finite geometry.* This is a finite geometry of 10 points and 10 lines with 3 points on a line and 3 lines on a point. It can be defined by the following postulates:

P1. *There exists at least one point.*

DEFINITIONS. A line *p* is called a *polar* of a point *P* if there is no line on both *P* and a point on *p*. A point *P* is called a *pole* of a line *p* if there is no point on both *p* and a line on *P*.

P2. *Every point has a polar.*

P3. *Every line has at most one pole.*

P4. *There is at most one line on any two distinct points.*

P5. *There are at least three distinct points on every line.*

P6. *If line* p *is not on point* Q, *then there is a point on both* p *and any polar* q *of* Q.

Two lines without a point in common are said to be *parallel lines*; two points without a line in common are said to be *parallel points*. This geometry is interesting in that it is Desarguesian, has duality, has polarity, and is Lobachevskian in the sense that a line may have as many as three lines parallel to it through a given point not on it. The reader may care to try to establish the following theorems.

T1. *A point has exactly one polar and a line has exactly one pole.*

T2. *If point* P *is the pole of line* p, *then line* p *is the polar of point* P.

T3. *If point* P *is on the polar of point* Q, *then point* Q *is on the polar of point* P.

T4. *Two lines parallel to the same line have a point in common.*

T5. *There are exactly three points on every line and three lines on every point.*

T6. *There are exactly ten points and ten lines in the geometry.*

T7. *The principle of duality holds.*

T8. *Desargues' Theorem holds.*

The geometry may be represented by the following symbolic diagram or by the graphical diagram of Figure 8.9b.

1	2	3	2	4	3	6	1	4	7
4	5	6	3	6	1	4	2	5	8
10	10	10	7	7	8	8	9	9	9

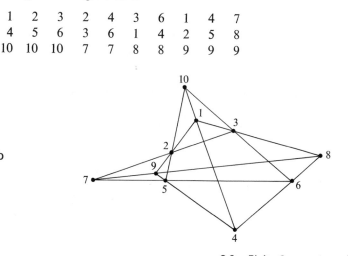

Figure 8.9b

8.9.6 *Affine geometries.* An *affine geometry* can be defined by the following postulates:

P1. *There exists at least one line.*

P2. *There are at least two points on every line.*

P3. *Not all points are on the same line.*

P4. *There is exactly one line on any two distinct points.*

P5. *Given a line* l *and a point* P *not on* l, *there is exactly one line on* P *and not on any point on* l. (This may be called the *parallel postulate.*)

The following theorems can be established from the above postulates.

T1. *Every line contains either an infinite number of points or the same finite number.*

T2. *Every point has either an infinite number of lines on it, or the same finite number.*

T3. *If every line contains exactly* n *points, then every point has exactly* n + 1 *lines on it.*

T4. *If every line contains exactly* n *points, then there are exactly* n^2 *points and* n(n + 1) *lines in the geometry.*

DEFINITION 1. Two lines having no point in common are said to be *parallel.*

T5. *If a line is parallel to one of two intersecting lines, it intersects the other.*

T6. *If line* p *is parallel to line* q *and line* q *is parallel to line* r, *then line* p *is parallel to line* r.

DEFINITION 2. If *p* is any line, then the set of lines consisting of *p* and all lines parallel to *p* is called the *parallel class* of *p*, and is denoted by [*p*].

T7. *If line* q *is in* [p], *then* [q] = [p].

T8. *Two parallel classes either coincide or have no members in common.*

T9. *If every line contains exactly* n *points, then for each line* p *class* [p] *has exactly* n *members.*

T10. *If a common ideal point is assigned to all the lines of a parallel class, and if the set of all such ideal points is called an ideal line, then the augmented plane containing all ordinary and ideal points and lines is a projective plane.*

8.9.7 *An interesting finite affine geometry.* Young's geometry of 8.9.3 is clearly a finite affine geometry. As another example of a finite affine geometry consider the following model.

The 25 letters *A*, *B*, *C*,..., *Y* (all the letters of the English alphabet except *Z*) shall be called *points*. The 30 sets of five letters which occur together in any row or any column of the following three blocks shall be called *lines*.

A	B	C	D	E		A	I	L	T	W		A	X	Q	O	H
F	G	H	I	J		S	V	E	H	K		R	K	I	B	Y
K	L	M	N	O		G	O	R	U	D		J	C	U	S	L
P	Q	R	S	T		Y	C	F	N	Q		V	T	M	F	D
U	V	W	X	Y		M	P	X	B	J		N	G	E	W	P

The notion of *a point on a line* and *a line on a point* will have the obvious interpretation.

We leave to the reader the straightforward task of showing that the model is indeed an example of a finite affine geometry, and now formulate the following definitions.

DEFINITION 1. By the *distance* $Z_1 Z_2$ between two points Z_1 and Z_2 on a line p is meant the least number of steps along the line p from one point to the other, where the first letter of the line is considered as following the last letter of the line. (Thus on line *ABCDE*, distance $DB = 2$ and distance $AE = 1$.)

DEFINITION 2. Two point pairs Z_1, Z_2 and Z_3, Z_4 are *congruent*, and we write $Z_1 Z_2 = Z_3 Z_4$, if they both lie on row lines or both lie on column lines and if distance $Z_1 Z_2 =$ distance $Z_3 Z_4$. (Thus $AC = RD$ and $DS = KG$.)

DEFINITION 3. A line q is *perpendicular* to a line p if there exist two points Z_1 and Z_2 on p such that for each point Z on q, $ZZ_1 = ZZ_2$. (Thus line *AFKPU* is perpendicular to line *ABCDE*, since we may take $Z_1 = B, Z_2 = E$.)

A surprising number of theorems of Euclidean plane geometry can be shown to hold in this finite geometry. For example we have:

T1. *Through any given point there is a unique perpendicular to a given line.*

T2. *The three altitudes of a triangle are concurrent in a point* α.

T3. *The perpendicular bisectors of the three sides of a triangle are concurrent in a point* β.

T4. *The three medians of a triangle are concurrent in a point* γ.

T5. *The three points* α, β, γ *of T2, T3, T4 are collinear and such that* γ *divides the segment* $\alpha\beta$ *in the ratio* $2:1$.

An extensive theory of parallelograms and quadrilaterals can be developed. There are counterparts of such loci as circles, parabolas, ellipses, and hyperbolas. As an example, the theorem *The locus of the midpoints of a system of parallel chords of a parabola is a line perpendicular to the directrix of the parabola*, continues to hold. A theory of areas can be formulated.

PROBLEMS

1. Consider the following postulates:

    P1. *There is exactly one point on any two distinct lines.*

    P2. *Every point is on exactly one pair of lines.*

    P3. *There are exactly four lines.*

  (a) Establish the absolute consistency of the above postulate set.

  (b) Establish the independence of the above postulate set.

  (c) Deduce the following theorems from the above postulate set:

    T1. *There are exactly six points.*

    T2. *There are exactly three points on each line.*

    DEFINITION. Two points which have no line in common are called *parallel points.*

    T3. *Each point has exactly one point parallel to it.*

    T4. *On a line not containing a given point there is one and only one point which is parallel to the given point.*

2. (a) Show that the postulate set of 8.9.3 is independent.

  (b) Try to establish the theorems of 8.9.3.

3. (a) Show that the postulate set of 8.9.4 is independent.

  (b) Try to establish the theorems of 8.9.4.

4. (a) Show that the postulates of 8.9.5 are independent.

  (b) Try to establish the theorems of 8.9.5.

5. (a) Show that the postulates of 8.9.6 are independent.

  (b) Try to establish the theorems of 8.9.6.

6. (a) Give an example in the model of 8.9.7 of theorems T1 through T5.

  (b) Is line *ABCDE* perpendicular to line *AFKPU*?

  (c) Prove that every segment has a midpoint.

  (d) Find the points on the circle with center *A* and passing through point *B*.

  (e) Find the points on the circle with center *A* and passing through point *C*.

  (f) Show that triangle *ABJ* is isosceles, with vertex *B*, and that the altitude from *B* bisects the base *AJ*.

  (g) Check, by examples in this model, the following theorems:

    (1) If one pair of sides of a quadrilateral are congruent and parallel, so are the other pair of sides.

    (2) The diagonals of a parallelogram bisect each other.

    (3) The diagonals of a rhombus are perpendicular.

    (4) There is a unique tangent to a circle at each point of the circle.

BIBLIOGRAPHY

BLUMENTHAL, L. M., *A Modern View of Geometry*. San Francisco: W. H. Freeman and Company, 1961.

BIRKHOFF, G. D., and RALPH BEATLEY, *Basic Geometry*. Chicago: Scott, Foresman and Company, 1940. Reprinted by Chelsea Publishing, Inc., 1959.

BORSUK, KAROL and WANDA SZMIELEW, *Foundations of Geometry*, tr. by Erwin Marquit. Amsterdam: North-Holland Publishing Company, 1960.

BRUMFIEL, C. F., R. E. EICHOLZ, and M. E. SHANKS, *Geometry*. Reading, Mass.: Addison-Wesley Publishing Company, Inc., 1960.

CHOQUET, GUSTAVE, *Geometry in a Modern Setting*. Boston, Mass.: Houghton Mifflin Company, 1969.

CURTIS, C. W., P. H. DAUS, and R. J. WALKER, *Euclidean Geometry Based on Ruler and Protractor Axioms*. New Haven, Conn.: School Mathematics Study Group, 1959.

EVES, HOWARD, and C. V. NEWSOM, *An Introduction to the Foundations and Fundamental Concepts of Mathematics*. New York: Holt, Rinehart and Winston, revised ed., 1965.

FORDER, H. G., *Geometry*. New York: Hutchinson's University Library, 1950.

———, *The Foundations of Euclidean Geometry*. New York: Cambridge University Press, 1927. Reprinted by Dover Publications, Inc.

GOODSTEIN, R. L., and E. J. F. PRIMROSE, *Axiomatic Projective Geometry*. Leicester: University College, 1953.

HALSTED, G. B., *Rational Geometry*. New York: John Wiley and Sons, Inc., 1904.

HEATH, T. L., *The Thirteen Books of Euclid's Elements*, 2d ed., 3 vols. New York: Cambridge University Press, 1926. Reprinted by Dover Publications, Inc., 1956.

HILBERT, DAVID, *The Foundations of Geometry*, 2nd ed. tr. by E. J. Townsend. Chicago: The Open Court Publishing Company, 1902. 10th ed. tr. by Leo Unger, Open Court Publishing Company, 1971.

IVINS, W. M., JR., *Art and Geometry, a Study in Space Intuitions*. Cambridge, Mass.: Harvard University Press, 1946.

JOLLY, R. F., *Synthetic Geometry*. New York: Holt, Rinehart and Winston, Inc., 1969.

KEYSER, C. J., *Mathematical Philosophy*. New York: E. P. Dutton and Company, Inc., 1922.

———, *Mathematics and the Question of Cosmic Mind, with Other Essays*. The Scripta Mathematica Library, No. 2. New York: Scripta Mathematica, 1935.

LEVI, HOWARD, *Foundations of Geometry and Trigonometry*. Englewood Cliffs, N.J.: Prentice-Hall, Inc., 1960.

MACH, ERNST, *Space and Geometry*, tr. by T. J. McCormack. Chicago: The Open Court Publishing Company, 1906.

MESERVE, B. E., *Fundamental Concepts of Geometry*. Reading, Mass.: Addison-Wesley Publishing Company, Inc., 1955.

MOISE, E. E., *Elementary Geometry from an Advanced Standpoint*. Reading, Mass.: Addison-Wesley Publishing Company, Inc., 1963.

PRENOWITZ, WALTER, *A Contemporary Approach to Classical Geometry*. Herbert Ellsworth Slaught Memorial Paper No. 9. Buffalo, N.Y.: Mathematical Association of America, 1961.

RAINICH, G. Y., and S. M. DOWDY, *Geometry for Teachers*. New York: John Wiley and Sons, Inc., 1968.

REDEI, L., *Foundations of Euclidean and Non-Euclidean Geometries According to F. Klein*. New York; Pergamon Press, 1968.

ROBINSON, G. DE B., *The Foundations of Geometry*. Toronto: University of Toronto Press, 1946.

RUSSELL, B. A. W., *An Essay on the Foundations of Geometry*. New York: Dover Publications, Inc., 1956.

SCHOOL MATHEMATICS STUDY GROUP, *Mathematics for High School, Geometry (Parts 1 and 2)*. New Haven, Conn.: School Mathematics Study Group, 1959.

ŚNIATYCKI, A., *An Axiomatics of Non-Desarguean Geometry Based on the Half-Plane as the Primitive Notion*. Warsaw, Poland: Państwowe Wydawnictwo Naukowe, 1961.

STABLER, E. R., *An Introduction to Mathematical Thought*. Reading, Mass.: Addison-Wesley Publishing Company, Inc., 1953.

TULLER, ANNITA, *A Modern Introduction to Geometries*. Princeton, N.J.: D. Van Nostrand Company, Inc., 1967.

VEBLEN, OSWALD, and J. W. YOUNG, *Projective Geometry*, 2 vols. Boston: Ginn and Company, 1910, 1918.

WYLIE, C. R., JR., *Foundations of Geometry*. New York: McGraw-Hill Book Company, 1964.

YOUNG, J. W., *Lectures on Fundamental Concepts of Algebra and Geometry*. New York: The Macmillan Company, 1911.

YOUNG, J. W. A., ed., *Monographs on Topics of Modern Mathematics Relevant to the Elementary Field*. New York: Longmans, Green and Company, 1911. Reprinted by Dover Publications, Inc., 1955.

Appendix

1

Euclid's First Principles and the Statement of the Propositions of Book I[*]

The Initial Explanations and Definitions

    1. A *point* is that which has no part.

    2. A *line* is length without breadth.

    3. The extremities of a line are points.

    4. A *straight line* is a line which lies evenly with the points on itself.

    5. A *surface* is that which has only length and breadth.

    6. The extremities of a surface are lines.

    7. A *plane surface* is a surface which lies evenly with the straight lines on itself.

    8. A *plane angle* is the inclination to one another of two lines in a plane if the lines meet and do not lie in a straight line.

    9. When the lines containing the angle are straight lines, the angle is called a *rectilinear angle*.

    10. When a straight line erected on a straight line makes the adjacent angles equal to one another, each of the equal angles is called a *right angle*, and the straight line standing on the other is called a *perpendicular* to that on which it stands.

* Taken, with permission, from T. L. Heath, *The Thirteen Books of Euclid's Elements*, 3 vols., 2 ed. New York: The Cambridge University Press, 1926. Reprint, New York: Dover Publications, Inc., 1956.

11. An *obtuse angle* is an angle greater than a right angle.

12. An *acute angle* is an angle less than a right angle.

13. A *boundary* is that which is an extremity of anything.

14. A *figure* is that which is contained by any boundary or boundaries.

15. A *circle* is a plane figure contained by one line such that all the straight lines falling upon it from one particular point among those lying within the figure are equal.

16. The particular point (of Definition 15) is called the *center* of the circle.

17. A *diameter* of a circle is any straight line drawn through the center and terminated in both directions by the circumference of the circle. Such a straight line also bisects the circle.

18. A *semicircle* is the figure contained by a diameter and the circumference cut off by it. The center of the semicircle is the same as that of the circle.

19. *Rectilinear figures* are those which are contained by straight lines, *trilateral* figures being those contained by three, *quadrilateral* those contained by four, and *multilateral* those contained by more than four straight lines.

20. Of the trilateral figures, an *equilateral triangle* is one which has its three sides equal, an *isosceles triangle* has two of its sides equal, and a *scalene triangle* has its three sides unequal.

21. Furthermore, of the trilateral figures, a *right-angled triangle* is one which has a right angle, an *obtuse-angled triangle* has an obtuse angle, and an *acute-angled triangle* has its three angles acute.

22. Of the quadrilateral figures, a *square* is one which is both equilateral and right-angled; an *oblong* is right-angled but not equilateral; a *rhombus* is equilateral but not right-angled; and a *rhomboid* has its opposite sides and angles equal to one another but is neither equilateral nor right-angled. Quadrilaterals other than these are called *trapezia*.

23. *Parallel* straight lines are straight lines which, being in the same plane and being produced indefinitely in both directions, do not meet one another in either direction.

The Postulates

Let the following be postulated:

1. A straight line can be drawn from any point to any point.

2. A finite straight line can be produced continuously in a straight line.

3. A circle may be described with any center and distance.

4. All right angles are equal to one another.

5. If a straight line falling on two straight lines makes the interior angles on the same side together less than two right angles, the two straight lines, if produced indefinitely, meet on that side on which the angles are together less than two right angles.

The Axioms or Common Notions

1. Things which are equal to the same thing are also equal to one another.
2. If equals be added to equals, the wholes are equal.
3. If equals be subtracted from equals, the remainders are equal.
4. Things which coincide with one another are equal to one another.
5. The whole is greater than the part.

The Forty-Eight Propositions of Book I

1. On a given finite straight line to construct an equilateral triangle.
2. To place at a given point (as an extremity) a straight line equal to a given straight line.
3. Given two unequal straight lines, to cut off from the greater a straight line equal to the less.
4. If two triangles have the two sides equal to two sides respectively, and have the angles contained by the equal straight lines equal, they will also have the base equal to the base, the triangle will be equal to the triangle, and the remaining angles will be equal to the remaining angles respectively, namely those which the equal sides subtend.
5. In isosceles triangles the angles at the base are equal to one another, and, if the equal straight lines be produced further, the angles under the base will be equal to one another.
6. If in a triangle two angles be equal to one another, the sides which subtend the equal angles will also be equal to one another.
7. Given two straight lines constructed on a straight line (from its extremities) and meeting in a point, there cannot be constructed on the same straight line (from its extremities), and on the same side of it, two other straight lines meeting in another point and equal to the former respectively, namely each to that which has the same extremity with it.
8. If two triangles have the two sides equal to two sides respectively, and have also the base equal to the base, they will also have the angles equal which are contained by the equal straight lines.
9. To bisect a given rectilinear angle.
10. To bisect a given finite straight line.
11. To draw a straight line at right angles to a given straight line from a given point on it.
12. To a given infinite straight line, from a given point which is not on it, to draw a perpendicular straight line.
13. If a straight line set up on a straight line make angles, it will make either two right angles or angles equal to two right angles.
14. If with any straight line, and at a point on it, two straight lines not lying on the same side make the adjacent angles equal to two right angles, the two straight lines will be in a straight line with one another.
15. If two straight lines cut one another, they make the vertical angles equal to one another.

16. In any triangle, if one of the sides be produced, the exterior angle is greater than either of the interior and opposite angles.

17. In any triangle two angles taken together in any manner are less than two right angles.

18. In any triangle the greater side subtends the greater angle.

19. In any triangle the greater angle is subtended by the greater side.

20. In any triangle two sides taken together in any manner are greater than the remaining one.

21. If on one of the sides of a triangle, from its extremities, there be constructed two straight lines meeting within the triangle, the straight lines so constructed will be less than the remaining two sides of the triangle, but will contain a greater angle.

22. Out of three straight lines, which are equal to three given straight lines, to construct a triangle; thus it is necessary that two of the straight lines taken together in any manner should be greater than the remaining one.

23. On a given straight line and at a point on it to construct a rectilinear angle equal to a given rectilinear angle.

24. If two triangles have the two sides equal to two sides respectively, but have the one of the angles contained by the equal straight lines greater than the other, they will also have the base greater than the base.

25. If two triangles have the two sides equal to two sides respectively, but have the base greater than the base, they will also have the one of the angles contained by the equal straight lines greater than the other.

26. If two triangles have the two angles equal to two angles respectively, and one side equal to one side, namely, either the side adjoining the equal angles, or that subtending one of the equal angles, they will also have the remaining sides equal to the remaining sides and the remaining angle equal to the remaining angle.

27. If a straight line falling on two straight lines make the alternate angles equal to one another, the straight lines will be parallel to one another.

28. If a straight line falling on two straight lines make the exterior angle equal to the interior and opposite angle on the same side, or the interior angles on the same side equal to two right angles, the straight lines will be parallel to one another.

*   *   *   *   *   *

29. A straight line falling on parallel straight lines makes the alternate angles equal to one another, the exterior angle equal to the interior and opposite angle, and the interior angles on the same side equal to two right angles.

30. Straight lines parallel to the same straight line are also parallel to one another.

31. Through a given point to draw a straight line parallel to a given straight line.

32. In any triangle, if one of the sides be produced, the exterior angle is equal to the two interior and opposite angles, and the three interior angles of the triangle are equal to two right angles.

33. The straight lines joining equal and parallel straight lines (at the extremities which are) in the same directions (respectively) are themselves also equal and parallel.

34. In parallelogrammic areas the opposite sides and angles are equal to one another, and the diameter bisects the areas.

35. Parallelograms which are on the same base and in the same parallels are equal to one another.

36. Parallelograms which are on equal bases and in the same parallels are equal to one another.

37. Triangles which are on the same base and in the same parallels are equal to one another.

38. Triangles which are on equal bases and in the same parallels are equal to one another.

39. Equal triangles which are on the same base and on the same side are also in the same parallels.

40. Equal triangles which are on equal bases and on the same side are also in the same parallels.

41. If a parallelogram have the same base with a triangle and be in the same parallels, the parallelogram is double of the triangle.

42. To construct, in a given rectilinear angle, a parallelogram equal to a given triangle.

43. In any parallelogram the complements of the parallelograms about the diameter are equal to one another.

44. To a given straight line to apply, in a given rectilinear angle, a parallelogram equal to a given triangle.

45. To construct, in a given rectilinear angle, a parallelogram equal to a given rectilinear figure.

46. On a given straight line to describe a square.

47. In right-angled triangles the square on the side subtending the right angle is equal to the squares on the sides containing the right angle.

48. If in a triangle the square on one of the sides be equal to the squares on the remaining two sides of the triangle, the angle contained by the remaining two sides of the triangle is right.

Appendix

2

Hilbert's Postulates for Plane Euclidean Geometry

PRIMITIVE TERMS

point, line, on, between, congruent

Group I: Postulates of Connection

I–1. *There is one and only one line passing through any two given distinct points.*

I–2. *Every line contains at least two distinct points, and for any given line there is at least one point not on the line.*

Group II: Postulates of Order

II–1. *If point* C *is between points* A *and* B, *then* A, B, C *are all on the same line, and* C *is between* B *and* A, *and* B *is not between* C *and* A, *and* A *is not between* C *and* B.

II–2. *For any two distinct points* A *and* B *there is always a point* C *which is between* A *and* B, *and a point* D *which is such that* B *is between* A *and* D.

II–3. *If* A, B, C *are any three distinct points on the same line, then one of the points is between the other two.*

DEFINITIONS. By the *segment AB* is meant the points *A* and *B* and all points which are between *A* and *B*. Points *A* and *B* are called the *end points* of the segment. A point C is said to be *on* the segment AB if it is A or B or some point between A and B.

DEFINITION. Two lines, a line and a segment, or two segments, are said to *intersect* if there is a point which is on both of them.

DEFINITIONS. Let A, B, C be three points not on the same line. Then by the *triangle ABC* is meant the three segments AB, BC, CA. The segments AB, BC, CA are called the *sides* of the triangle, and the points A, B, C are called the *vertices* of the triangle.

II–4. (Pasch's Postulate) *A line which intersects one side of a triangle but does not pass through any of the vertices of the triangle must also intersect another side of the triangle.*

Group III : Postulates of Congruence

III–1. *If* A *and* B *are distinct points and if* A′ *is a point on a line* m, *then there are two and only two distinct points* B′ *and* B″ *on* m *such that the pair of points* A′, B′ *is congruent to the pair* A, B *and the pair of points* A′, B″ *is congruent to the pair* A, B; *moreover,* A′ *is between* B′ *and* B″.

III–2. *If two pairs of points are congruent to the same pair of points, then they are congruent to each other.*

III–3. *If point* C *is between points* A *and* B *and point* C′ *is between points* A′ *and* B′, *and if the pair of points* A, C *is congruent to the pair* A′, C′, *and the pair of points* C, B *is congruent to the pair* C′, B′, *then the pair of points* A, B *is congruent to the pair* A′, B′.

DEFINITION. Two segments are said to be *congruent* if the end points of the segments are congruent pairs of points.

DEFINITIONS. By the *ray* AB is meant the set of all points consisting of those which are between A and B, the point B itself, and all points C such that B is between A and C. The ray AB is said to *emanate from* point A.

THEOREM. *If* B′ *is any point on the ray* AB, *then the rays* AB′ *and* AB *are identical.*

DEFINITIONS. By an *angle* is meant a point (called the *vertex* of the angle) and two rays (called the *sides* of the angle) emanating from the point. By virtue of the above theorem, if the vertex of the angle is point A and if B and C are any two points other than A on the two sides of the angle, we may unambiguously speak of the angle BAC (or CAB).

DEFINITIONS. If ABC is a triangle, then the three angles BAC, CBA, ACB are called the *angles* of the triangle. Angle BAC is said to be *included* by the sides AB and AC of the triangle.

III–4. *If* BAC *is an angle whose sides do not lie in the same line, and if* A′ *and* B′ *are two distinct points, then there are two and only two distinct rays,* A′C′ *and* A′C″, *such that angle* B′A′C′ *is congruent to angle* BAC *and angle* B′A′C″ *is congruent to angle* BAC; *moreover, if* D′ *is any point on the ray* A′C′ *and* D″ *is any point on the ray* A′C″, *then the segment* D′D″ *intersects the line determined by* A′ *and* B′.

III–5. *Every angle is congruent to itself.*

III–6. *If two sides and the included angle of one triangle are congruent, respectively, to two sides and the included angle of another triangle, then each of the remaining angles of the first triangle is congruent to the corresponding angle of the second triangle.*

Group IV: Postulate of Parallels

IV–1. (Playfair's Postulate) *Through a given point A not on a given line* m *there passes at most one line which does not intersect* m.

Group V: Postulates of Continuity

V–1. (Postulate of Archimedes) *If* A, B, C, D *are four distinct points, then there is, on the ray* AB, *a finite set of distinct points,* $A_1, A_2, \ldots, A_n$ *such that* (1) *each of the pairs* A, A_1; A_1, A_2; $\ldots$; A_{n-1}, A_n *is congruent to the pair* C, D, *and* (2) B *is between* A *and* A_n.

V–2. (Postulate of Completeness) *The points of a line constitute a system of points such that no new points can be assigned to the line without causing the line to violate at least one of the nine postulates* I–1, I–2, II–1, II–2, II–3, II–4, III–1, III–2, V–1.

Alternative Group V

DEFINITIONS. Consider a segment *AB*. Let us call one end point, say *A*, the *origin* of the segment, and the other point, *B*, the *extremity* of the segment. Given two distinct points *M* and *N* of *AB*, we say that *M precedes N* (or *N follows M*) if *M* coincides with the origin *A* or lies between *A* and *N*. A segment *AB*, considered in this way, is called an *ordered segment.*

V′–1. (Dedekind's Postulate) *If the points of an ordered segment of origin* A *and extremity* B *are separated into two classes in such a way that*

(1) *each point of* AB *belongs to one and only one of the classes,*

(2) *the points* A *and* B *belong to different classes (which we shall respectively call the* first *class and the* second *class),*

(3) *each point of the first class precedes each point of the second class,*

then there exists a point C *on* AB *such that every point of* AB *which precedes* C *belongs to the first class and every point of* AB *which follows* C *belongs to the second class.*

Suggestions for Solutions
of Selected Problems

Section 1.2

1. Let the quadrilateral be $ABCD$, with $AB = a$, $BC = b$, $CD = c$, $DA = d$. Show that $2K = ad \sin A + bc \sin C \leq ad + bc$; similarly, $2K \leq ab + cd$; therefore $4K \leq ad + bc + ab + cd = (a + c)(b + d)$. This formula illustrates the prevalent idea, in early empirical geometry, of averaging.

2. Take $\pi = 3$.

6. (b) Consider a frustum having B, B', h, V for bottom base, top base, altitude, and volume. Let a, a' represent the altitudes of the complete and cut-off pyramids. Then $V = (aB - a'B')/3 = [(h + a')B - (a - h)B']/3 = (hB + hB' + a'B - aB')/3$. But $a/a' = \sqrt{B}/\sqrt{B'}$, whence $a'\sqrt{B} = a\sqrt{B'}$. Therefore $a'B - aB' = a\sqrt{BB'} - a'\sqrt{BB'} = h\sqrt{BB'}$, etc.

9. (a) If r is the radius of the circle, θ half the central angle subtended by the chord, and A the sought area, then $r = (4s^2 + c^2)/8s$, $\theta = \sin^{-1}(c/2r)$, $A = r^2\theta + c(s - r)/2$.
   (b) 12 ft.

Section 1.3

1. Set up a vertical stick of length s near the pyramid. Let S_1, P_1 and S_2, P_2 be the points marking the shadows of the top of the stick and the apex of the pyramid at two different times of the day. Then, if x is the sought height of the pyramid, $x = s(P_1P_2)/(S_1S_2)$.

2. (c) $T_n = n(n + 1)/2$, $S_n = n^2$, $P_n = n(3n - 1)/2$.

4. A convex polyhedral angle must contain at least three faces, and the sum of its face angles must be less than 360°.

5. (a) Suppose that $\sqrt{2} = p/q$, where p and q are relatively prime integers. Then $p^2 = 2q^2$ and it follows that p^2, and hence p, is even. Set $p = 2r$. Then $2r^2 = q^2$ and it follows that q^2, and hence q, is even. This contradicts the hypothesis that p and q are relatively prime.

(b) Suppose that the diagonal d and the side s of a square are commensurable. Then $d = pt$, $s = qt$, where p and q are integers and t is some segment. It follows that $\sqrt{2} = d/s = p/q$, a rational number.

(c) Suppose that the line passes through the lattice point (q,p) and let m denote the slope of the line. Then $\sqrt{2} = m = p/q$, a rational number.

9. THEOREM 1. Let a be a member of S. By P1 there exists a club A to which a belongs. By P3 there exists a club B conjugate to club A. Since B is nonempty, it has at least one member b, and $b \neq a$. By P2 there exists a club C containing a and b. By P3 there exists a club D conjugate to club C. Since D is nonempty, it has a member c, and $c \neq a$, $c \neq b$. By P2 there exists a club E containing a and c. Since c belongs to E but not to C, clubs C and E are distinct. Thus a belongs to two distinct clubs, C and E.

THEOREM 2. Let A be a club. Since A is nonempty, it has at least one member a. Suppose a is the only member of A. By T1, there exists a club B distinct from club A and containing a as a member. Now B must contain a second member $b \neq a$, for otherwise clubs A and B could not be distinct. By P3 there exists a club C such that B is conjugate to C. It then follows that A is also conjugate to C. But this contradicts P3. Hence the theorem by *reductio ad absurdum*.

THEOREM 3. In the proof of T1 we showed the existence in S of at least two distinct people a and b. By P2 there exists a club A to which a and b belong. By P3 there exists a club B conjugate to club A. But, by T2, B must contain at least two distinct members, c and d. Since A and B are conjugate, it follows that a, b, c, d are four distinct people of S.

Section 1.4

4. (a), (b) Use Venn diagrams.

5. (b) See L. S. Shively, *An Introduction to Modern Geometry* (John Wiley and Sons, Inc.), p. 141, or Nathan Altshiller-Court, *College Geometry* (Barnes and Noble, Inc.), pp. 72–73.

6. (c) A definition of an entity (such as an angle bisector) does not prove that the entity exists; existence is established by actual construction of the entity.

7. Take the angle in the law of cosines equal to 90°.

9. (a) We have $(x - r)(x - s) \equiv x^2 - px + q^2$.
(h) Denote the parts by x and $a - x$. Then $x^2 - (a - x)^2 = x(a - x)$, or $x^2 + ax - a^2 = 0$.

10. (b) First trisect the diagonal BD by points E and F. Then the broken lines AEC and AFC divide the figure into three equivalent parts. Transform these parts so as to fulfill the conditions by drawing parallels to AC through E and F.

(c) Draw the line through P and the midpoint of the median of the trapezoid.

(d) Through B draw BD parallel to MN to cut AC in D. Then, if the required triangle is $AB'C'$, AC' is a mean proportional between AC and AD.

(e) Let ABC be the given triangle. Draw AB' making the given vertex angle with AC and let it cut the parallel to AC through B in B'. Now use 10 (d), Section 1.4.

Section 1.5

1. (c) $h_c = b \sin A$.

   (f) $h_a = t_a \cos[(B - C)/2]$.

   (g) $4h_a^2 + (b_a - c_a)^2 = 4m_a^2$.

   (h) $b_a - c_a = 2R \sin(B - C)$.

   (i) $4R(r_a - r) = (r_a - r)^2 + a^2$. If M and N are the midpoints of side BC and arc BC, then $MN = (r_a - r)/2$; clearly any two of R, a, MN determine the third.

   (j) $h_a = 2rr_a/(r_a - r)$.

2. (b) See Problem 3336, *The American Mathematical Monthly*, Aug. 1929.

   (c) See Problem E 1447, *The American Mathematical Monthly*, Sept. 1961. The solution given in this reference is a singularly fine application of the method of data.

3. (b) Let M be the midpoint of BC. The broken line EMA bisects the area. Through M draw MN parallel to AE to cut a side of triangle ABC in N. Then EN is the sought line.

5. Take the radius of C as one unit and set $p_k = 2k \sin(\pi/k)$, $P_k = 2k \tan(\pi/k)$, $a_k = (1/2)p_k \cos(\pi/k)$, $A_k = (1/2)P_k$.

6. See, e.g., G. A. Wentworth, *Plane and Solid Geometry*, revised edition, 1899, Ginn and Co.

7. (a) $(GC)^2 = (TW)^2 = 4r_1r_2$.

8. (e) The volume of the segment is equal to the volume of a spherical sector minus the volume of a cone. Also, $a^2 = h(2R - h)$.

   (f) The segment is the difference of two segments, each of one base, and having, say, altitudes u and v. Then

$$V = \pi R(u^2 - v^2) - \pi(u^3 - v^3)/3$$
$$= \pi h[(Ru + Rv) - (u^2 + uv + v^2)/3].$$

But $u^2 + uv + v^2 = h^2 + 3uv$, $(2R - u)u = a^2$, $(2R - v)v = b^2$. Therefore

$$V = \pi h[(a^2 + b^2)/2 + (u^2 + v^2)/2 - h^2/3 - uv]$$
$$= \pi h[(a^2 + b^2)/2 + h^2/2 + uv - h^2/3 - uv], \text{ etc.}$$

Section 1.6

1. (b) Let line AB cut line C in point D. Find T and T' on line C such that $DT = DT' = $ (mean proportional between DA and DB). Draw circles ABT and ABT'.

   (c) Reflect the point in a bisector of the angles formed by the two lines.

   (d) Let F be the focus and d the directrix. Find the reflection F' of F in m. Draw the circles through F and F' and tangent to d.

   (e) Two.

2. (a) Mark off a chord of the given length in the given circle; draw a concentric circle tangent to the chord; draw tangents from the given point to the second circle.

(b) Denote AB by a, AC by b, BC by c, and $\angle ADB$ by θ. Then, by the law of sines, applied first to triangle BCD and then to triangle ABD,

$$\sin 30°/\sin \theta = a/c, \quad \sin \theta/\sin 120° = a/(b + a).$$

Consequently, $1/\sqrt{3} = \tan 30° = a^2/c(b + a)$. Squaring both sides and recalling that $c^2 = b^2 - a^2$, we find $2a^3(2a + b) = b^3(2a + b)$, or $b^3 = 2a^3$.

(c) Use the fact that triangles DCO and COB are isosceles, and the fact that an exterior angle of a triangle is equal to the sum of the two remote interior angles.

3. (b) Let I and E divide AB internally and externally in the given ratio. If P is a point on the desired locus, then PI and PE are the internal and external bisectors of angle APB. It follows that P lies on the circle on IE as diameter.

   For the second part of this problem, let A and B be the given points, O the midpoint of AB, s^2 the given sum of squares, and P any point on the sought locus. By the law of cosines: $(PA)^2 = (PO)^2 + (OA)^2 - 2(PO)(OA)\cos(POA)$, $(PB)^2 = (PO)^2 + (OB)^2 - 2(PO)(OB)\cos(POB)$. Therefore $s^2 = (PA)^2 + (PB)^2 = 2(PO)^2 + 2(OA)^2$, whence PO is constant.

6. (a) Take the radius of the circumscribed circle as one unit. Then the side of the regular heptagon is $2 \sin(\pi/7)$, and the apothem of the regular inscribed hexagon is $\sqrt{3}/2$.

(b) Let a light ray, emanating from a point A, reflect at point C on mirror m, and then pass through a point B. If B' is the image of B in m, path ACB is a minimum when path ACB' is a minimum.

9. (b) Let $ABCD$ be the cyclic quadrilateral and let E on AC be such that $\angle CDE = \angle ADB$. From similar triangles CDE and ADB it follows that $(EC)(BD) = (AB)(DC)$. Likewise, from similar triangles ADE and BCD it follows that $(AE)(BD) = (BC)(AD)$. Therefore $(BC)(AD) + (AB)(DC) = (BD)(AE + EC) = (BD)(AC)$.

(c) In part (b) take AC as a diameter, $BC = a$, and $CD = b$.

(d) In part (b) take AB as a diameter, $BD = a$, and $BC = b$.

(e) In part (b) take AC as a diameter, $BD = t$ and perpendicular to AC.

(h) (1) Taking the side of the triangle as one unit, apply Ptolemy's Theorem to quadrilateral $PACB$. (2) Taking the side of the square as one unit, apply Ptolemy's Theorem to quadrilaterals $PBCD$ and $PCDA$. (3) Taking the side of the pentagon as one unit, apply Ptolemy's Theorem to quadrilaterals $PCDE$, $PCDA$, $PBCD$. (4) Taking the side of the hexagon as one unit, apply Ptolemy's Theorem to quadrilaterals $PBCD$, $PEFA$, $PBCF$, $PCFA$.

10. (a) From the similar triangles DFB and DBO, $FD/DB = DB/OD$. Therefore $FD = (DB)^2/OD = 2(AB)(BC)/(AB + BC)$.

(b) From similar right triangles, $OA/OB = AF/BD = AF/BE = AC/CB = (OC - OA)/(OB - OC)$. Now solve for OC.

(f) We have $ax \sin 60° + bx \sin 60° = ab \sin 120°$.

11. For a very neat application of this theorem to the establishment of the Erdös-Mordell inequality: "If P is any point inside or on the perimeter of triangle

ABC and if p_a, p_b, p_c are the distances of P from the sides of the triangle, then $PA + PB + PC \geq 2(p_a + p_b + p_c)$," see N. D. Kazarinoff, *Geometrical Inequalities*, Random House (1961), pp. 84–87.

12. (a) $V = 2\pi^2 R r^2$, $S = 4\pi^2 R r$.

(b) A distance of $2r/\pi$ from the diameter of the semicircle.

(c) A distance of $4r/3\pi$ from the diameter of the semicircle.

Section 1.7

1. (a) $x = hd/(2h + d)$.

(b) 8 cubits and 10 cubits.

(c) 20 cubits.

2. (a) Draw the circumdiameter through the vertex through which the altitude passes, and use similar triangles.

(b) Apply Problem 2 (a), Section 1.7 to triangles DAB and DCB.

(c) Use the result of Problem 2 (b), Section 1.7 along with Ptolemy's relation, $mn = ac + bd$.

(d) Here $\theta = 0°$ and $\cos \theta = 1$. Now use Problem 2 (b), (c), Section 1.7.

3. (b) Since the quadrilateral has an incircle we have $a + c = b + d = s$. Therefore $s - a = c$, $s - b = d$, $s - c = a$, $s - d = b$.

(c) In Figure S1 we have

$$a^2 + c^2 = r^2 + s^2 + m^2 + n^2 - 2(rn + sm) \cos \theta,$$
$$b^2 + d^2 = r^2 + s^2 + m^2 + n^2 + 2(sn + rm) \cos \theta.$$

Therefore $a^2 + c^2 = b^2 + d^2$ if and only if $\cos \theta = 0$, or $\theta = 90°$.

(d) Use Problem 3 (c), Section 1.7.

(e) The consecutive sides of the quadrilateral are 39, 60, 52, 25; the diagonals are 56 and 63; the circumdiameter is 65; the area is 1764.

5. (b) $x = 2.3696$.

(c) $x = 4.4934$.

7. (a) Find z such that $b/a = a/z$, then m such that $n/z = a/m$.

8. (a) Draw any circle Σ on the sphere and mark any three points A, B, C on its circumference. On a plane construct a triangle congruent to triangle ABC, find its circumcircle, and thus obtain the radius of Σ. Construct a right triangle having the radius of Σ as one leg and the polar chord of Σ as hypotenuse. It is now easy to find the diameter of the given sphere.

(b) If d is the diameter of the sphere and e the edge of the inscribed cube, then $e = (d\sqrt{3})/3$, whence e is one-third the altitude of an equilateral triangle of side $2d$.

Figure S.1

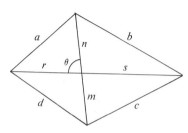

(c) If d is the diameter of the sphere and e the edge of the inscribed regular tetrahedron, then $e = (d\sqrt{6})/3$, whence e is the hypotenuse of a right isosceles triangle with leg equal to the edge of the inscribed cube. See Problem 8 (b), Section 1.7.

9. (a) Let the legs, hypotenuse, and area of the triangle be a, b, c, K, $a \geq b$. Then $a^2 + b^2 = c^2$ and $ab = 2K$. Solving for a and b we find

$$a = \frac{\sqrt{c^2 + 4K} + \sqrt{c^2 - 4K}}{2}, \quad b = \frac{\sqrt{c^2 + 4K} - \sqrt{c^2 - 4K}}{2}.$$

11. (a) Following is essentially the solution given by Regiomontanus. We are given (see Figure S2) $p = b - c$, h, $q = m - n$. Now $b^2 - m^2 = h^2 = c^2 - n^2$, or $b^2 - c^2 = m^2 - n^2$, or $b + c = qa/p$. Therefore

$$b = (qa + p^2)/2p \quad \text{and} \quad m = (a + q)/2.$$

Substituting these expressions in the relation $b^2 - m^2 = h^2$ yields a quadratic equation in the unknown a.

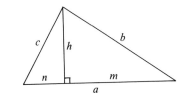

Figure S.2

(b) Following is essentially the solution given by Regiomontanus. Here we are given (see Figure S2) a, h, $k = c/b$. Set $2x = m - n$. Then

$$4n^2 = (a - 2x)^2, \quad 4c^2 = 4h^2 + (a - 2x)^2,$$
$$4m^2 = (a + 2x)^2, \quad 4b^2 = 4h^2 + (a + 2x)^2.$$

Then $k^2[4h^2 + (a + 2x)^2] = 4h^2 + (a - 2x)^2$. Solving this quadratic we obtain x, and thence b and c.

(c) On AD produced (see Figure S3) take $DE = bc/a$, the fourth proportional to the given segments a, b, c. Then triangles DCE and BAC are similar and $CA/CE = a/c$. Thus C is located as the intersection of two loci, a circle of Apollonius and a circle with center D and radius c.

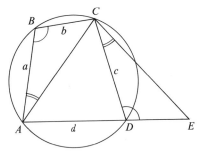

Figure S.3

12. Using standard notation we have

$$(rs)^2 = s(s - a)(s - b)(s - c),$$

or $16s^2 = s(s - 14)(6)(8)$, and $s = 21$. The required sides are then $21 - 6 = 15$ and $21 - 8 = 13$. This is not Pacioli's method of solving the problem; his solution is needlessly involved.

Section 1.8

1. No, but induction may be employed to conjecture the proposition to which the process is applied.

2. Fold the vertices onto the incenter of the triangle, or fold the vertices onto the foot of one of the altitudes.

7. No. Only for so-called *orthocentric tetrahedra* are the four altitudes concurrent. An orthocentric tetrahedron is a tetrahedron each edge of which is perpendicular to the opposite edge.

9. In the first three cases the binomial coefficients appear. Therefore one might expect the pentatope to have 5 zero-dimensional, 10 one-dimensional, 10 two-dimensional, and 5 three-dimensional bounding elements.

11. (a) Parallelepiped, rectangular parallelepiped (box), sphere.
    (b) Prism, right prism, sphere.

13. Instead of isogonal conjugate lines of a plane angle, consider *isogonal conjugate planes* of a dihedral angle.

14. It holds for all convex polyhedra and, more generally, for all polyhedra continuously deformable into a sphere.

15. The list cannot be continued: there is no convex polyhedron all of whose faces are hexagons. In fact, it can be shown that any convex polyhedron must have some face with less than six sides.

17. (b) Let D be a point on the surface of P such that the distance CD is a minimum. Show that D can be neither a vertex of P nor lie on an edge of P, and that CD is perpendicular to the face F of P on which D lies.

18. The point is known as the *isogonic center* of the triangle; from it each side of the triangle subtends an angle of $120°$. It may be found mechanically by three equally-weighted strings radiating from a common point and hanging over pulleys located at the vertices of a horizontal triangular table. See H. Steinhaus, *Mathematical Snapshots* (New York: Oxford University Press, 1950).

19. Draw, on cardboard, $a_1, a_2, \ldots, a_n$ as successive chords in a circle so large that the polygonal line made up of the chords does not close or cross itself. Cut out the polygon bounded by the polygonal line and the radii to the endpoints of the polygonal line. Fold the cardboard along the $n - 1$ other radii and paste together the two extreme radii, obtaining an open polyhedral surface. Push the n free edges gently against a plane, thus forming an isosceles pyramid.

20. See Problem E 753. *The American Mathematical Monthly*, Aug.-Sept. 1947.

Section 2.1

1. Use Theorem 2.1.3 along with mathematical induction.

2. Start with $\overline{AM} = \overline{MB}$ and then insert an origin at P.

3. Start with $\overline{AB} = \overline{OB} - \overline{OA}$ and then square both sides.

4. Insert an origin at P.

5. Set $\overline{AA'} = \overline{OA'} - \overline{OA} = (\overline{O'A'} - \overline{O'O}) - \overline{OA}$, etc.

6. Insert an origin O and let M and N denote the midpoints of CR and PQ. Then $4\overline{OM} = 2\overline{OR} + 2\overline{OC} = \overline{OA} + \overline{OB} + 2\overline{OC} = \overline{OB} + \overline{OC} + 2\overline{OQ} = 2\overline{OP} + 2\overline{OQ} = 4\overline{ON}$. Or, M and N clearly coincide if A, B, C are not collinear; now let C approach collinearity with A and B.

9. Insert an origin at O.

10. Use Theorem 2.1.9.

11. Use Theorem 2.1.9.

13. First consider the case where P is on the line and take P as origin; then let P' be the foot of the perpendicular dropped from P to the line.

14. Use Stewart's Theorem.

15. Use Stewart's Theorem.

16. By Problem 15, Section 2.1, $t_a^2 = bc[1 - a^2/(b + c)^2]$, $t_b^2 = ca[1 - b^2/(c + a)^2]$. Show that $t_a^2 - t_b^2 = (b - a)f(a,b,c)$ where $f(a,b,c)$ contains only positive terms.

17. Use Stewart's Theorem.

18. This problem is a generalization of Stewart's Theorem in that the point P is replaced by a circle.

19. (a) Let a line cut OA, OB, OC, OD in A', B', C', D'. Apply Euler's Theorem to A', B', C', D'. Multiply through by p^2, where p is the perpendicular from O onto line $A'B'C'D'$. Replace $p\overline{A'B'}$ by $(OA')(OB') \sin \overline{A'OB'}$, etc.
   (b) $2\triangle\overline{AOB} = (OA)(OB) \sin \overline{AOB}$, etc. Now use Problem 19 (a), Section 2.1.

20. Take O at the center of the circle.

21. Consider three cases: first where O is within triangle ABC, next where O is within $\angle BAC$ but on the other side of BC from A, then where O is within the vertical angle of $\angle BAC$.

23. Consider the equation where A is replaced by a variable point X on the line. Take any origin O on the line and obtain a quadratic equation in $x = OX$. Take $X = B$, C, D in turn, obtaining three identities. Then the quadratic equation has three distinct roots, and therefore is an identity. That is, X can be any point on the line, e.g. A.

Section 2.2

3. Each follows from Theorem 2.2.5 or Convention 2.2.4. Thus Theorem (a) becomes: "Through a given point there is one and only one plane containing the ideal line of a given plane not passing through the given point," and this is an immediate consequence of Theorem 2.2.5. Theorem (b) becomes: "Two lines which intersect a third line in its ideal point, intersect each other in their ideal points," and this follows from Convention 2.2.4.

5. Multiply the identity of Problem 23, Section 2.1 by $\overline{AD}$ and then take D as an ideal point.

6. Divide the identity of Problem 23, Section 2.1 by $\overline{AQ}$ and then take Q as an ideal point.

Section 2.3

2. $\overline{BD}/\overline{DC} = \overline{BF}/\overline{FL}$, etc.

3. If A', B', C' are the feet of the perpendiculars and if DEF cuts $A'B'C'$ in T, then $\overline{BD}/\overline{DC} = \overline{B'T}/\overline{TC'}$, etc.

5. Use Menelaus' Theorem.

6. $PD/AD = \triangle BPC/\triangle BAC$, etc.

7. $AP/AD = (AD - PD)/AD = 1 - PD/AD$, etc. Now use Problem 6, Section 2.3.

8. $\overline{AF}/\overline{FB} = \triangle\overline{CAP}/\triangle\overline{BCP}$, $\overline{AE}/\overline{EC} = \triangle\overline{PAB}/\triangle\overline{BCP}$.

9. Apply Menelaus' Theorem to triangles ABD and BCD.

10. Use mathematical induction. The converse is not true. A correct converse is: If the relation holds and if $n - 1$ of the points A', B', C', D', ... are collinear, then all n of them are collinear.

11. Apply Menelaus' Theorem to triangles ABD and BCD.

12. Let the plane cut line DB in point P. Use Menelaus' Theorem on triangle ABD with transversal $D'A'P$ and on triangle CBD with transversal $C'B'P$.

13. Use Theorem 2.1.9.

14. In the figure for Problem 13, Section 2.3, draw a sphere with center O to cut OA', OB', OC', OD', OE', OF' in A, B, C, D, E, F.

15. $\overline{AA'}/\overline{A'B} = (OA \sin \overline{AOA'})/(OB \sin \overline{A'OB})$, etc.

Section 2.4

1. (b) Use the trigonometric form of Ceva's Theorem.
   (c) Use the trigonometric form of Ceva's Theorem.

2. Use Ceva's Theorem.

3. Use Ceva's Theorem.

4. Use Ceva's Theorem.

6. Use Menelaus' Theorem.

7. Use the trigonometric form of Ceva's Theorem.

8. Use the trigonometric form of Menelaus' Theorem.

9. Use Ceva's Theorem along with the fact that $(AF)(AF') = (AE)(AE')$, etc.

10. Let D, E, F be the points of intersection of the tangents at A, B, C with the opposite sides. Then triangles ABD and CAD are similar, and $\overline{BD}/\overline{DC} = -BD/CD = -(BD)^2/(\overline{BD})(\overline{CD}) = -(BD)^2/(AD)^2 = -c^2/b^2$, etc. Or obtain the result as a special case of Pascal's " mystic hexagram " theorem, by regarding ABC as a hexagon with two vertices coinciding at A, two at B, and two at C. Or use the fact that, by Problem 2, Section 2.4, the given triangle and the triangle formed by the tangents are copolar.

11. Use Desargues' two-triangle theorem.

12. Use the trigonometric form of Menelaus' Theorem, or use Problem 12, Section 2.1.

13. Use the trigonometric form of Menelaus' Theorem.

14. Let DD', $A'B$ intersect in E and $A'D'$, BD in G. Then, using Menelaus' Theorem on triangle $GA'B$ and transversal $D'DE$ we find

$$(\overline{BD}/\overline{DG})(\overline{GD'}/\overline{D'A'})(\overline{A'E}/\overline{EB}) = -1.$$

But $\overline{BD} = \overline{CA}$, $\overline{GD'} = \overline{BB'}$, $\overline{DG} = \overline{AA'}$, $\overline{D'A'} = \overline{B'C}$. Therefore

$$(\overline{CA}/\overline{AA'})(\overline{A'E}/\overline{EB})(\overline{BB'}/\overline{B'C}) = -1.$$

15. Apply Menelaus' Theorem to triangle ABD with transversal EG and to triangle CBD with transversal HF.

16. Triangles CAC' and $B'AB$ are congruent and therefore sin ACC'/sin $B'BA$ = sin ACC'/sin $CC'A$ = AC'/CA = AB/CA, etc.

18. Let CE cut AB in T, AD cut CB in S, and the altitude from B cut AC in R. Then $AT/TB = EA/BC = AB/BC$, $BS/SC = AB/CD = AB/BC$, $CR/RA = (BC/AB)^2$.

19. Let t_A, t_B, t_C be the tangent lengths from D, E, F. Then $t_A^2 = (\overline{BD})(\overline{CD})$, $t_B^2 = (\overline{CE})(\overline{AE})$, $t_C^2 = (\overline{AF})(\overline{BF})$. But, by Menelaus' Theorem,

$$-1 = (\overline{BD}/\overline{DC})(\overline{CE}/\overline{EA})(\overline{AF}/\overline{FB}).$$

It now follows that $-(t_A t_B t_C)^2 = -[(AF)(BD)(CE)]^2$.

20. Multiply together results obtained by applying Menelaus' Theorem to triangles ABC, FBD, ECD, ECD, AFE, AFE, FBD with transversals DEF, AX, AX, BY, BY, CZ, CZ respectively.

21. Let AX, BY, CZ, cut BC, CA, AB in A', B', C'; let PO, QO, RO cut QR, RP, PQ in P', Q', R'. Then we have

$$(\overline{QP'}/\overline{P'R})(\overline{RQ'}/\overline{Q'P})(\overline{PR'}/\overline{R'Q}) = 1,$$

$$(\overline{QX}/\overline{XR})(\overline{RY}/\overline{YP})(\overline{PZ}/\overline{ZQ}) = 1,$$

$$(\overline{CP}/\overline{PB})(\overline{BR}/\overline{RA})(\overline{AQ}/\overline{QC}) = 1.$$

But, by Theorem 2.1.9,

$$(\overline{QP'}/\overline{P'R})/(\overline{QX}/\overline{XR}) = (\overline{CP}/\overline{PB})/(\overline{CA'}/\overline{A'B}),$$

$$(\overline{RQ'}/\overline{Q'P})/(\overline{RY}/\overline{YP}) = (\overline{AQ}/\overline{QC})/(\overline{AB'}/\overline{B'C}),$$

$$(\overline{PR'}/\overline{R'Q})/(\overline{PZ}/\overline{ZQ}) = (\overline{BR}/\overline{RA})/(\overline{BC'}/\overline{C'A}).$$

Setting the product of the left members of the above three equations equal to the product of the right members, it now follows that

$$(\overline{CA'}/\overline{A'B})(\overline{BC'}/\overline{C'A})(\overline{AB'}/\overline{B'C}) = 1.$$

22. We have $(\overline{B1}/\overline{1C})(\overline{CB'}/\overline{B'A})(\overline{A2}/\overline{2B}) = -1$, $(\overline{BA'}/\overline{A'C})(\overline{C3}/\overline{A3})(\overline{A2}/\overline{2B}) = -1$, $(\overline{AC'}/\overline{C'B})(\overline{B4}/\overline{4C})(\overline{C3}/\overline{3A}) = -1$, $(\overline{CB'}/\overline{B'A})(\overline{A5}/\overline{5B})(\overline{B4}/\overline{4C}) = -1$, $(\overline{BA'}/\overline{A'C})(\overline{C6}/\overline{6A})(\overline{A5}/\overline{5B}) = -1$, $(\overline{AC'}/\overline{C'B})(\overline{B7}/\overline{7C})(\overline{C6}/\overline{6A}) = -1$. Multiply together the first, fifth, and third of these equations; divide by the product of the other three; obtain $(\overline{B1}/\overline{C1})(\overline{C7}/\overline{B7}) = 1$, whence $1 \equiv 7$.

Section 2.5

3. On a line other than l through C construct A' and B' such that $\overline{CA'}/\overline{CB'} = r$. Let AA' and BB' intersect in D'. Through D' draw the parallel to CB' to intersect l in D.

4. (a) Expand.
   (b) Expand.

5. Insert an origin at O in the expansions of the two cross ratios. Now show that
$$(\overline{OB} - \overline{OA})(\overline{OB'} - \overline{OC})(\overline{OA'} - \overline{OC'}) = (\overline{OB'} - \overline{OA'})(\overline{OB} - \overline{OC'})(\overline{OC} - \overline{OA}).$$

6. By Theorem 2.5.2 (3), $(AC,BP) = 1 - m$, $(AC,BQ) = 1 - n$.

8. If A, B and C, D separate each other then $(AB,CD) = r$, where $-\infty < r < 0$. Define θ such that $r = -\cot^2 \theta$. Let the semicircles on AB and CD intersect in V. Show that the angle of intersection of the two semicircles is 2θ.

Section 2.6

3. Let A', B', C', D', P' be the points of contact of the tangents a, b, c, d, p, respectively. Let O be the center of the circle and K any fixed point on the circle. Then $\angle AOP' = (\angle A'OP')/2 = \angle A'KP'$, etc. It follows that $(AB,CD) = O(AB,CD) = K(A'B',C'D')$.

4. Let FA and DC intersect in T, BC and ED in U, BC and FE in V, and AF and DE in R. Consider the four tangents DE, AB, EF, CD. By Problem 3, Section 2.6, $(RA,FT) = (UB,VC)$, whence $D(EA,FC) = D(RA,FT) = E(UB,VC) = E(DB,FC)$. It now follows, by Corollary 2.6.3, that DA, EB, FC are concurrent.

5. Expand and use Menelaus' Theorem.

6. Set $(MN, AA') = k$. Expand and insert an origin at M. Solve for $\overline{MA'}$ to obtain $\overline{MA'} = (\overline{MA})(\overline{MN})/[k\overline{MN} + (1 - k)\overline{MA}]$, with similar expressions for $\overline{MB'}$, $\overline{MC'}$, $\overline{MD'}$. Now expand $(A'B', C'D')$; insert an origin at M; substitute the expressions for $\overline{MA'}$, $\overline{MB'}$, $\overline{MC'}$, $\overline{MD'}$; simplify to obtain (AB,CD).

7. The proof of Theorem 2.6.9 as given in the text applies here.

8. Let CC', $A'B$, $B'A$ be concurrent in P; let BB', $C'A$, $A'C$ be concurrent in Q. Let CB' and $B'C$ intersect in O, $PB'A$ cut $OC'B$ in X, PCC' cut QBB' in Y, and $PA'B$ cut $QC'A$ in Z. Then $A(OX,C'B) = B'(OX,C'Q) = B'(CP,C'Y) = Q(A'P,ZB) = A(A'X,C'B)$. It now follows that O, A, A' are collinear.

Section 2.7

2. Apply Theorem 2.7.3 (2) to the homographic pencils $A(X')$ and $B(X)$.

3. Apply Theorem 2.7.3 (2) to the homographic pencils $R(A)$ and $Q(A)$.

4. Apply Theorem 2.7.3 (1) to the homographic ranges (B) and (C).

5. (L) and (M) are homographic ranges. $F(L)$ and $H(M)$ are homographic pencils having FH as a common line.

6. $B(Y)$ and $C(X)$ are homographic pencils having BC as a common line.

7. (B) and (R) are homographic ranges having E as a common point.

8. Let Ω and Ω' be the infinite points on ranges (A) and (A') respectively. Then $(AX,I\Omega) = (A'X',\Omega'J')$. That is $\overline{AI}/\overline{IX} = \overline{J'X'}/\overline{A'J'}$. Therefore $(\overline{IX})(\overline{J'X'}) = (\overline{IA})(\overline{J'A'})$, a constant.

9. Let A and A' be any positions of X and X'. Then $(\overline{IX})(\overline{J'X'}) = (\overline{IA})(\overline{J'A'})$. Hence, reversing the steps in the proof of Problem 8, Section 2.7, we get $(AX,I\Omega) = (A'X',\Omega'J')$. Or, since $(\overline{IX})(\overline{J'X'}) =$ constant, it immediately follows (from Theorem 2.7.4) that $(X) = (X')$. (This is the converse of Problem 8, Section 2.7.)

10. (a) Pencils $V(P)$ and $V(P')$ are congruent.
    (b) Apply Problem 8, Section 2.7.

11. Let the polygon by $A_1A_2 \ldots A_{n-1}A_n$, and let the sides A_1A_2, A_2A_3, $\ldots$, $A_{n-1}A_n$ each pass through a fixed point. Then $(A_1) = (A_n)$.

12. Parallel the proof of Problem 11, Section 2.7.

13. Use Theorem 2.6.7 and the facts that $AC/A'C' = VA/VC'$, $DB/D'B' = VB/VD'$, $CB/C'B' = VB/VC'$, $AD/A'D' = VA/VD'$.

Section 2.8

1. (a) $AC/CB = AP/MB = AP/BN = AD/DB$.
   (b) If O is the center of the semicircle, $(\overline{OC})(\overline{OD}) = (OT)^2 = (OB)^2$.
   (c) Apply Theorem 2.4.1.

2. $(OB)^2 = (\overline{OC})(\overline{OD}) = (\overline{O'C} - \overline{O'O})(\overline{O'D} - \overline{O'O}) = (\overline{OO'} + \overline{O'C})(\overline{OO'} - \overline{O'C})$.

3. (a) The angles in the pencil are multiples of $45°$.
   (b) Let AB be the diameter, CD the chord, and P a point on the circle. Then $P(AB,CD) = A(TB,CD)$, where AT is the tangent at A. But $A(TB,CD) = (\sin \overline{TAC}/\sin \overline{CAB})/(\sin \overline{TAD}/\sin \overline{DAB})$.
   (c) $(AC,LM) = B(AC,LM) = B(AC,DE)$. Now use part (b).
   (d) Use part (b).

4. (a) Use Theorem 2.8.3.
   (b) Show that PA bisects $\angle QPR$.
   (c) We have $AU/UT = [AB \cos(A/2)]/[BT \sin TBU]$ and also $AV/VT = [AC \cos(A/2)]/[TC \sin TCV]$, $AB/AC = BT/TC$, $\angle TBU = \angle TVC$.

5. Use Problem 3(b), Section 2.8.

6. By Theorem 2.1.9 show that $AC/CB = PA/PB = AD/DB$.

7. Use Theorem 2.8.11 and Problem 6, Section 2.8.

8. Draw any circle S cutting the circles drawn on AB and CD as diameters and let O be the intersection of the common chords of S and each of the other two circles. The circle with center O and radius equal to the tangent from O to either of the circles on AB and CD as diameters cuts the line $ABCD$ in the required points P and Q. The construction fails if the pairs A,B and C,D separate each other.

9. Use the theorems of Ceva and Menelaus.

10. $(TA)(TC)/(SA)(SC) = (\triangle ATC)/(\triangle ASC) = TU/SU = TV/SV = (\triangle BTD)/(\triangle BSD) = (TB)(TD)/(SB)(SD)$.

11. By Theorem 2.8.4, $2/\overline{AB} = 1/\overline{AC} + 1/\overline{AD} = (\overline{AD} + \overline{AC})/(\overline{AC} \cdot \overline{AD}) = 2\overline{AO}/(\overline{AC} \cdot \overline{AD})$.

12. $(\overline{OP})(\overline{OP'}) = (OB)^2 = (\overline{OQ})(\overline{OQ'})$.

13. Let AB cut PQ in O. Then O is the midpoint of PQ and $(\overline{OL})(\overline{OM}) = (\overline{OQ})^2$.

14. Let R and r be the radii of the semicircle and Σ respectively. Draw a concentric semicircle of radius $R - r$ and note that r is the geometric mean of $AC - r$ and $CB - r$.

15. Use Corollary 2.6.1.

16. We have $\overline{AC}/\overline{CB} = -\overline{AD}/\overline{DB}$, consequently $(AC)^2/(AD)^2 = (CB)^2/(DB)^2 = (\overline{OB} - \overline{OC})^2/(\overline{OB} - \overline{OD})^2 = (\overline{OD} \cdot \overline{OB} - \overline{OD} \cdot \overline{OC})^2/(OD)^2(\overline{OB} - \overline{OD})^2 = (\overline{OD} \cdot \overline{OB} - \overline{OB} \cdot \overline{OB})^2/(OD)^2(\overline{OB} - \overline{OD})^2 = (OB)^2/(OD)^2 = (\overline{OC})(\overline{OD})/(OD)^2 = \overline{OC}/\overline{OD}$.

17. We have $(DB, PV) = -1$, whence (by Problem 16, Section 2.8) $\overline{XP}/\overline{XV} = (DP)^2/(DV)^2$, with similar relations for Y and Z. Now apply Menelaus' Theorem to triangle PVU.

18. Draw the diametral secant AMN and let it cut the chord of contact $T'T$ of the tangents from A in R. Now, if O is the center of the given circle, $(\overline{OR})(\overline{OA}) = (OT)^2$. It follows that if circle Σ on AB as diameter cuts the given circle in P and Q, then OP is tangent to Σ. Thus, if S is the center of Σ, SP is tangent to the given circle. Therefore $(\overline{SC})(\overline{SD}) = (SP)^2 = (SB)^2$, and $(AB,CD) = -1$.

19. Take O, not on line ABC, and draw OA, OB, OC, OP, OQ, OR. Through A draw AM parallel to OQ, and let OB, OC, OP, OR cut AM in B', C', P', R'. Now B' is the midpoint of AC', P' is the first trisection point of $B'C'$, R' is the second trisection point of AB'. It follows that R' is the midpoint of AP', whence $O(QR,PA) = -1$.

20. The circle is a circle of Apollonius for PP' and for QQ'. It follows that if V is any point on this circle, then VB bisects angles PVP' and QVQ'.

22. Since $(AB,PQ) = -1$, we have $2/\overline{AB} = 1/\overline{AP} + 1/\overline{AQ}$. Therefore $2\overline{OB}/\overline{AB} = 2(\overline{OA} + \overline{AB})/\overline{AB} = 2\overline{OA}/\overline{AB} + 2 = (\overline{OA}/\overline{AP} + 1) + (\overline{OA}/\overline{AQ} + 1) = (\overline{OA} + \overline{AP})/\overline{AP} + (\overline{OA} + \overline{AQ})/\overline{AQ} = \overline{OP}/\overline{AP} + \overline{OQ}/\overline{AQ}$.

23. Take E and E' as a fixed pair of harmonic conjugates. Now ranges (F) and (F') are homographic. Therefore pencils $E(F)$ and $E'(F')$ are homographic. But EE' is a common line of these two pencils. It follows that the intersections of EF and $E'F'$ lie on a line. But B and C are two of these intersections. Therefore EF and $E'F'$ intersect on line BC.

24. Let R be the harmonic conjugate of O with respect to P and Q. Then the locus of R is a straight line n through the intersection of the two given lines, and $2/\overline{OR} = 1/\overline{OP} + 1/\overline{OQ}$. Now draw line n' parallel to n and through the midpoint R' of OR. Then $1/\overline{OR'} = 1/\overline{OP} + 1/\overline{OQ}$, and n' is the sought line.

25. Replace $1/\overline{OP_1} + 1/\overline{OP_2}$ by $1/\overline{OX_1}$, $1/\overline{OX_1} + 1/\overline{OP_3}$ by $1/\overline{OX_2}$, and so on, using Problem 24, Section 2.8.

Section 2.9

2. Use Theorem 2.8.5.

3. Let the circles have centers O and O' and intersect in P. The circles are orthogonal if and only if triangle OPO' is a right triangle—that is, if and only if $[(c/2)d]/2 = (rr')/2$.

4. Use Theorem 2.9.2 (3).

5. Let M be the center of the given circle and N the center of circle CPQ. Now $\measuredangle PCQ = \measuredangle PNM$ and $\measuredangle PAQ = \measuredangle PMN$. Therefore $\measuredangle PNM + \measuredangle PMN = \measuredangle PCQ + \measuredangle PAQ = 180° - \measuredangle CQA = 90°$. It follows that triangle NPM is a right triangle and the two circles are orthogonal.

6. This is a converse of Problem 5, Section 2.9, and may be established by reversing the proof of Problem 5.

7. Use Problem 5, Section 2.9.

8. By symmetry the center of the circle through the four points must be the midpoint of the line of centers of the two given circles.

9. Let R and S be the centers of circles AMO and BMO respectively. Then $\measuredangle ROM = (\measuredangle AOM)/2$ and $\measuredangle SOM = (\measuredangle BOM)/2$. Therefore $\measuredangle ROS = 90°$.

10. $(BH)(BE) = (BD)(BC)$ and $(CH)(CF) = (CD)(CB)$. Therefore $(BH)(BE) + (CH)(CF) = (BC)^2$.

Section 2.10

1. The tangents drawn from the radical center to the three circles are all equal in length.

3. Use Theorems 2.10.6 and 2.10.7.

4. The midpoint of a common tangent to two circles has equal powers with respect to the two circles.

5. P is the radical center of the three circles.

6. (a) The altitudes are common chords of pairs of the circles.
   (b) Let A', B', C' be the feet of the altitudes and H the orthocenter. Then AA', BB', CC' are chords of the three circles and $(\overline{AH})(\overline{HA'}) = (\overline{BH})(\overline{HB'}) = (\overline{CH})(\overline{HC'})$.

7. (a) Draw the diameter of C_1 through the center of C_2.
   (b) Draw the chord of C_1 perpendicular to the diameter of C_1 through P.
   (c) Use part (b) and the fact that P has equal powers with respect to the three circles.
   (d) If circle $M(m)$ bisects the given circles $A(a)$ and $B(b)$, then $(MA)^2 + a^2 = m^2 = (MB)^2 + b^2$, whence $(MA)^2 - (MB)^2 = b^2 - a^2$.
   (e) The center of the given circle has equal powers with respect to all the bisecting circles.

8. Use Theorem 2.10.6 (1).

9. (a) The center of the sought circle is the radical center of the two given circles and the given point.
   (b) If the two given circles are intersecting, use part (a) and Theorem 2.10.11.

10. The required circle is the radical circle of the three circles having the given points as centers and the given tangent lengths as radii.

11. See Art. 113 in Johnson's *Modern Geometry*, or Art. 471 in Altshiller-Court's *College Geometry* (listed in bibliography of Chapter 2).

12. (a) A circle concentric with the given circle.
   (b) A circle whose center is midway between the centers of the given circles.
   (c) A straight line parallel to the radical axis of the given circles. Use Problem 11, Section 2.10.
   (d) A circle coaxial with the given circles. Use Problem 11, Section 2.10.

13. (a) Use Theorem 2.10.9 (3).

    (b) Use Theorem 2.10.11.

14. The point of concurrence is the radical center of the given circle and any two circles of the coaxial pencil.

15. If the three circles are distinct, their radical axes must be concurrent.

Section 3.1

1. (a) Onto and one-to-one.

   (b) Not onto; 3 is not the image of any element of A.

   (c) Not onto; 2 is not the image of any element of A.

   (d) Not onto; no even integer is the image of any element of A.

   (e) Onto and one-to-one.

   (f) Onto and one-to-one.

3. Let r' be any real number and let r be a real root of $x^3 - x = r'$. Then $r \to r'$, whence the mapping is onto. The mapping is not one-to-one since $0 \to 0$ and $1 \to 0$.

4. (b) $n \to 4n^2 + 12n + 9$, $n \to 2n^2 + 3$, $n \to n^4$, $n \to 4n + 9$, $n \to 4n^4 + 12n^2 + 9$, $n \to 4n^4 + 12n^2 + 9$.

7. $S = IS = (T^{-1}T)S = T^{-1}(TS) = T^{-1}I = T^{-1}$.

8. Since $(T_2 T_1)(T_1^{-1}T_2^{-1}) = I$ it follows (by Theorem 3.1.11) that $T_1^{-1}T_2^{-1} = (T_2 T_1)^{-1}$.

9. $T^{-1}T = I$. Therefore, by Theorem 3.1.11, $T = (T^{-1})^{-1}$.

11. $(T_3 T_2 T_1)^{-1} = [(T_3 T_2)T_1]^{-1} = T_1^{-1}(T_3 T_2)^{-1} = T_1^{-1}(T_2^{-1}T_3^{-1}) = T_1^{-1}T_2^{-1}T_3^{-1}$.

12. (a) $(STS^{-1})^{-1} = (S^{-1})^{-1}T^{-1}S^{-1} = ST^{-1}S^{-1}$.

    (b) $(ST)(TS)(ST)^{-1} = (ST)(ST)(T^{-1}S^{-1}) = ST$.

    (c) $(ST_2 S^{-1})(ST_1 S^1) = S(T_2 T_1)S^{-1}$.

13. G1 is satisfied by Definition 3.1.5, G2 by Theorem 3.1.9, G3 by Theorem 3.1.10 (1), and G4 by Theorem 3.1.10(2).

Section 3.2

1. (a) (2,1), (b) $(-1,1)$, (c) $(1,-1)$, (d) $(1,-1)$, (e) $(-1,-1)$, (f) $(2,-2)$ (g), $(4,-2)$, (h) $(1,1/2)$, (i) $(2,-1)$, (j) $(-2,2)$, (k) $(-2,2)$, (l) $(2-2\sqrt{2},0)$.

2. No, for all parts.

3. If $A = A'$, then $O = A$. If $B = B'$, then $O = B$. If $A \neq A'$, $B \neq B'$, and AA' is not parallel to BB', then O is the point of intersection of the perpendicular bisectors of AA' and BB'. If $A \neq A'$, $B \neq B'$, AA' is parallel to BB', and AB is not collinear with $A'B'$, then O is the point of intersection of lines AB and $A'B'$. If AB is collinear with $A'B'$, then O is the common midpoint of AA' and BB'.

4. Obvious from a figure.

5. Place an x axis on BC with origin at the midpoint of BC.

8. Obvious from a figure.

9. (a), (b), (c) Obvious from a figure.

10. (a) The midpoints of the sides of a quadrilateral form a parallelogram.
    (b) Use part (a).

11. Obvious from a figure.

12. Let $R(O,\theta)$ carry P and Q into P'' and Q''; let $R(O',-\theta)$ carry P'' and Q'' into P' and Q'. Now the angle from directed line PQ to directed line $P''Q''$ is θ, and the angle from directed line $P''Q''$ to directed line $P'Q'$ is $-\theta$. It follows that $P'Q'$ is parallel to PQ. Also, $P'Q' = PQ$.

13. On line O_1O_2 such that $\overline{O_2 P} = (1 - k_1)\overline{O_1O_2}/(k_1 - k_2)$.

14. Let $T(AB)$ carry P into P_1, $R(l_1)$ carry P_1 into P_2, $T(CD)$ carry P_2 into P_3, $R(l_2)$ carry P_3 into P', $R(l_2)$ carry P_2 into P_4. Then $T(CD)$ carries P_4 into P'. From the figure it is easily seen that $P_2 P_3 P'P_4$ is a rectangle and that $\angle P_2 OP_4 = 2\theta$.

Section 3.3

2. Let a common tangent to circles $A(a)$ and $B(b)$ cut the line of centers AB in point S. Draw the radii to the points of contact of the common tangent. Then, from similar triangles, $SA/SB = a/b$, etc.

5. Use Theorem 3.3.9.

6. Let P be any point on the circle of similitude of the two given circles $A(a)$, $B(b)$. Then $PA/PB = a/b$, or $(PA^2 - a^2)/(PB^2 - b^2) = a^2/b^2$. Now use Problem 12 (d), Section 2.10. For a direct proof not depending upon Casey's Power Theorem, see Daus, *College Geometry*, p. 91.

7. Use the theorem of Menelaus.

8. This is a special case of Problem 7, Section 3.2.

9. Its distance from A is $k^2c/(k^2 - 1)$, where $k = a/b$.

10. Use Theorem 2.9.3.
    (b) We have, power of $A_1 = (\overline{A_1M_2})(\overline{A_1H_2}) = (\overline{A_1A_3})(\overline{A_1H_2})/2$. Similarly, power of $A_1 = (\overline{A_1A_2})(\overline{A_1H_3})/2$. It follows that the power of A_1 is $[(\overline{A_1A_3})(\overline{A_1H_2}) + (\overline{A_1A_2})(\overline{A_1H_3})]/4$.

12. Let the given triangle ABC have altitudes AD, BE, CF, orthocenter H, circumcenter O, and nine-point center N. Let the tangential triangle $A'B'C'$ have circumcenter O'. Now triangles DEF and $A'B'C'$ are homothetic. But H is the incenter (or an excenter) of triangle DEF and O is the incenter (or corresponding excenter) of triangle $A'B'C'$. N is the circumcenter of triangle DEF and O' is the circumcenter of triangle $A'B'C'$. It follows that H, O, N, O' are collinear.

13. Let Y_1, Y_2, Y_3 be the images of X_1, X_2, X_3 under the homothety $H(H,2)$, which carries the nine-point circle into the circumcircle. Show that triangle $Y_1 Y_2 Y_3$ is equilateral.

14. If the Euler line is parallel to side $A_2 A_3$ we have $OM_1 = (A_1 H_1)/3$, or $R \cos A_1 = (2R \sin A_2 \sin A_3)/3$, R the circumradius. That is, $\cos A_1/\sin A_2 \sin A_3 = 2/3$. But $\cos A_1 = -\cos(A_2 + A_3) = \sin A_2 \sin A_3 - \cos A_2 \cos A_3$, etc. Also see Problem E 259, *The American Mathematical Monthly*, Oct. 1937.

15. The trilinear polar of the orthocenter is the radical axis of the circumcircle and the nine-point circle.

Section 3.4

2. (a), (b) Use Theorem 3.4.7.

3. (a) Use Theorems 3.4.4 and 3.4.5.
   (b) Use Theorem 3.4.9.
   (c) Use Theorems 3.4.7 and 3.4.9.

4. (a) Obvious from a figure.

5. Use Theorem 3.4.9 and Lemma 3.4.8.

6. Obvious from a figure.

7. Easily shown from a figure.

8. Use Lemma 3.4.8.

9. (a) Use Theorem 3.4.9.
   (b) For brevity represent $R(l_1)$, $R(l_2)$, $R(l_3)$, by R_1, R_2, R_3 respectively. By Theorem 3.4.10, $R_1 R_3 R_2$ and $R_3 R_2 R_1$ are glide-reflections, whence $(R_1 R_3 R_2)^2$ and $(R_3 R_2 R_1)^2$ are translations. Since translations commute we have

   $$(R_1 R_3 R_2)^2 (R_3 R_2 R_1)^2 = (R_3 R_2 R_1)^2 (R_1 R_3 R_2)^2,$$

   or

   $$R_1 R_3 R_2 R_1 R_3 R_2 R_3 R_2 R_1 R_3 R_2 R_1 = R_3 R_2 R_1 R_3 R_2 R_1 R_1 R_3 R_2 R_1 R_3 R_2,$$

   or, since $R_1 R_1 = I$,

   $$R_1(R_3 R_2 R_1 R_3 R_2)^2 R_1 = (R_3 R_2 R_1 R_3 R_2)^2,$$

   or

   $$R_1(R_3 R_2 R_1 R_3 R_2)^2 = (R_3 R_2 R_1 R_3 R_2)^2 R_1.$$

   Since R_1 and $(R_3 R_2 R_1 R_3 R_2)^2$ commute, it follows that $(R_3 R_2 R_1 R_3 R_2)^2$, which must be a translation or a rotation, is a translation along l_1.

   (c) $G(l_3, A_1 A_2)G(l_2, A_3 A_1)G(l_1, A_2 A_3)$
   $$= R_3 T(A_1 A_2)R_2 T(A_3 A_1)R_1 T(A_2 A_3)$$
   $$= R_3 T(A_1 A_2)T(A_3 A_1)R_2 R_1 T(A_2 A_3)$$
   $$= R_3 T(A_3 A_2)R_2 R_1 T(A_2 A_3)$$
   $$= R_3 T(A_3 A_2)R(A_3, 2 \nleftrightarrow A_3)T(A_2 A_3)$$
   $$= R_3 R(A_2, 2 \nleftrightarrow A_3).$$

   (d) Use Theorem 3.4.5.
   (e) $R_3 R_2 R_1$, being an opposite isometry, is a glide-reflection. To find which glide-reflection it is, consider the fates of H_1 and H_3.

10. $R(l_3)R(l_2)R(l_1)$ is an opposite isometry, and therefore (by Theorem 3.4.9) a reflection in a line or a glide-reflection. But O is an invariant point.

Section 3.5

2. (a), (b) Use Theorems 3.5.3 and 3.5.4.

3. (a) Use Theorem 3.4.9.
   (b) Use Theorem 3.5.4.

4. (a), (b) Obvious from a figure.

5. If the isometry is opposite, it is a reflection or a glide-reflection; in either case the midpoints of segments PP' lie on the axis of the transformation. If the isometry is direct, the result follows from Theorem 3.5.6, the locus degenerating to a point if the isometry is a half-turn.

6. The maps are related by a direct nonisometric similarity, that is, by a homology.

7. (a) A translation or a homology.
   (b) A glide-reflection or a stretch-reflection.

8. Consider a triangle ABC; it is carried into a triangle $A'B'C'$ whose sides are parallel to the corresponding sides of triangle ABC. It follows that the two triangles are coaxial (on the line at infinity), and thus (by Desargues' two-triangle theorem) also copolar at a point P. If P is an ideal point, the similarity is the translation $T(AA')$; if P is an ordinary point, the similarity is the homothety $H(\overline{P,A'B'}/AB)$.

10. Use Theorem 3.5.6.

Section 3.6

1. (a) The figure consists of certain parts of four concurrent circles.
   (b) The figure consists of certain parts of two circles and of two lines.

2. (a) A system of coaxial intersecting circles.
   (b) A system of coaxial tangent circles.

4. (a) Let OB cut the given circle again in M. Then $(\overline{OM})(\overline{OB}) = (\overline{OC})(\overline{OC'})$, or $\overline{OB} = (\overline{OC})(\overline{OC'})/2(\overline{OC}) = \overline{OC'}/2$.
   (b) Denote the center of inversion by O and the center of K' by M, and let OM cut K in S and T and K' in S' and T', where S and S', and T and T', are corresponding points under the inversion. Now $(\overline{OC})(\overline{OC'}) = (\overline{OS})(\overline{OS'}) = (\overline{OT})(\overline{OT'})$ and

$$(\overline{MO})(\overline{MC'}) = \overline{OM}(\overline{OM} - \overline{OC'}) = \overline{OM}[(\overline{OS'} + \overline{OT'})/2 - (\overline{OT})(\overline{OT'})/\overline{OC}]$$
$$= \overline{OM}[(\overline{OS'} + \overline{OT'})/2 - 2(\overline{OT})(\overline{OT'})/(\overline{OS} + \overline{OT})]$$
$$= \overline{OM}[(\overline{OS'} + \overline{OT'})(\overline{OS} + \overline{OT}) - 4(\overline{OT})(\overline{OT'})]/2(\overline{OS} + \overline{OT})$$
$$= \overline{OM}(\overline{OT} - \overline{OS})(\overline{OS'} - \overline{OT'})/4\overline{OC} = \overline{OM}(2\overline{CT})(2\overline{MS'})/4\overline{OC}$$
$$= (\overline{MS'}/\overline{CT})(\overline{CT})(\overline{MS'}) = (MS')^2.$$

There are much easier ways to solve this problem; see, e.g., Problem 12, Section 3.7.
   (d) Let the centers of the circles be C and C' and let their common chord ST cut CC' in M. Then $(\overline{CM})(\overline{CC'}) = (CS)^2$.

Section 3.7

1. Invert with respect to O.

2. Invert with respect to A.

3. (a) If C' is any point on a semicircle having diameter $A'OB'$, the circles $A'OC'$ and $B'OC'$ are orthogonal.

(b) If B', C', D' are three collinear points and A is a point not collinear with them, the circles $AB'D'$ and $AC'D'$ intersect the lines AB' and AD', respectively, in equal or supplementary angles.

4. Invert with respect to A.

5. Invert with respect to M.

6. Invert with respect to any one of the four points.

7. Invert with respect to T.

8. Invert with respect to a point of intersection of K and K_2.

9. Invert with respect to A.

10. Invert with respect to the given point and discover that the sought locus is the circle through the given point and orthogonal to the system of coaxial circles; its center lies on the common tangent of the system and passes through the common point of the system.

11. Invert with respect to C.

Section 3.8

1. Consider two circles orthogonal to C_1 and C_2.

2. Invert with respect to O, then use Menelaus' Theorem and Theorem 3.7.5.

3. Invert with respect to a point on C, and show that $t_{ij}^2/r_i r_j$ is invariant.

4. (a) Let P be any point. Let P' be the inverse of P for circle $A(a)$ and let P'' be the inverse of P' for circle $B(b)$. Let Q be the inverse of P for circle $B(b)$. Now invert the figure with respect to circle $B(b)$. Since circle $A(a)$ inverts into itself (Theorem 3.6.4) and P and P' are inverse points for circle $A(a)$, it follows (by Theorem 3.7.4) that Q and P'' are also inverse points for circle $A(a)$.

(b) Invert the figure in circle K_2 and use Theorem 3.7.4.

5. Invert with respect to D. Then $A'C' + C'B' = A'B'$. But, if p' is the perpendicular distance from D to line $A'B'C'$, we have $A'C'/p' = AC/r$. Similarly, $C'B'/p' = CB/q$, $A'B'/p' = AB/p$, etc.

6. Let a secant through O cut C in A and B and cut C' in A' and B', where A,A' and B,B' are pairs of inverse points. Then $pp' = (\overline{OA})(\overline{OB})(\overline{OA'})(\overline{OB'}) = r^4$.

7. Invert the coaxial system into a system of concentric circles, a system of concurrent lines, or a system of parallel lines.

8. (b) Let A and A' be a pair of antinverse points for circle K and let J be any circle through A and A'. Let O be the center of K and let P and Q be the points of intersection of K and J. Let PO cut J again in R. Then $(\overline{OP})(\overline{OR}) = (\overline{OA})(\overline{OA'}) = -r^2$, where r is the radius of K. It follows that $P \equiv Q$.

9. Let triangles PQR and UVW, inscribed in a circle K, be copolar at C. Let U', V', W' be the inverse of U, V, W for C as center of inversion and r^2 as power of inversion. Let p denote the power of C with respect to circle K. Then $\overline{CU'}/\overline{CP} = r^2/(\overline{CU})(\overline{CP}) = r^2/p$.

10. (a) Such a point is the center of a circle orthogonal to both of the given circles. Invert with respect to this circle.

(b) When the radical center is outside all three circles.

11. Use Theorem 3.7.7.

12. Invert with respect to any point on their radical circle.

13. Use Theorem 3.7.5 and some circles of Apollonius.

14. Use Problem 13, Section 3.8.

15. Use Theorem 3.7.4 (2).

16. Use Ptolemy's Theorem.

17. Subject the figure to the inversion $I(A,1)$.

18. Invert with respect to D and then apply Stewart's Theorem.

19. (b) Let C_1 be the circle orthogonal to circle $ABCD$ and passing through A and C; let C_2 be the circle orthogonal to circle $ABCD$ and passing through B and D. Let C_1 and C_2 intersect in X and Y. Invert with respect to X.

22. We have $OP = (BD)(AO)/AB$ and $OP' = (AC)(OB)/AB$, whence $(OP)(OP') = (BD)(AC)(AO)(OB)/(AB)^2$. But $(BD)(AC) = (AD)^2 - (AB)^2$.

23. If V is outside the circle, invert the circle into itself with respect to V as center of inversion. If V is inside the circle, invert the circle into its reflection in V.

Section 3.9

3. Let O and r be the center and radius of the circle and let Q' be the inverse of Q. Then $(PQ)^2 = (OP)^2 + (OQ)^2 - 2(OQ)(OQ') = (OP)^2 + (OQ)^2 - 2r^2$.

4. By Theorem 3.6.7, P and Q are inverse points for circle K_2.

5. (a), (b) Use Theorems 3.9.7, 3.6.7, and 2.9.3.

6. (a) Let Q be diametrically opposite P on circle R. Then (by Problem 5 (b), Section 3.9) P, Q are conjugate points for K_1, K_2, K_3. Therefore the poles of P for K_1, K_2, K_3 all pass through Q.
   (b) Draw the circle on PQ as diameter. Prove that this circle is orthogonal to the coaxial system. Then use Problem 5 (b), Section 3.9.
   (c) Use part (b).
   (d) Use part (b).
   (e) Use Problem 5 (a) and (b), Section 3.9.

7. Let P', Q' be the inverses of P, Q. Show that $OP'YQ$ is similar to $OQ'XP$.

Section 3.10

1. Line t is the polar of point T.

2. Let T be the point of contact of the tangent to the incircle. Then the poles, for the incircle, of AP, BQ, CR are the feet of the perpendiculars from T on the sides of the triangle determined by the points of contact of the incircle with the sides of triangle ABC.

7. (a) Use Problem 6, Section 3.10.
   (b) Use Problem 6, Section 3.10.
   (c) The inverse of each vertex of the triangle is the foot of the altitude through that vertex.

8. (a) Let AB and CD intersect in M and let the tangents to the circle at A and B intersect in N. Then $-1 = (NM,CD) = B(NM,CD) = B(BA,CD) = (BA,CD) = (AB,CD)$.

(b) By Theorem 2.6.7, $(AB,CD) = e(AC/CB)/(AD/DB)$. Therefore $(AC)(BD) = (BC)(AD)$. The rest follows by Ptolemy's Theorem.

9. (a) Use Theorem 3.10.2.

(b) Use part (a) and Problem 5 (a), Section 3.9.

11. (a) The pole, with respect to the incircle, of YZ is A. Let P' be the pole of AA'. Then, since YZ and AA' are conjugate lines, $A(BC,A'P') = -1$, whence AP' is parallel to BC. It now follows that the poles of AA', PX, YZ are collinear.

(b) Let the line through P parallel to BC cut YZ in S. Since YZ and PM are conjugate lines, $P(ZY,MS) = -1$. Therefore $(RQ,M\infty) = -1$ and M is the midpoint of RQ.

Section 3.11

2. (a), (b), (c) Yes.

4. There is a unique isometry that carries a given noncoplanar tetrad of points A, B, C, D into a given congruent tetrad A', B', C', D'.

5. (a) See the proof of Theorem 3.4.6.

(b) See the proof of Theorem 3.4.5.

7. Use Theorem 3.11.4.

8. Use Problem 7, Section 3, 11.

11. Generalize the proof of Theorem 3.6.4.

13. For the first part, generalize the proofs of Theorems 3.6.10, 3.6.11, 3.6.12, 3.6.13. The second part follows since the intersection of two "spheres" is a "circle."

15. Invert with respect to the point of contact of S_1 and S_2.

16. Generalize the proof of Theorem 3.9.3.

Section 4.1

2. (b) Let A be the given point and BC the given line segment. Construct, by Proposition 1, an equilateral triangle ABD. Draw circle $B(C)$, and let DB produced cut this circle in G. Now draw circle $D(G)$ to cut DA produced in L. Then AL is the sought segment.

3. It is a matter of existence; there exists a greatest triangle inscribed in a circle, but there does not exist a greatest natural number. To complete argument I, we must prove that a maximum triangle inscribed in a circle *exists*. The problem illustrates the importance in mathematics of existence theorems.

Section 4.2

1. (8) Let A and B be the two given points, P any point on the locus, and M the foot of the perpendicular from P on AB. Then $(MA)^2 - (MB)^2 = [(PM)^2 + (MA)^2] - [(PM)^2 + (MB)^2] = (PA)^2 - (PB)^2 = $ a constant, whence M is fixed. (9) Let A and B be the two given points, P any point on the locus, and O the midpoint of AB. Then, by the law of cosines,

$$(PA)^2 = (PO)^2 + (AO)^2 - 2(PO)(AO)\cos(AOP),$$
$$(PB)^2 = (PO)^2 + (BO)^2 - 2(PO)(BO)\cos(BOP).$$

It follows that $(PA)^2 + (PB)^2 = 2(PO)^2 + (AB)^2/2 = a$ constant, whence $PO =$ constant.

3. Use loci (1) and (2).

4. Let r be half the distance between the two parallel lines. Draw the circle of radius r with center at the given point, to intersect the line midway between the two parallel lines in the centers of the sought circles.

5. Use loci (5) and (1).

6. Locate the point as an intersection of some circles of Apollonius.

7. Denote the initial positions of the balls by A and B, and the center of the table by O. Construct the circle of Apollonius of A and B for the ratio $\overline{AO}/\overline{OB}$.

8. Let the given points be A and B and let the required chords be AC and BD. Let E, F be the midpoints of AB, CD; let O be the center of the circle. Then F lies on $O(E)$ and on $E(s/2)$, where $s = AC + BD$.

9. Let R and r be the radii of the given and required circles. Draw circles of radii $R \pm r$ concentric with the given circle.

10. Locate the vertex of the right angle by the method of loci.

11. Use loci (1) and (2).

12. Draw a line parallel to the given line at a distance from the given line equal to the radius of the given circle. Let P be the foot of the perpendicular from the given point to the drawn line. Find the center of the required circle as the intersection of the perpendicular to the given line at the given point, and the perpendicular bisector of PC, where C is the center of the given circle.

13. Draw a circle concentric with the given circle and having a radius equal to the other leg of a right triangle of hypotenuse equal to the tangent length and leg equal to the radius of the given circle.

14. In cyclic quadrilateral $ABCD$, let us be given angle A, AB, BD, AC. First construct triangle ABD and then find C by loci (1) and (6).

15. Locate the midpoint of the chord cut off by the circle on the required line by loci (2) and (5).

16. Use two circles of Apollonius.

17. It is a rectangle having the given lines as diagonal lines.

18. (a) Use the locus of Problem 17, Section 4.2.
    (b) Draw a rectangle having the given lines as diagonal lines and having one side tangent to the given circle.

Section 4.3

1. Translate one circle through a vector determined by the given line segment.

2. Use Problem 1, Section 4.3.

3. Reflect two adjacent sides of the quadrilateral in the given point.

4. Reflect one of the curves in the given line.

5. Use Problem 4, Section 4.3.

6. Let ABC be the sought triangle and let D, E, F be the given points on the sides BC, CA, AB respectively. Let $\overline{BD}/\overline{DC} = m/n$, $\overline{CE}/\overline{EA} = p/q$, $\overline{AF}/\overline{EB} = r/s$. Find F_1 on line FE such that $\overline{FE}/\overline{EF_1} = q/p$. Next find F_2 on line F_1D such that $\overline{F_1D}/\overline{DF_2} = n/m$. Then $F_2 F$ lies along the line AB, etc.

7. Use the method of similitude.

8. Let the given line and the given directrix meet in a point O; draw the line determined by O and the given focus F. Take any point on the given line as center and draw a circle tangent to the directrix. Now use O as a center of homothety.

9. Use the method of similitude.

10. Take any point D' on BA. Then take E'' on CA such that $CE'' = BD'$. Let circle $D'(B)$ cut the parallel to BC through E'' in E'. Draw a line through E' parallel to AC to cut BA in A' and BC in C'. We now have a figure homothetic to the desired figure, with B as center of homothety.

11. Use Problem 10, Section 4.3.

12. Subject C_2 to the homothety $H(O,k)$, where $k = OP_1/OP_2$.

13. Subject the given circle to the homothety $H(O, -k)$, where k is the given ratio.

14. Use Problem 12, Section 4.3.

15. Use the method of similitude.

16. Use the method of similitude.

17. Use General Problem 4.3.4.

18. Take any point O on one of the circles and let C_1 and C_2 denote the other two circles. Now use General Problem 4.3.4.

19. Let $(OP_1)(OP_2) = a^2$. Invert C_2 into C'_2 in the circle $O(a)$.

20. Use Problem 19, Section 4.3.

21. Use Problem 19, Section 4.3.

22. Invert with respect to the given point.

23. This is a special case of Problem 22, Section 4.3.

24. Increase the radius of each circle by half the distance between two of them, obtaining in this way three circles of which two are externally tangent to one another. Invert with respect to the point of contact of these two circles.

Section 4.4

1. Let A be an arbitrary point on m and let A' be the corresponding point on m'. Let PA cut m in A''. Then range (A'') is homographic to range (A'). If D is a double point of these two coaxial homographic ranges, PD is a desired line.

2. Find the double points of the homographic ranges cut by the two given pencils on line m.

3. Join the vertices V and V' of the pencils, and on VV' describe the circular arc containing the given angle. Let corresponding rays of the two homographic pencils cut this arc in points P and P'. We seek the double points of the concyclic homographic ranges (P) and (P').

4. Draw a line m parallel to the join of the vertices V and V' of the given pencils. Let corresponding lines of the two pencils cut m in A and A'. Mark off A'' on m such that $\overline{AA''} = \overline{VV'}$. Find the common points of the two coaxial homographic ranges (A') and (A'').

5. Let the two given lines be m and m' and the two given points be V and V'. Let A be any point on m and find A' and A'' on m' such that angles AVA' and $AV'A''$ are equal to the given angles. Now find the double points of the two coaxial homographic ranges (A') and (A'').

6. Let ABC be the given triangle, m the given line, and P any point on m. Let P' and P'' be points on m such that AP and AP' are isogonal lines for vertex A and BP and BP'' are isogonal lines for vertex B. Now find the double points of the two coaxial homographic ranges (P') and (P'').

7. Take any point A' on BC, draw $A'C'$ in the assigned direction to cut BA in C', draw $C'B'$ in the assigned direction to cut AC in B'. Draw $A'B'$ in the assigned direction to cut AC in B''. Now find the double points of the two coaxial homographic ranges (B') and (B'').

8. Take any point A on p and find the feet R and S of the perpendiculars from A on r and s. Find R' and S' on r and s such that RR' and SS' have the given projected lengths. Let the perpendicular to r at R' and the perpendicular to s at S' cut q in B and B' respectively. Now find the double points of the two coaxial homographic ranges (B) and (B').

9. Let P be the given point and m and n the given lines. Let A be any point on m and let PA cut n in A'. Mark off the given lengths AB and $A'B''$ on m and n respectively, and let PB cut n in B'. We seek the double points of the coaxial homographic ranges (B') and (B'').

10. Take any point A on m and let AP cut m' in A'. Find A'' on m' such that $OA/O'A''$ is the given constant. Now find the double points of the two coaxial homographic ranges (A') and (A'').

11. Parallel the solution suggested for Problem 10, Section 4.4.

12. Let m and m' be the two given lines, O their point of intersection, and P the given point. Take any point A on m and let AP cut m' in A'. Find A'' on m' such that triangle AOA'' has the given area. Now find the double points of the two coaxial homographic ranges (A') and (A'').

13. Let ABC be the given triangle and P, Q, R the given points. Let A' be any point on BC. Find B' on CA such that $\measuredangle A'RB'$ is equal to the first of the given angles; next find C' on AB such that $\measuredangle B'PC'$ is equal to the second of the given angles; then find A'' on BC such that $\measuredangle C'QA''$ is equal to the third of the given angles. We seek the double points of the coaxial homographic ranges (A') and (A'').

14. Let P, Q, R be the three given points and let A be any point on the circle. Draw AP to cut the circle again in B; next draw BQ to cut the circle again in C; then draw CR to cut the circle again in A'. We seek the double points of the concyclic homographic ranges (A) and (A'). See Problems 13, Section 2.7, and 23, Section 3.8.

15. As in Problem 14, Section 4.4, inscribe in the circle a triangle whose sides shall pass through the poles of the given lines, and then draw the tangents to the circle at the vertices of this triangle.

Section 4.5

1. (b) Use successive applications of part (a).

   (d) Case 2. Find N on OM such that $ON = n(OM) > (OD)/2$. By Case 1 find N', the inverse of N in $O(D)$. Finally find M' such that $OM' = n(ON')$.

   (e) See Problem 4 (a), Section 3.6.

   (f) See Problem 4 (b), Section 3.6.

   (g) From the points A, B, C, D one can, with a Euclidean compass alone, obtain a circle k whose center is not on AB or $C(D)$. Under inversion in k, line AB and circle $C(D)$ become circles whose centers are constructible (by parts (e) and (f)), and points on which are constructible (by part (d)). These circles can then be drawn, and their intersections found. The inverses in k of these intersections are the sought points X and Y.

   (h) From the points A, B, C, D one can, with a Euclidean compass alone, obtain a circle k whose center is not on AB or CD. Under inversion in k, lines AB and CD become circles through the center O of inversion. The centers of these circles are constructible (by parts (e) and (f)), and, since they pass through O, the circles can be drawn. The inverse in k of the other point of intersection of these circles is the sought point X.

2. Circle ABC is the inverse of line BC' in circle $A(B)$. Hence use Problem 4 (a), Section 3.6.

3. Find C such that $\overline{AC} = 2\overline{AB}$. Let $A(B)$ and $C(A)$ intersect in X and Y. Draw $X(A)$ and $Y(A)$ to intersect in the sought midpoint M.

4. We suppose the center O of the circle is given. Draw $A(C)$ and $D(B)$ to intersect in M. Draw $A(OM)$ to cut the given circle in X, Y. Then A, X, D, Y are vertices of an inscribed square. The proof is easy.

Section 4.6

2. Case 1, P not on k. Draw PAB, PCD cutting k in A, B and C, D. Draw AD, BC to intersect in M. Draw AC, BD to intersect in N. Then MN is the sought polar.

   Case 2, P on k. Draw any secant m through P and let R and S be any two points on m but not on k. Find the polars r and s of R and S. Then r and s intersect in M, the pole of m. PM is the sought polar.

3. (a) Find, by Problem 2, Section 4.6, the polar p of P, and let p cut k in S and T. Then PS and PT are the sought tangents.

   (b) Find the polar p of P by Problem 2, Section 4.6. Or inscribe a hexagon 123456 in k, where $1 \equiv 2 \equiv P$, and use Pascal's mystic hexagram theorem.

4. (a) Draw (by Problem 4.6.1) a line m parallel to line ABC. Choose a point V not on m or ABC and let VA, VB, VC cut m in A', B', C'. Let BA' and CB' intersect in V'. Draw $V'C'$ to cut ABC in D; draw DV to cut m in D'; draw $V'D'$ to cut ABC in E; etc.

   (b) We illustrate with $n = 5$. In the figure of the solution for part (a), let $A'A$ and $F'B$ intersect in U. Then $B'U$ cuts AB in a point Y such that $AY = AB/5$.

Section 4.7

1. Consider an oblique cone with a circular base and vertex V. There exist circular sections of the cone which are not parallel to the base; let c be one of these circular sections. A straightedge construction of the center of the circular base of the cone would lead, by projection from V, to a straightedge construction of the center of c. But the center of c is not the projection from V of the center of the base.

2. (a) By Problem 4.6.1 draw through P lines parallel to the diagonals of the given parallelogram to cut AB in C and D. By Problem 4.6.1 draw CE parallel to PD and DE parallel to PC, and let PE cut CD in M. Now, by Problem 4.6.1, draw through P a line parallel to AB.
   (b) By part (a) draw three chords parallel to one diagonal of the parallelogram. Now, by connecting opposite ends of pairs of these chords, obtain two points on the diameter of k that is perpendicular to the three chords; draw this diameter. Carry out a similar construction with respect to the other diagonal of the parallelogram.

3. (a) Take M and N, any two points on the inner circle and such that MN is not a diameter of the circle. By Problem 3 (b), Section 4.6, draw the tangents to the inner circle at M and N. We now have, in the outer circle, two bisected chords in two different directions. This allows us (by Problem 4.6.1) to construct a parallelogram with sides parallel to these bisected chords. Now use Problem 2 (b), Section 4.7.
   (b) Let P be the point of contact. Draw any three secants AA', BB', CC' through P, where A, B, C are on one circle and A', B', C' are on the other. Then BA is parallel to $B'A'$, and BC is parallel to $B'C'$. Now use Problem 2 (b), Section 4.7.
   (c) Let the two circles intersect in P and Q. Draw any two secants MQT and UQS, where M and U are on one circle and T and S are on the other. Next draw secants TPV, SPN, where V and N are on the other circle. Then MN is parallel to VU. Similarly obtain another pair of parallel lines, and use Problem 2 (b), Section 4.7.

4. Draw any two secant lines through the center of similitude and by connecting pairs of corresponding points of intersection of these secant lines with the circles obtain two pairs of parallel lines. Now use Problem 2 (b), Section 4.7.

5. (a) Take any point P on AB and with the parallel ruler draw any pair of parallel lines PP', MM' the width of the ruler apart. Let MM' cut AB in M. Then draw NN' parallel to PP' and the width of the ruler from PP'. Let NN' cut AB in N. Then P is the midpoint of segment MN.
   (b) Use Problem 4.6.1.
   (c) Suppose P is on AB. Draw any line $P'P$ through P and then draw $M'M$ and $N'N$ parallel to $P'P$ the width of the ruler from $P'P$. Then draw NN'' not parallel to $P'P$ and the width of the ruler from P. Let NN'' meet $M'M$ in R. RP is the required perpendicular.
      Suppose P is not on AB. Take any point Q on AB and (by the above case) draw QR perpendicular to AB. By part (b) draw PT parallel to RQ. Then PT is the required perpendicular.

(d) Let k denote the width of the ruler. Take any point C' and draw $C'D'$ parallel to CD. Take D' such that $C'D' = k$; this can be done by erecting a perpendicular to $C'D'$ at C', and then using the ruler to obtain D'. Draw CC' and DD' to meet in S. Draw $C'A'$ parallel to CA to meet SA in A'. Then draw $A'B'$ parallel to AB. We have now reduced the problem to that where $CD = k$. We proceed, then, with this special case.

Take any point P on AB outside the circle $C(D)$. Draw PT at distance k from C; PT is then a tangent to $C(D)$. Draw CT perpendicular to PT. Draw CR perpendicular to AB, TR perpendicular to PC. Through R draw RX and RY at distance k from C to cut AB in the required points X and Y. The proof is easy. TR is the polar of P, hence AB is the polar of R. Hence, since RX and RY are tangent to $C(D)$, X and Y are the required points of intersection of AB and $C(D)$.

(e) We may proceed exactly as in Problems 4.6.5 and 4.6.6.

9. PROOF I. Let C be a circle inside S and such that G', A', R' lie outside C. Invert G', A' in C, obtaining G'', A'' inside C. Find R'' (also inside C) and then invert R'' in C to obtain R'.

PROOF II. Take point P inside S as center of homothety and construct, from G', A', loci G'', A'' homothetic to G, A, but lying entirely inside S. Now construct R'' (also inside S) and then obtain R'.

12. (a) Use the fact that the sum of the infinite geometric series $\frac{1}{2} - \frac{1}{4} + \frac{1}{8} - \frac{1}{16}$ $+ \cdots$ is $\frac{1}{3}$. For another asymptotic Euclidean solution of the trisection problem see Problem 4134, *The American Mathematical Monthly*, Dec. 1945.

14. See Problem 8, Section 1.7.

15. (a) Note that $\angle OLP = \angle ORL + \angle LOR$ and $\angle OPQ = \angle OPL + \angle LPQ$. But $\angle OLP = \angle OPQ$ and $\angle ORL = \angle OPL$. Therefore $\angle LOR = \angle LPQ$ $= \angle LOQ = 30°$ and $\angle LKQ = 2(\angle LOQ) = 60°$. It follows that $RL = QL$ $= KL$.

Section 5.2

1. In its movement, line s can pass over only one vertex of P at a time.

3. A vertex V of a polyhedron P is a *projecting vertex* if there exists a plane which separates V from all the other vertices of P.

4. Devise a space analog of the proof suggested in Problem 1, Section 5.2.

Section 5.3

27. Let AB and $A'B'$ be a corresponding pair of congruent arcs on the boundary of P. We may connect the ends A and B of arc AB by a polygonal path p such that the region K bounded by arc AB and path p contains nothing but interior points of P, while the congruent region K' bounded by arc $A'B'$ and the corresponding path p' contains nothing but exterior points of P. Let us cut off region K and place it on K'. Doing this for all corresponding pairs of congruent arcs on the boundary of P converts P by dissection into a polygon. Q may be treated similarly. We then apply Theorems 5.3.6 and 5.1.3.

Section 5.4

4. Note that the proof usually given in high school solid geometry texts is incomplete.

Section 5.5

1. Connect the midpoint of one edge of the regular tetrahedron to the ends of the opposite edge. If we take an edge of the tetrahedron as 2, the isosceles triangle formed has base 2 and sides $\sqrt{3}$. The vertex angle θ of this triangle measures the dihedral angles of the tetrahedron. But $\sin(\theta/2) = 1/\sqrt{3}$, whence $\cos\theta = 1 - 2\sin^2(\theta/2) = 1/3$.

2. See a high school solid geometry text.

3. (a), (b), (c) See a high school solid geometry text.

Section 5.6

2. Such a dissection is indicated in the proof of Theorem 5.6.5.

3. See the proof of Theorem 5.6.5.

4. This follows from Theorem 5.6.5, since P and Q are equivalent.

6. See a high school solid geometry text.

Section 5.7

2. If the points coincide, then $i - j = 2k\pi$ for some integer k. But, since π is irrational, this is impossible.

3. A good question.

5. Let M_1 be the midpoint of AB, M_2 the midpoint of M_1B, M_3 the midpoint of M_2B, etc. Denote by E the set of all points on $[AB]$ with the exception of points A, B, M_1, M_2, M_3, ... Then we have

$$[AB] = E \cup A \cup B \cup M_1 \cup M_2 \cup M_3 \cup \ldots,$$
$$(AB] = E \cup B \cup M_1 \cup M_2 \cup M_3 \cup \ldots,$$
$$[AB) = E \cup A \cup M_1 \cup M_2 \cup M_3 \cup \ldots,$$
$$(AB) = E \cup M_1 \cup M_2 \cup M_3 \cup \ldots.$$

It is now apparent how we may put the points of any one of the four segments in one-to-one correspondence with the points of any other one of the four segments.

6. (a) In fact, $R(O,\theta)$ carries z into $z(\cos\theta + i\sin\theta) = ze^{i\theta}$.
   (b) Use part (a) and mathematical induction.
   (c) If $E_1 \cap E_2 \neq \emptyset$, then there exist polynomials P and Q in e^i, having integral coefficients, such that $Pe^i = Q + 1$.
   (d) Assuming the contrary we have, by part (c), that e^i satisfies a polynomial equation with integral coefficients. But this is impossible, since e^i is transcendental.

7. Yes, by the Banach-Tarski paradox.

Section 5.8

1. Since (see Figure 5.8a) right triangles DJE and GKF are congruent, we have $DJ = GK$ and $JE = KF$. This is ample to guarantee that the reassembled figure is a rectangle R. But R must be a square since one side is the square root of the area of R.

2. In Figure 5.8a, take E' and F' as the midpoints of $A'B'$ and $D'C'$. Mark off $E'G' = 3^{-1/4}(A'B')$. Mark off $F'H' = H'K' = E'G'$.

3. Draw the regular octagon and the equivalent square (see Figure 5.8c). With radius equal to the apothem of the octagon, draw a circle concentric with the square, and take alternate intersections of the circle with the square as the cut points A, B, C, D. The corresponding set of points in the octagon will be the set of midpoints A', B', C', D' of alternate sides. Now rotate the boundary $A'B'$ so that A' and B' fall on B and A, etc. The sides of the octagon become cut lines of the square, and the sides of the square are the cut lines of the octagon.

6. Dissect the smallest square of the rectangle into squares as shown in Figure 5.8e. Repeat the process.

7. Start with successive squares of sides 36, 25, 16, 28, 33 around the perimeter of the rectangle.

8. A pair of diagonally opposite corner squares are of the same color, say red. It follows that in the remaining piece there are more black squares than red ones. But a domino requires one square of each color.

9. See, e.g., pp. 199-207 of Maurice Kraitchik, *Mathematical Recreations*.

10. The figure has axial symmetry and contains 6 congruent rhombuses of one shape, 6 congruent rhombuses of a second shape, and 3 congruent rhombuses of a third shape.

11. Employ mathematical induction. See Problem E 468, *The American Mathematical Monthly*, Feb. 1942.

12. (b) One obtains Perigal's dissection of Figure 5.1c.

13. $[\sqrt{3}(a^2 + b^2 + c^2)/4 + 3ab/2]/2$.

14. (a) See Problem E 1406, *The American Mathematical Monthly*, Nov. 1960.
    (b) See V. E. Hoggatt, Jr. and Russ Denman, "Acute isosceles dissection of an obtuse triangle," *The American Mathematical Monthly*, Nov. 1961, pp. 912–913.

15. See Item 5.8.16.

16. (a) Divide the cube into 8 smaller cubes by the three planes midway between the pairs of opposite faces; divide one of the smaller cubes similarly into 8 still smaller cubes. Carry out the dividing operation q times. Finally divide one of the smallest cubes into k subcubes.
    (b) Note that $55 = 20 + 5(7)$, $56 = 49 + 7$, $57 = 1 + 8(7)$, $58 = 51 + 7$, $59 = 38 + 3(7)$, $60 = 39 + 3(7)$, $61 = 61 + 0(7)$.

Section 6.1

2. (a) The sides of the angle which the vanishing points of m and n subtend at V are parallel to m' and n'.
   (b) Use part (a).

3. (a) Denote the required center of perspectivity by V. Then (by Problem 2 (a), Section 6.1) $\angle AVC = \angle A'B'C' = \alpha$ and $\angle DVF = \angle D'E'F' = \beta$. To find V draw on AC an arc of a circle containing angle α and on DF (on the same side of l) an arc of a circle containing angle β. Since A, C, D, F are in the order A, D, C, F, these arcs must intersect; let X be such an intersection. Now rotate X about line $ADCF$ out of plane π to a position V. Then if we project plane π from center V onto a plane parallel to the plane of V and l the problem is solved.

(b) Not necessarily—only so long as the circular arcs described in the solution of part (a) intersect one another.

4. Let AB and CD intersect in U, AD and BC in V, AC and BD in W. By Problem 3, Section 6.1, project line UV to infinity and angles VAU and LWM into right angles.

5. Draw any line cutting the rays of the pencil $U(AB,CD)$ in (ab,cd). Let U' $(A'B', C'D')$ be the projection, under any perspectivity, of the pencil $U(AB,CD)$, and let $(a'b',c'd')$ be the projection of (ab,cd). Then $U(AB,CD) = (ab,cd)$ $= (a'b',c'd') = U'(A'B',C'D')$.

6. Let V be the center of perspectivity and let X and Y' be the feet of the perpendiculars from V on π and π'. Then the bisectors of $\angle XVY'$ cut π and π' in the isocenters of the perspectivity. Suppose, for example, the internal bisector of $\angle XVY'$ cuts π and π' in E and E'. Then XE and $E'Y'$ intersect on the axis of perspectivity in a point K, and $EK = E'K$. Let the sides of an angle at E cut the axis of perspectivity in L and M. Then $\angle LEM$ maps into $\angle LE'M$. But triangles LEM and $LE'M$ are congruent, etc.

7. The isolines are the reflections of the axis of perspectivity in the vanishing lines of the two planes.

Section 6.2

1. Project $ABB'A'$ into a square (by Problem 4, Section 6.1).

2. (b) The line at infinity.
   (c) Itself; in fact each point of line b maps into itself.
   (d) A line parallel to lines a and b.
   (e) Itself.
   (f) A line parallel to lines a and b.
   (g) Let m be a line not parallel to lines a and b. Then m cuts a and b in points U and V. m' is the line through V parallel to OU.
   (h) At O.
   (i) Let c be any line through the intersection of lines a and b. Let OA cut c in Q and QB cut PO in P'. La Hire's mapping may be obtained from this generalization by projecting line c to infinity.

3. Project line PQ to infinity and then use Problem 14, Section 2.4.

4. The expression is an h-expression.

5. The expression

$$(\overline{MN})(\overline{MN'})(\overline{M'P})(\overline{M'P'})/(\overline{MP})(\overline{MP'})(\overline{M'N})(\overline{M'N'})$$

is an h-expression. Now project $ABCD$ into a parallelogram.

Section 6.3

1. (b) Suppose there are two parallel tangents a and b, touching the parabola at A and B respectively. Then the pole of AB is the ideal point C of intersection of a and b. Now the line s at infinity touches the parabola and passes through C. It follows that a, b, s are three distinct tangents to the parabola through C. This is impossible.

(c) T is the pole of PQ. Therefore PQ passes through the pole U of TV and $(PQ,VU) = -1$. Since V is the midpoint of PQ, U is the point at infinity on PQ. Let TV cut the parabola in R and S. Then the tangents at R and S intersect in U. It follows, by part (b), that S must be an ideal point. But $(TV,RS) = -1$, whence R is the midpoint of TV.

(d) Let T be any point and t its polar; let S be the ideal point on the parabola and let ST cut the parabola in R and cut t in V. Then, since $(VT,RS) = -1$, it follows that R is the midpoint of VT. Now t, the tangent at R, and the tangent at S (that is, the line at infinity) all pass through the pole U of VT. It follows that U is a point at infinity, and the tangent at R is parallel to t. But the tangent at R is the line midway between T and t.

(e) This is a consequence of part (d).

2. (a), (b), (c) The line at infinity touches, intersects, and fails to intersect a parabola, a hyperbola, and an ellipse respectively.

(e) For the polar of the center of the parallelogram is the line at infinity.

(f) Let W be the other point of intersection of CT with the conic. Then $(UW,TV) = -1$.

3. (a) For a diameter is the polar of the point at infinity in the direction of the family of parallel chords.

(b) The diagonals of the quadrilateral formed by the points of contact are diameters (since their poles are at infinity) and hence bisect each other; the quadrilateral is thus a parallelogram. It follows that the poles of the diagonals of the circumscribed parallelogram are at infinity, and these diagonals are then diameters.

(c) Let P, Q be any two points on the conic and let V be the midpoint of PQ. If C is the center of the conic, CV is the diameter bisecting chords parallel to PQ. Thus CV and PQ are parallel to a pair of conjugate diameters, and are thus perpendicular to one another. Since $PV = VQ$, it follows that $CP = CQ$, and all radii of the conic are equal.

5. (a) Use Theorem 6.3.12.

(b) By part (a), the diameter through the midpoint of QQ' also bisects RR'.

(c) The tangent is parallel to the diameter conjugate to the diameter through the point of contact of the tangent. Now use part (a).

Section 6.4

1. See Theorem 2.8.4.

2. Let u be the polar of U and let PQ and RR' cut u in V and W respectively. Since $(PQ,UV) = -1$, u is the exterior bisector of angle PTQ, and is therefore perpendicular to TU. But $(RR',UW) = -1$. It follows that TU and u are the bisectors of angle RTR'.

Suggestions for Solutions of Selected Problems **413**

3. Let AB, CT intersect in R. Then the polar of R passes through C and through the harmonic conjugate of R for AB; and hence is CD. Since R is on AB, it follows that CD passes through the pole of AB.

4. TN and TR are conjugate lines. Now use Theorem 6.3.12.

5. If R is the pole of PQ, then the polar of U is the line through R parallel to AB.

6. Let TR cut PQ in V and let U be the point at infinity on PQ. Then $(PQ,VU) = -1$, whence also $(MN,RU) = -1$.

7. R, S, T lie on the polar of the point of intersection of BC and AD.

8. (a), (b) See Problem 3, Section 4.6.

9. (a), (b), (c) These are special cases of Pascal's Theorem where certain of the six vertices of the inscribed hexagon coalesce.

10. (a), (b), (c) These are special cases of Brianchon's Theorem where certain of the six sides of the circumscribed hexagon coalesce.

11. Parallel the proof of Theorem 6.4.5.

12. Apply Carnot's Theorem to triangle ABC; or project into a circle and employ the cross ratios (BC,A_1A_2) and (AC,B_1B_2).

13. Use Ceva's Theorem for triangle ABC and points X, Y, Z, and then apply Carnot's Theorem.

14. Let AD, BE, CF intersect in G. Project (see Theorem 6.1.7) the figure so that G becomes the centroid of ABC. Then D, E, F are the midpoints of BC, CA, AB respectively, and $A'D$, $B'E$, $C'F$ are diameters of the conic.

15. Let L be the pole of MN and let ML and NL cut ON and OM respectively in S and T. Then $(OP,TM) = -1$ and $(OQ,NS) = -1$. Therefore PQ, TN, MS are concurrent. That is, PQ passes through L.

Section 6.5

2. Using the lettering of Figure 2.6b we have $A(EB,DF) = (EL,DH) = (EM,KF) = C(EB,DF)$. Since no three of the points A, B, C, D, E, F are collinear, $A(C)$ does not correspond to $C(A)$. It follows, by Theorem 6.5.3, that E, B, D, F lie on a proper conic passing through A and C.

   Or, denote the vertices of the hexagon by 1, 2, 3, 4, 5, 6, and let p be the intersection of lines 61 and 34, q the intersection of lines 12 and 45, r the intersection of lines 23 and 56. By Theorem 6.5.7 there is a unique proper conic c passing through 1, 2, 3, 4, 5. Let $6'$ be the other point of intersection of c with $5r$. Then, by Pascal's Theorem applied to hexagon $123456'$, $6'1$ must pass through p. It follows that $6 \equiv 6'$.

3. Let the conic determined by A_1, A_2, B_1, B_2, C_1 cut AB in C_2'. Apply Carnot's Theorem and show that $C_2' \equiv C_2$.

4. Let ABC be the triangle. Let the parallel to BC through a point G cut AB in C_2 and AC in B_1; let the parallel to CA through G cut BC in A_2 and BA in C_1; let the parallel to AB through G cut CA in B_2 and CB in A_1. Consider the h-expression

$$(\overline{AC_1})(\overline{AC_2})(\overline{BA_1})(\overline{BA_2})(\overline{CB_1})(\overline{CB_2})/$$
$$(\overline{AB_1})(\overline{AB_2})(\overline{BC_1})(\overline{BC_2})(\overline{CA_1})(\overline{CA_2}).$$

Now (by Theorem 6.1.7) project ABC and point G into a triangle and its centroid. In the projected figure the h-expression has the value 1. Now apply Problem 3, Section 6.5.

5. (a) Let the five points be A, B, C, D, E. Use Pascal's Theorem on the hexagon $ABCDEF$, where $F \equiv A$, to obtain the tangent at A.

(b) Let the four points be A, D, E, F, where we are given the tangent at A. Draw any other line through A. To find the intersection C of this line with the conic, use Pascal's Theorem on hexagon $ABCDEF$, where $B \equiv A$.

(c) Let the three points be A, B, C. Use Pascal's Theorem on the hexagon $ADBECF$, where $D \equiv A$, $E \equiv B$, $F \equiv C$.

6. Use Carnot's Theorem and Problem 3, Section 6.5.

7. For $O(T) = (P) = S(P)$.

8. For $C(R') = C'(R') = C'(P) = C(R) = C'(R)$.

9. For $U(JK,LM) = U(AB,CD) = V(AB,CD) = V(JK,LM)$.

Section 6.6

1. Denote the sides of the hexagon by 1, 2, 3, 4, 5, 6, and let p be the join of points 61 and 34, q is the join of points 12 and 45, r the join of the points 23 and 56. By Theorem 6.3.14 there is a unique proper conic c touching 1, 2, 3, 4, 5. Let 6′ be the other tangent to c from point $5r$. Then, by Brianchon's Theorem applied to hexagon 123456′, point 6′1 must lie on p. It follows that $6 \equiv 6'$.

2. (a) Let the five tangents be a, b, c, d, e. Use Brianchon's Theorem on hexagon $abcdef$, where $f \equiv a$, to obtain the point of contact of a.

(b) Let the four tangents be a, d, e, f, where we are given the point of contact of a. Take any other point on a. To find the tangent c from this point to the conic, use Brianchon's Theorem on hexagon $abcdef$, where $b \equiv a$.

(c) Let the three tangents be a, b, c. Use Brianchon's Theorem on the hexagon $adbecf$, where $d \equiv a$, $e \equiv b$, $f \equiv c$.

3. See Problem 8, Section 2.7.

4. See Problem 8, Section 2.7.

5. Let S and S' be points of contact with c and c' of a common tangent to c and c'. Then $S(P) = (R) = S'(P)$. But $S(S')$ corresponds to $S'(S)$.

6. Let AR and BS intersect in T. Then $A(T) = A(R) = (R) = (S) = B(S) = B(T)$, and T traces a conic passing through A and B. Taking $R \equiv S \equiv O$, we see that $T \equiv O$, and O is also on the conic.

7. (a) Let the triangle be ABC, where BC, CA, AB pass through the fixed points D, E, F respectively, and B and C lie on fixed lines b and c respectively. Now $F(A) = F(B) = D(B) = D(C) = E(C) = E(A)$, and A lies on a conic through E and F.

(b) Let the triangle be ABC, where A, B, C lie on fixed lines a, b, c respectively, and AB and AC pass through fixed points F and E respectively. Now $(B) = F(B) = F(A) = E(A) = E(C) = (C)$, and BC touches a conic tangent to b and c.

8. Let $A'C'$ cut BA and BC in D and E, and let AC cut $B'A'$ and $B'C'$ in D' and E'. Then $(A'C',DE) = B(A'C',AC) = B'(A'C',AC) = (D'E',AC)$. Therefore $A'D'$, $C'E'$, DA, EC touch a conic tangent to $A'C'$ and AC.

Section 6.7

1. (a) Two points P and Q on a tangent m to a conic.
   (b) A line m cutting a conic in two points P and Q, two tangents p and q to the conic intersecting in a point N on m.

2. A complete quadrilateral.

3. If lines x, y, u are concurrent and if $x(ab,cu) = y(ab,cu)$, then lines a, b, c are concurrent.

4. abc is a triangle, l a fixed line through point ab, o a variable line through point cl. If p is the join of ao and cb, and q is the join of bo and ca, then the join of pq and ab is a fixed line m.
   Or: If ABC is a triangle, F is a fixed point on AB, and a variable line through F cuts BC in D and CA in E, then DA and BE intersect on a fixed line through C.

9. Let P be the pole of p for proper conic c, whose reciprocal is proper conic c'. From an arbitrary point Q of p, external to c, draw the two tangents t, u to c. Denote PQ by q. Then $(pq,tu) = -1$. Now if P', Q', T', U', q', p' are the reciprocals of p, q, t, u, Q, P, it follows that P', Q', T', U' lie on q' and $(P'Q',T'U') = -1$. But T', U' are points of c'. Hence Q' lies on the polar of P' for c'. But Q' lies on p'. Thus p' and the polar of P' with respect to c' both coincide with the locus of Q', and so are identical. It follows that p' is the polar of P' for conic c'.

10. Use Problem 9, Section 6.7.

11. A hexagon is inscribed in a conic if and only if the points of intersection of the three pairs of opposite sides are collinear.

12. (a) Let m be the straight line and let M and M' be the poles of m for the two conics. Let A be a variable point on m and let a and a' be the polars of A for the two conics. As A moves along m, p and p' rotate about M and M'. But $(a) = (A) = (a')$, whence the intersection of a and a' traces a conic through M and M'.
   (b) If a line rotates about a fixed point, the join of the poles of the line for two given proper conics envelops a conic.

13. (a) Let c denote the proper conic through A, B, C, D, E. By Pascal's Theorem find A', where PA cuts c again. Similarly find B', where PB cuts c again. Let AB', $A'B$ intersect in U, and let AB, $A'B'$ intersect in V. Then UV is the sought polar.
   (b) Dualize the construction of part (a).

14. (a) Let P be the moving point and let U and V be fixed points on the same conic. Let p, u, v be the polars of P, U, V for the second conic. Now $u(p) = U(P) = V(P) = v(p)$. It follows that p envelops a conic.
   (b) The pole with respect to a second conic of a variable tangent to a first conic traces a third conic.

15. (a) Let A travel along line l and let L be the pole of l for the first conic. Then a passes through L. Let l' be the polar of L for the second conic. Then A' lies on l'.
   (b) The pole of a line a with respect to one conic is A, and the polar of A with respect to a second conic is a'. As a rotates about a fixed point, a' also rotates about a fixed point.

16. (a) Use Ceva's Theorem and the converse of Carnot's Theorem (Problem 3, Section 6.5).

(b) The joins of the vertices of a triangle to the points of intersection with the opposite sides of two fixed transversals are all tangent to a conic.

17. (a) This is a special case of the converse of Brianchon's Theorem (Problem 1, Section 6.6).

(b) If L, M, N are three collinear points on the sides BC, CA, AB of a triangle ABC, there is a proper conic touching AL, BM, CN at A, B, C respectively.

Section 6.8

3. Angle α of Figure 6.8a can be continuously altered.

4. There are two spheres inscribed in the cone and passing through the given point, and then two planes tangent to these spheres at the given point.

5. The angles α and β of Figure 6.8a are the same for the two parallel sections.

6. Choose α of Figure 6.8a so that $\sin \alpha / \sin \beta = e$, the eccentricity of the given ellipse, etc.

7. The sum of the distances is (referring to Figure 6.8b) $VE_1 + VE_2$.

Section 6.9

1. Use Theorem 6.9.2.

2. Orthogonally project the ellipse into a circle.

3. $(3ab\sqrt{3})/4$.

4. $(4\pi K\sqrt{3})/9$.

5. Each of two perpendicular diameters of a circle bisects all chords parallel to the other.

6. Use Problem 5, Section 6.9.

7. Use Problem 5, Section 6.9.

8. Draw the diameter from the given point and find the trisection point of this diameter that lies nearer the other end. The side of the triangle opposite the given vertex passes through this trisection point and is parallel to the tangent to the ellipse at the given vertex.

9. Orthogonally project the ellipse into a circle.

10. Orthogonally project the ellipse into a circle. Note that in the projected figure $(OA)(OB)/(O'A')(O'B') = (OC)(OD)/(O'C')(O'D')$. This same relation holds (by Theorem 6.9.1 (5)) for the original figure. Etc.

11. Orthogonally project the two ellipses into two concentric circles.

12. Orthogonally project the ellipse into a circle and use Problem 5, Section 6.9.

13. Orthogonally project the ellipse into a circle, of center O', say. Then the circle on $O'T'$ as diameter passes through P', Q', V', whence $\measuredangle T'V'Q' = \measuredangle T'P'Q'$. Etc.

14. Let E be the ellipse inscribed in a given parallelogram and touching the sides at their midpoints. Orthogonally project E into a circle. The parallelogram then projects into a square. The circle is the maximum ellipse that can be inscribed in the square.

15. Orthogonally project the ellipse into a circle.

16. $ab(\sqrt{3} - \pi/2)$.

17. Lines drawn from any point on an ellipse parallel to the diameters of the ellipse which bisect the sides of an inscribed triangle cut the respective sides of the triangle in three collinear points.

18. The center of any ellipse inscribed in a quadrilateral is collinear with the midpoints of the diagonals of the quadrilateral.

19. A chord AQ of an ellipse cuts the diameter of the ellipse conjugate to the diameter through A in point R; CP is the radius of the ellipse parallel to AQ. Then $(AQ)(AR) = 2(CP)^2$.

20. First consider triangles whose bases lie on the line of intersection of the two planes.

Section 7.1

1. To deduce Euclid's fifth postulate, let AB and CD be cut by the transversal ST, and suppose $\angle BST + \angle DTS < 180°$. Through S draw QSR, making $\angle RST + \angle DTS = 180°$. Now apply, in turn, I 28, Playfair's Postulate, I 17.

2. (a), (b) It is easily shown that, given Playfair's Postulate, each of (a) and (b) follows, since the contradiction of either (a) or (b) contradicts Playfair's Postulate. Conversely, given either (a) or (b), the Playfair Postulate follows, since the contradiction of the Playfair Postulate contradicts each of (a) and (b).

3. See Wolfe, *Introduction to Non-Euclidean Geometry*, pp. 21–23.

4. Try the same experiment and reasoning on a spherical triangle, using a great circle arc in place of the straightedge.

5. Let P be above Q and let PM be a variable ray rotating counterclockwise about P, starting from PQ as initial position. Though there may be a first position of PM that fails to intersect QB, there is no final position of PM that intersects QB.

6. The proof assumes that the noncongruent triangles AEO and ADC are similar.

7. Each of Propositions I 27 and I 28 implies the other. Let P be a given point not on a given line m. Through P draw a line n cutting line m. Now through P draw a line p such that a pair of corresponding angles formed with p and m by n are equal. By Proposition I 28, p is parallel to m.

8. (a) Let $ABCD$ be the Saccheri quadrilateral, where angles A and B are right angles and $AD = BC$. Draw the diagonals AC and BD and show that triangles DAB and CBA are congruent (sas), and then that triangles ADC and BCD are congruent (sss).

   (b) Using the notation of the solution of part (a), let M and N be the midpoints of AB and CD. Draw AN and BN and show that triangles ADN and BCN are congruent (sas). Next show that triangles AMN and BMN are congruent (sss). This makes MN perpendicular to AB. Similarly prove MN is perpendicular to CD by drawing DM and CM.

   (c) Drop the perpendicular from the third vertex upon the line joining the midpoints of two sides, and prove some right triangles are congruent.

(d) Using the solution of part (b), let R and S be the midpoints of AD and BC. Then $ABSR$ is a Saccheri quadrilateral and MN perpendicularly bisects RS.

9. Using the notation introduced in the solution of Problem 8 (b), Section 7.1, $MBCN$ is a Lambert quadrilateral and is half the Saccheri quadrilateral $ABCD$.

10. Consult a high school solid geometry text.

12. (a) Angle A is less than 60°.

(b) Carrying out the construction n times we obtain a triangle containing the given triangle at least 2^n times, and therefore having a defect at least 2^n times the defect of the given triangle. By taking n sufficiently large we can then obtain a triangle whose defect is greater than 180°. But this is absurd.

13. (b) Let ABC be a triangle having the sum of its angles equal to two right angles. If ABC is not already right isosceles, draw the altitude BD. If neither of the resulting triangles is right isosceles, mark off on the longer leg of one of them a segment equal to the shorter leg. By part (a), the resulting isosceles right triangle has the sum of its angles equal to two right angles. By putting together two such congruent isosceles right triangles, a quadrilateral can be formed having all its sides equal and all its angles right angles. By putting together four such congruent quadrilaterals, a larger quadrilateral of the same kind can be formed. By repeating the last construction enough times, one can obtain a quadrilateral of the same kind having its sides greater in length than any given segment. A diagonal of this last quadrilateral will give an isosceles right triangle of the type desired.

(c) Let ABC be any right triangle, right-angled at C. By part (b), there exists an isosceles right triangle DEF, right-angled at E, having the sum of its angles equal to two right angles and its legs greater than either of the legs of triangle ABC. Now produce CA and CB to A' and B', respectively, so that $CA' = CB' = ED$. Then triangles DEF and $A'CB'$ are congruent. Draw $A'B$, and apply part (a) to triangle $A'CB'$. The extension to *any* triangle ABC is now easily accomplished by dividing ABC into two right triangles by one of its altitudes.

14. See Wolfe, *Introduction to Non-Euclidean Geometry*, pp. 23–24.

15. See Wolfe, *Introduction to Non-Euclidean Geometry*, p. 25.

Section 7.3

1. Draw PQ perpendicular to AB and use Theorem 7.2.4.

2. Draw the line through the midpoint of the finite side of an isosceles limit triangle and parallel to the other two sides. The isosceles limit triangle is divided into two congruent limit triangles (Theorem 7.3.3), and it follows that the drawn parallel is perpendicular to the finite side. Now consider two isosceles limit triangles having equal finite sides and divide each of the triangles into a pair of congruent right-angled limit triangles as above. Any one of the limit triangles of one pair is then congruent to any one of the other pair (Theorem 7.3.3), and the angles of one isosceles limit triangle are equal to those of the other.

3. Use Theorem 7.3.2.

4. The two lines do not intersect (by Proposition I 28), and they are not parallel (by Problem 3, Section 7.3).

5. Draw $A'E'$ parallel to $B'D'$ and Use Theorem 7.3.3.

6. Let AX and BY be the parallel sides of an isosceles limit triangle and let M be the midpoint of the finite side AB. In the solution of Problem 2, Section 7.3, it is shown that the perpendicular bisector MZ of AB is parallel to AX and BY. Let P be any point on MZ and let R and S be the feet of the perpendiculars from P on AX and BY respectively. Now triangles AMP and BMP are congruent (sas), whence $PA = PB$ and $\angle MAP = \angle MBP$. It follows that $\angle RAP = \angle SBP$ and right triangles PRA and PSB are congruent, whence $PR = PS$.

7. Use Theorem 7.3.3.

8. Let the two limit triangles be $XABY$ and $UCDV$, where $\angle B = \angle D$ and $AB > CD$. Mark off $BE = DC$ on BA and draw EZ parallel to AX and BY. Then limit triangles $ZEBY$ and $UCDV$ are congruent (Theorem 7.3.3) and $\angle UCD = \angle ZEB > \angle A$ (Theorem 7.3.2).

9. (a) $h = \log \cot [\Pi(h)/2]$.

10. (b) Absolute.

Section 7.4

1. Let $ABCD$ and $A'B'C'D'$ be two Saccheri quadrilaterals with equal bases AB and $A'B'$ and equal summits DC and $D'C'$. Suppose $AD > A'D'$. Mark off AD'' on AD equal to $A'D'$, and BC'' on BC equal to $B'C'$. Then $ABC''D''$ and $A'B'C'D'$ are congruent and $D''C'' = D'C'$. Let the perpendicular to AB at its midpoint M cut $D''C''$ in P and DC in Q. Then (Theorem 7.4.3) PQ perpendicularly bisects both $D''C''$ and DC. It follows that $PQCC''$ is a Saccheri quadrilateral with base PQ and summit $C''C$, and angles QCC'' and $PC''C$ must be equal acute angles (Theorem 7.4.2). But this is impossible since angle $BC''P$ is acute (Theorem 7.4.2).

2. Let $ABCD$ and $A'B'C'D'$ be two Saccheri quadrilaterals with equal bases AB and $A'B'$ and equal summit angles. Suppose $AD > A'D'$. Mark off AD'' on AD equal $A'D'$, and BC'' on BC equal to $B'C'$. Then $ABC''D''$ and $A'B'C'D'$ are congruent, whence angles ADC, $AD''C''$, BCD, $BC''D''$ are all equal. But then the sum of the angles of quadrilateral $D''C''CD$ is 360°, which is impossible.

3. In quadrilateral $ABCD$, let $\angle A = \angle B = 90°$ and $\angle D = \angle C$. Suppose $AD > BC$. Mark off AD' on AD equal to BC. Then $ABCD'$ is a Saccheri quadrilateral and $\angle AD'C = \angle BCD' < \angle BCD = \angle ADC$. But this is impossible.

4. Otherwise there would be a contradiction of Theorem 7.4.5.

5. Dissect the convex polygon into $n - 2$ triangles by diagonals through a chosen vertex.

Section 7.5

3. If $AD > BC$, mark off $AE = BC$ on AD. Then $\angle BCD > \angle BCE = \angle AEC > \angle ADC$. Etc.

4. (a) Put two congruent Lambert quadrilaterals together to form a Saccheri quadrilateral.
   (b) Use Problem 3, Section 7.5.

5. By drawing the line connecting the midpoints of the summit and base, and applying Problem 3, Section 7.5, we see that the summit is greater than the base.

6. Let $ABCD$ be the given Saccheri quadrilateral, with base AB. Let E and F be the midpoints of AD and BC and let G and H be the midpoints of AB and DC. Then $ABFE$ is a Saccheri quadrilateral. Therefore GH, which is perpendicular to AB at G, must be perpendicular to EF. The second part of the problem now follows from Theorem 7.5.6.

7. In Figure 7.4d, $2LM = DE$. But, by Problem 5, Section 7.5, $DE < AB$.

Section 7.6

1. In the model, each line must pass through the pole of the other.

3. In the model, two lines are conjugate to the same third line if and only if they pass through the pole of the third line.

4. Use Theorem 3.10.6.

5. In the model, we must project m on n from the pole of n.

7. In the model, let the hyperparallel lines a and b cut K in A and B and in C and D respectively. Consider the lines AD, BC, AC, BD.

12. By Pascal's mystic hexagram theorem.

13. This is an easy consequence of Pascal's mystic hexagram theorem.

Section 7.8

3. (a) Apply transformation S to the segment AB, taking O as center and $\angle \overline{BOC}$ as angle. Then B will map into C and A into D', where $\angle \overline{AOD'} = \angle \overline{BOC}$ and AD' is perpendicular to OD'. Thus segment AB maps into segment $D'C$, and segment OA into segment OD'. But the right angle OAB, one side of which passes through the center O, must map into a right angle one side of which passes through O. It follows that $\angle OD'C = 90°$. Etc.

3. (b) $OABC$ is concyclic and $\angle OAC = \angle OBC$.

4. (a) Let ABC and DEF be plane angles of the dihedral angle and let O be the midpoint of BE. Construct $\angle A'EC'$, the image of $\angle ABC$ under the reflection $R(O)$. Then $\angle A'EC' = \angle ABC$. Now $\angle A'EO$ is the image of $\angle ABO$ under $R(O)$, whence $A'E$ is perpendicular to EO. It follows that A', E, D colline. Similarly, C', E, F colline, and $\angle A'EC' = \angle DEF$.

Section 8.1

1. (a) It is tacitly assumed, in Figure 8.1c, that E lies inside the triangle.
   (b) It is tacitly assumed, in Figure 8.1d, that PE lies between PB and PF.
   (c) It is tacitly assumed, in Figure 8.1e, that CD does not pass through B.

2. (b) Let m be a line and P a point not on m. Associate with each line through P its point of intersection with m.
   (c) Let m be a line and P a point not on m. Let c be the semicircle having P as center and any convenient diameter parallel to m. Associate with each point Q of c the point Q' of m where PQ cuts m.

3. Verification of the first four postulates presents little difficulty. To verify the fifth postulate it suffices to show that two ordinarily intersecting lines, each determined by a pair of restricted points, intersect in a restricted point. This may be accomplished by showing that the equation of a straight line determined by two points having rational coordinates has rational coefficients, and that two such lines, if they intersect, must intersect in a point having rational coordinates. For the last part of the problem, consider the unit circle with center at the origin, and the line through the origin having slope one.

5. (a) Let the line enter the triangle through vertex A. Take any point U on the line and lying inside the triangle. Let V be any point on the segment AC, and draw line VU. By Pasch's Postulate, VU will (1) cut AB, or (2) cut BC, or (3) pass through B. If VU cuts AB, denote the point of intersection by W and draw WC; now apply Pasch's Postulate, in turn, to triangles VWC and BWC. If VU cuts BC, denote the point of intersection by R; now apply Pasch's Postulate to triangle VRC. If VU passes through B, apply Pasch's Postulate to triangle VBC.

Section 8.2

1. (b) The janitor repaired the chairs.

4. (b) 1. Let p be a dabba. By P2 there exists another dabba q. By P3 there is an abba L containing both p and q. By P4 there is a dabba r not in L. By P3 there is an abba M containing p and r. L and M are distinct since q is in L but not in M.

2. Let L be an abba. By P1, L contains a dabba p. Suppose p is the only dabba in L. By Theorem (1) there exists an abba M distinct from L and containing p. Now M must contain a second dabba $q \neq p$, for otherwise L and M would not be distinct. By P4 there exists a dabba r not in M. By P5 there exists an abba N containing r but not containing any dabba of M. Since N does not contain p, N contains no dabba of L. We thus have two distinct abbas, L and M, containing dabba p but containing no dabba of N. But this contradicts P5, and the theorem follows by *reductio ad absurdum*.

3. By P2 there exist at least two dabbas p and q. By P3 there is an abba L containing p and q. By P4 there exists a dabba r not in L. By P5 there exists an abba M containing r but containing no dabba of L. By Theorem (2) there is a dabba $s \neq r$ in M. Now p, q, r, s are four distinct dabbas.

4. Consider the four distinct dabbas p, q, r, s of Theorem (3). Then, by P3, there are abbas L, M, N, P, Q, R containing p and q, r and s, p and r, p and s, q and r, q and s respectively, and these abbas are easily shown to be distinct.

(d) A plane.

5. (a)—2 Let the three points be A, B, C, and let T be a motion mapping A into A', B into B', C into C'. Designate the inverse of T by T^{-1}. Let R be any effective motion which leaves A and B fixed. Now consider the resultant of the three motions T^{-1}, R, T, made in this order.

(b)—1 Let the sphere have center A and pass through B, and let P be a point on the sphere. Let T be a motion mapping A into A', B into B', P into P'. Designate the inverse of T by T^{-1}. Let S be a motion which leaves A fixed but maps B into P. Now consider the resultant of the three motions T^{-1}, S, T, made in this order.

(c) Pieri says AC is *perpendicular* to AB if there exists a motion which leaves A and B fixed but maps C into another point of the straight line CA.

6. Look the matter up in Hilbert's *The Foundations of Geometry*.

7. (a) Let A and B be two distinct points. By Postulate II-2 there is a point C_1 between A and B, a point C_2 between C_1 and B, a point C_3 between C_2 and B, etc. It follows that $C_1, C_2, C_3, \ldots$ are distinct points between A and B.
   (b) Otherwise we would have a contradiction of Postulate IV-1.
   (c) Employ *reductio ad absurdum*.
   (d) Employ *reductio ad absurdum*.

9. (a) Let h be a horn angle for which $\theta = 0$, and let h' be any horn angle for which $\theta > 0$. Then, for every positive integer n, we have $nh < h'$.
   (d) If $a = a'$ and $k = k'$, set $M = M'$; if $a > a'$, set $M > M'$; if $a = a'$ but $k > k'$, set $M > M'$. Define $M' = nM$ if and only if $a' = na$, $k' = nk$.
   (f) See part (e).

10. Huntington makes the following definitions. A sphere which does not include any other sphere is called a *point*. If A and B are two points, the *segment* $[AB]$ is the class of all points X such that every sphere which includes A and B also includes X. The *extension* of $[AB]$ *beyond A* is the class of all points X such that $[BX]$ contains A; similarly, the *extension* of $[AB]$ *beyond B* is the class of all points X such that $[AX]$ contains B. The *ray AB* is the class of all points belonging to $[AB]$ or to the extension of $[AB]$ beyond B. The *line AB* is the class of all points belonging to $[AB]$ or to one of its extensions.

Section 8.3

2. T1: The number of bees is the number of combinations of four things taken two at a time.
   T2: Each hive shares one and only one bee with each of the other three hives.
   T3: By P3, each bee belongs to two and only two hives, say m and n. By T2, m contains three bees and n contains three bees, whence m and n together contain five bees, leaving, by T1, one bee not in either m or n.

3. See Eves and Newsom, *An Introduction to the Foundations and Fundamental Concepts of Mathematics*, Sec. 6.2.

4. (a) Since $a\ G\ c$, there exists m such that $a\ F\ m$ and $m\ F\ c$. Since $b\ G\ c$, there exists n such that $b\ F\ n$ and $n\ F\ c$. By T6, $m = n$ and then $a = b$.
   (b) Suppose we have $a\ R\ b$ and $b\ R\ c$. Then, by T1, we also have $a\ R\ c$. It follows that $B(acb)$ is precluded. Suppose we have $c\ R\ b$ and $b\ R\ a$. Then, by T1, we also have $c\ R\ a$. It follows that $B(acb)$ is again precluded.

5. See Problem 2895, *The American Mathematical Monthly*, 29 (1922), 357–358.

6. See Problem E1098, *The American Mathematical Monthly*, 61 (1954), 474.

Section 8.4

1. (4) $\rightarrow q$ is true $\rightarrow r$ is false, which contradicts (3).

3. (a) Interpret the bees as six people, A, B, C, D, E, F, and the four hives as the four committees (A,B,C), (A,D,E), (B,F,E), and (C,F,D). Or, interpret the bees and the hives as six trees and four rows of trees, respectively, forming the vertices and sides of a complete quadrilateral.

(b) To show independence of P2, interpret the bees and the hives as four trees and four rows of trees forming the vertices and sides of a square. To show independence of P3, interpret the bees as four trees located at the vertices and the foot of an altitude of an equilateral triangle, and the hives as the four rows of trees along the sides and the altitude of the triangle. To show independence of P4, interpret the bees and the hives as three trees and three rows of trees forming the vertices and sides of a triangle.

(c) Let us designate the four hives by *I, II, III, IV* and the three bees in hive *I* by *A, B, C*. Let hive *II* have bee *A*, and only bee *A*, in common with hive *I*. Then we may designate the bees in hive *II* by *A, D, E*. Now let hive *III* be that hive which has bee *B* in common with hive *I*. Then we may designate the bees of hive *III* by either *B, D, F* or *B, E, F*. In the first case, hive *IV* must contain bees *C, E, F*, and, in the second case, hive *IV* must contain bees *C, D, F*. There are, then, the following two ways of designating the hives and the bees:

I	*A, B, C*		*I*	*A, B, C*	
II	*A, D, E*	or	*II*	*A, D, E*	
III	*B, D, F*		*III*	*B, E, F*	
IV	*C, E, F*		*IV*	*C, D, F*	

But the second designation can be changed into the first by interchanging the two labels *D* and *E*. It follows, then, that any interpretation of the postulate set can be labeled as in the first designation, and the desired isomorphism is established.

6. In Problem 2, Section 8.3, T1 is derived from P1, P2, P3, P4. It suffices to show that P2 can be derived from P1, T1, P3, P4. This may be accomplished by an indirect argument. Suppose, on the contrary, that we have *a R b* and *a = b*. Then we also have *b R a*. But this contradicts T1.

7. Show that there is essentially only one way of labelling the four elements of *K*.

9. See Problem 2895, *The American Mathematical Monthly*, 29 (1922), 357–358.

10. Interpret the elements of *S* as a set of all rectangular Cartesian frames of reference which are parallel to one another and let *b F a* mean that the origin of frame *b* is in the first quadrant of frame *a*. Or, interpret the elements of *S* as the set of all ordered pairs of real numbers (x,y), and let $(m,n) F (u,v)$ mean $m > u$ and $n > v$.

Section 8.9

6. (a) Take, for example, triangle *ABJ*. The altitudes from *A, B, J* are the lines *ASGYM, RKIBY, EJOTY*, which are concurrent in *Y*. The midpoints of the sides *AB, BJ, AJ* are *D, P, R* respectively. The three perpendicular bisectors of the sides are *DINSX, IVOCP, RKIBY*, which are concurrent at *I*. The three medians are *WKDQJ, AFKPU, RKIBY*, which are concurrent at *K*. Note that *K* divides median *AP* in the ratio 2:1 because *AF* is congruent to *FK* and *KP*, and *K* divides median *BR* in the ratio of 2:1 because *BI* is congruent to *IK* and *KR*. Also note that points *Y, I, K* are collinear, on line *RKIBY*, and that *K* divides *YI* in the ratio 2:1.

Index

Construction(s) *(Cont.)*
  simplicity of, 182
  Swale's, 184
  symbol of, 182
  well-defined, 180
  with compass alone, 169*ff*
  with straightedge alone, 174*ff*
Contraparallelogram (Hart), 138
Contrapositive statement, 19
Converse statement, 19
Convex polygon, 203*n*
Convex polyhedron, 212*n*
Coolidge, J. L. (*The Mathematics of Great Amateurs*), 44
Copolar triangles, 68
Court, N. A. (*see* Altshiller-Court, Nathan)
Coxeter, H. S. M. (*Regular Polytopes*), 45
Cremona, Luigi, 241
Cross ratio, 73*ff*
  of cyclic range, 77
  harmonic, 82
  of pencil, 74, 75
  of range, 73
Cubic equations, geometrical solution, 43
Curve(s):
  inverse, 123
  line, 263
  orthogonal, 88
  point, 263
Cut lines, 207
Cutting a cube into cubes, 236
Cutting the cheese, 235
Cycle, 32
Cycloid, 46*n*

Dabbas and abbas, 335
Dandelin, Germinal, 274
Data (Euclid), 22
Datum, 22
Daus, P. H. (*College Geometry*), 398
Decomposition, von Neumann's, 223
Dedekind, Richard, 17, 325
  continuity postulate, 325, 332, 333, 382
Deductive geometry, 8
Defect of a triangle, 300
Definition, explicit and implicit, 323, 327
Degenerate conic, first and second type, 269
Dehn, Max, 195, 288
  theorem, 217
De Moivre's theorem, 216*n*
Demonstrative geometry, 8
Denman, Russ, 411
Der barycentrische Calcul (Möbius), 54, 73
Desargues, Gérard, 59, 70, 139, 240, 241,

242, 251
Brouillon project, 59, 70
  finite geometry, 368
  two-triangle theorem, 68, 77, 245, 249, 362, 364
Desarguesian projective plane, 365
Descartes, René, 54, 241, 345
Descriptive property, 241
De triangulis omnimodis (Regiomontanus), 45
Diagonal lines of a complete quadrilateral, 84
Diagonal points of a complete quadrangle, 85
Diagonal 3-line of a complete quadrilateral, 84
Diagonal 3-point of a complete quadrangle, 85
Diameter(s) of a proper conic, 255
  conjugate, 255
Die Theorie der Parallellinien (Lambert), 285
Dilatation, 105
Directed angle, 56
Directed area, 57
Directed segment, 55
Directrix of a proper conic, 273
Direct statement, 19
Displacement, 113
Dissection:
  checkerboard, 231
  cutting a cube into cubes, 236
  cutting the cheese, 235
  into dominoes, 231
  duplicating the cube, 236
  equilateral triangle into a square, 226
  Fibonacci numbers, 232
  Kürschák Prize Competition problem, 235
  law of cosines, 233
  minimal problem, 195, 228
  order of, 229
  parpolygon into parallelograms, 232
  passing a cube through a cube, 237
  perfect, 229
  regular octagon into a square, 227
  simple perfect, 231
  squaring the square, 229
  tessellations, 231
  translation, 231
  volume, 237
Division, internal and external, 56
D'Ocagne, Maurice (trisection of an angle), 49

Inversion (*Cont.*)
  center of, 121
  circle of, 121
  in a line, 123
  method of, 165
  power of, 121, 148
  properties of, 126*ff*
  radius of, 121, 148
  sphere of, 148
Inversive plane, 121
Inversive space, 149
Involutoric transformation, 102
Isocenter, 247
Isodynamic points, 72
Isogonal conjugate lines, 50
Isogonal conjugate points, 71
Isogonal lines, 71
Isogonic centers, 72, 389
Isolines, 247
Isometry, 113, 148
  direct, 148
  opposite, 148
Isomorphic interpretations, 344, 345
Isosceles limit triangle, 298
Isotomic conjugate points, 71
Isotomic points, 71

Jackson, W. H., 200
Johnson, R. A. (*Modern Geometry*), 41*n*, 396
Journal de l'École Polytechnique, 87

Kanayama, R., 136
Kantian theory of space, 282
Kazarinoff, N. D. (*Geometrical Inequalities*), 387
Kepler, Johann, 44, 59
Khayyam, Omar (solution of cubic equations), 43
Kirkman points, 70
Klamkin, M. S., 99*n*
Klein, Felix, 287, 330, 353
  Erlanger Programm, 330
  model, 353
Kraitchik, Maurice (*Mathematical Recreations*), 411
K'ui-ch'ang Suan-Shu, 7
Kurple, 335
Kürschák Prize Competition, 235, 277

La Hire, Philippe de, 139, 251
  mapping, 251

Lambert, J. H., 285
  Die Theorie der Parallellinien, 285
  quadrilateral, 285, 289, 304
Latus rectum, 29
Law of cosines, 17, 19, 233
Leacock, Stephen, 121*n*
Legendre, Adrien-Marie, 286, 290
  Éléments de géométrie, 286
  first theorem, 290
  second theorem, 290
Lehmus, D. C., 58
Lemoine, Émile, 54, 181
Lennes, N. J., 211, 212
  polyhedron, 195, 211, 212, 213
Leon, 10
Lewis Carroll:
  Alice in Wonderland, 233
  paradox, 233
L'Huilier, Simon A. J. (*Éléments d'analyse géométrique et d'analyse algébrique appliquées à la recherche des lieux géométriques*) 120
Liber assumptorum (Archimedes), 23, 27
Limit triangle, 295
  angles of, 295
  finite side of, 295
  isosceles, 298
Lindgren, Harry, 231, 238
Line(s):
  boundlessness of, 287
  Cayley, 70
  cevian, 65
  concurrent, 57
  conjugate, 140, 254
  curve, 263
  cut, 207
  Euler, 109, 113
  hyperparallel, 292
  ideal, 60, 61
  infiniteness of, 287
  at infinity, 60, 61
  isogonal, 50, 71
  parallel, 367, 368, 369
  parallel (Lobachevskian sense), 292
  Pascal, 69, 70
  pencil of, 57
  Plücker, 70
  Simson, 146
  vanishing, 243
Linkages, 136
Liouville, Joseph, 121
Lipkin, 135
Lobachevskian geometry, 287

Transformations (*Cont.*)
   orthogonal projection, 275
   perspective, 243
   perspectivity, 243
   product of, 102
   projective, 243
   projectivity, 243
   by reciprocal radii, 121
   reciprocation, 269
   reflection in a circle, 121
   reflection in a line, 104, 147
   reflection in a plane, 147
   reflection in a point, 105, 147
   reversal, 113
   rotation, 104
   rotation about an axis, 147
   rotatory-reflection, 148
   screw displacement, 147
   similarity, 113
   similitude, 113
   space, 147*ff*
   space homology, 148
   space inversion, 148
   stereographic projection, 150
   stretch, 105
   stretch-reflection, 105
   stretch-rotation, 105
   theory of, 100*ff*
   translation, 104, 147
Transform-discover-invert procedure, 100
Transform-solve-invert procedure, 100
   130, 276
Transitivity property of parallelism, 294
Translation, 104, 147
   dissection, 231
Transmissability property of parallelism, 292
Transversal of a pencil, 57
Transversals, reciprocal, 71
Travers, James, 227, 228, 238
Treatise on Plane Trigonometry, A
   (Hobson), 41*n*
Trial and error, method of, 166, 168
Triangle(s):
   area of, 7
   area of (in Lobachevskian plane), 301*ff*
   coaxial, 68
   conjugate, 145, 257
   copolar, 68
   defect of, 300
   isosceles limit, 298
   limit, 295
   polar, 145, 257
   self-conjugate, 145, 257

   self-polar, 145, 257
Triangular numbers, 12
Trilinear pole and polar, 72
Trisection of an angle, 33, 156, 183, 184
   approximate, 48, 49
   asymptotic, 184
   d'Ocagne's approximate construction,
      49
Tucker, Robert, 54
Tutte, W. T., 229, 230, 231

*Über die Entwickelung der Elementar-
      geometrie im XIX Jahrhundert*
      (Simon), 53
*Über die Hypothesen welche der
      Geometrie zu Grunde liegen*
      (Riemann), 322
Ultra-ideal points, 305, 306
Unger, Leo, 330*n*
University of Alexandria, 22, 28

Vanishing line, 243
Vanishing point, 243
Veblen, Oswald, 332, 332*n*, 333, 333*n*, 336
Vergings (Apollonius), 30, 31, 32
Verriest, Gustave (*Introduction à la géo-
      métrie non-Euclidienne par la
      méthode élémentaire*), 347*n*
Vertex:
   of an angle, 56
   of a cone, 252
   of a pencil, 57
   projecting (of a polygon), 201
   projecting (of a polyhedron), 409
Vieta, François, 120
   duplication of the cube, 33, 183
Vinci, Leonardo da, 174, 201
Viviani, Vincenzo, 72
Volume:
   frustum of a pyramid, 5, 6, 7, 48
   moment of, 47*n*
   prismatoid, 34
   pyramid, 7
   sphere, 7, 28
   spherical segment, 28
   tetrahedron, 7
   torus, 37
Von Neumann's decomposition, 223

Well-defined construction, 180
Wentworth, G. A. (*Plane and Solid
      Geometry*), 385
Wheeler, A. H., 236